STAT CITY

Understanding Statistics through Realistic Applications

The Irwin Series in Quantitative Analysis for Business

Consulting Editor   Robert B. Fetter   Yale University

STAT CITY

Understanding Statistics through Realistic Applications

HOWARD S. GITLOW
University of Miami

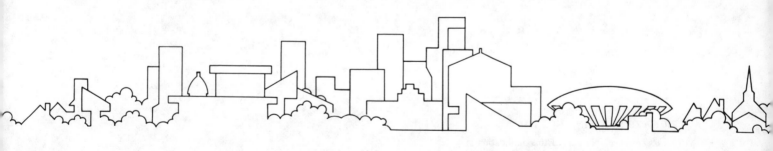

1982

RICHARD D. IRWIN, INC.
Homewood, Illinois 60430

© RICHARD D. IRWIN, INC., 1982

All rights reserved. No part of this publication may be
reproduced, stored in a retrieval system, or transmitted,
in any form or by any means, electronic, mechanical,
photocopying, recording, or otherwise, without the prior
written permission of the publisher.

ISBN 0-256-02654-8

Library of Congress Catalog Card No. 81–84833

Printed in the United States of America

1 2 3 4 5 6 7 8 9 0 ML 9 8 7 6 5 4 3 2

To Shelly—the rainbow in my sky

PREFACE

Stat City is a supplementary textbook which emphasizes the interpretation and communication of statistics. The objectives of *Stat City* are to:

1. Provide readers with *complete* statistical problems so that they can grasp the totality of statistical studies, from inception through memorandum.
2. Provide readers with *unified* statistical problems; all problems in *Stat City* are related to a Stat City data base and are performed for Stat City interest groups (e.g., the mayor, the telephone company, the electric company, and so on).
3. Provide readers with a role model for the performance of statistical studies.

Stat City is comprised of seven parts. Part One introduces Stat City, a fictitious town in the United States. Stat City is comprised of families, businesses, schools, parks, nonprofit organizations, and places of worship. Each dwelling unit in Stat City has five identifiers and 22 pieces of descriptive information. It is critical that Part One be mastered by the reader before he/she proceeds onto Parts Two through Seven; understanding Part One (Chapter 1) is the key to successfully using *Stat City*. Part Two introduces descriptive statistics, the basics of inferential statistics, and the fundamentals of statistical memorandum writing. The third part of the text introduces probability theory, probability distributions, and expands the fundamentals of statistical memorandum writing. Parts Four, Five, and Six formally introduce inferential statistics and complete the basics of statistical memorandum writing. Part Seven presents how computers (in particular SPSS—Statistical Package for the Social Sciences) can be used to facilitate statistical studies. All of the above parts are linked together through the Stat City theme.

Most chapters end with a brief summary and a series of problems which are designed to provide a vehicle for the reader to determine if he/she has grasped the essential concepts in the chapter.

Due to the nature of the Stat City Data Base, three topics frequently covered in basic statistics textbooks have been omitted—index numbers, time series, and decision theory. I believe that these ommissions are not a limitation due to the supplementary nature of this text and because these topics are adequately covered in most basic texts. Further, some readers may feel that Chapter 5, "Bivariate Statistics," is out of place. If this is the case, Chapter 5 can be introduced as a prelude to Chapter 12, "Regression and Correlation Analysis."

Stat City has been classroom tested for six semesters in basic and advanced courses, at both the undergraduate and graduate levels. The students' responses to the text have been extremely enthusiastic. A frequent comment is that the complete and unified examples greatly facilitate the learning of statistics.

I would like to express thanks to my colleagues and students at the University of Miami; especially Paul Sugrue, who was instrumental in the conception of this book; Paul Hertz, who tested the manuscript in his classroom; Margaret Updike and Mary McKenry, my graduate assistants who spent many hours searching for errors in the manuscript; Virginia Lee Fogle, my copy editor; and Deborah Jones, who typed and retyped the manuscript.

I am indebted to professors Paul D. Berger (Boston University), James F. Horrell (University of Oklahoma), Ernest Kurnow (New York University), Donald S. Miller (Emporia State University), Barbara Price (Wayne State University), Alan Oppenheim (Montclair State College), Rosa Oppenheim (Rutgers University), John C. Shannon (Suffolk University), and Howard J. Williams (Wilkes College) for their reviews and comments on the manuscript.

I am grateful to the Literary Executor of the late Sir Ronald A. Fisher, F.R.S., to Dr. Frank Yates, F.R.S., and to Longman Group Ltd. London, for permission to reprint Table III from their book *Statistical Tables for Biological, Agricultural and Medical Research* (6th edition, 1974).

Most important, I want to thank Shelly Gitlow for her encouragement and patience. One final acknowledgment is to my nieces, Jessica and Elissa, who are "dying" to see their names in print.

Howard S. Gitlow

CONTENTS

Contents

xii

Contents

xiv

APPENDIXES

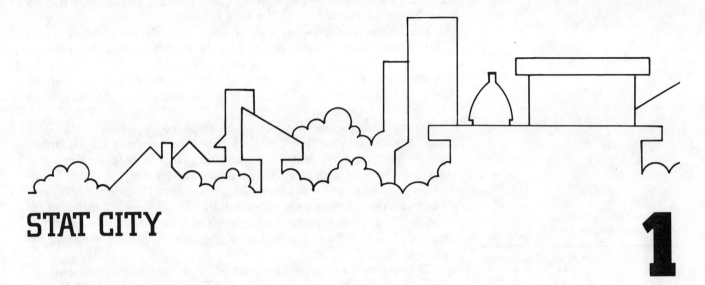

PART ONE

STAT CITY

1

Statistics involves methods and theory as they are
applied to numerical data or observations with
the objective of making rational decisions in
the face of uncertainty.[1]

Kurnow, Ottman, and Glasser

Stat City is a fictitious town in the United States. Stat City was created to help you learn how to *use* statistics, not to teach you about mathematical statistics. Once you have mastered using statistics to explain the characteristics of Stat City for decision-making purposes, you will be prepared to use statistics on any real-world problem.

Remember, statistics is a way of thinking and attacking tough problems; it does not have to be torture.

Stat City is comprised of families (dwelling units), businesses, schools, parks, nonprofit organizations, and places of worship. The map inside the back cover of this book gives you a pictorial view of Stat City. As you can see, the business district is located in approximately the center of town. The residential sections of Stat City surround the business district and are divided into four zones. Table 1–1 depicts the breakdown of Stat City dwelling units (families) into each residential zone as of January 1980.

Each zone in Stat City is subdivided into blocks. A block is a geographical area separated by streets and avenues. The subdivision of zones into blocks gives you an extremely detailed way of breaking down Stat City into component parts for analysis. For example, by grouping blocks you could compare the incomes of families in Zone 3 east of 10th Avenue with the incomes of families in Zone 3 west of 10th Avenue.

Each dwelling unit in Stat City has three identifiers: the last name of the head of the household, a street address, and an identification number. Each dwelling unit has a unique street address, except in the case of apartment buildings where families have the same street address. If more detail is needed to isolate a particular dwelling unit (family), their identification number can be used.

INTRODUCTION

Table 1–1   Dwelling units (families) in Stat City by residential zone (January 1980)

Zone	Number of units
1	130
2	157
3	338
4	748
Total	1,373

___

[1] *Statistics for Business Decisions* (Homewood, Ill.: Richard D. Irwin, 1959). © 1959 by Richard D. Irwin, Inc.

Aside from name, address, block, identification number, and zone, 22 other pieces of information were collected for each dwelling unit (family) in Stat City. Exhibit A–1 in Appendix A at the back of this book lists all the information collected on Stat City families.

The information listed in Exhibit A–1 was collected through a mail survey (questionnaire) sent out on January 30, 1980,—one questionnaire to each dwelling unit. The head of each household was instructed that he or she alone should complete the questionnaire and answer all questions relative to January 1980. A copy of the questionnaire can be seen in Exhibit A–2, Appendix A.

Once all 1,373 questionnaires were completed and checked, the data was compiled into the Stat City Data Base (see Appendix C). The information is ordered by identification number (variable 4). Zone 1 contains the dwelling units with identification numbers 1 through 130. Zone 2 contains the dwelling units with identification numbers 131 through 287. Zone 3 contains the dwelling units with identification numbers 288 through 625. Finally, Zone 4 contains the dwelling units with identification numbers 626 through 1,373. Table 1–2 depicts how the identification numbers are assigned.

To interpret the collected information for a particular dwelling unit (a line of data), simply read across the Stat City Data Base and use Exhibit A–1 as a reference guide for the descriptions of each piece of information. For example, an interpretation of the information collected for the Kilfoyle family as of January 1980 (the first line of data in the Stat City Data Base) appears in Exhibit 1–1.

Table 1–2 Identification number assignments

| Zone | Identification number | |
	From	To
1	1	130
2	131	287
3	288	625
4	626	1,373

Exhibit 1–1

NUMBER	NEUMONIC	DESCRIPTION	
1	NAME	Last name of the head of the household	Kilfoyle
2	ADDR	Street address	406 6th Street
3	BLOCK	Block location of dwelling unit	43
4	ID	Dwelling unit's identification code	0001
5	ZONE	Residential housing zone	1
6	DWELL	Type of dwelling unit 0 = apartment 1 = house	1
7	HCOST	Housing cost as of January 1980 Rent if apartment Mortgage if house	$502
8	ASST	Assessed value of home ($0.00 if apartment) as of January 1980	$88,264
9	ROOMS	Numbers of rooms in dwelling unit as of January 1980	10
10	HEAT	Average total yearly heating bill as of January 1980 (includes all types of heat—electric, gas, etc.)	$1,199
11	ELEC	Average monthly electric bill as of January 1980	$88
12	PHONE	Average monthly telephone bill as of January 1980	$90

Exhibit 1-1 (*continued*)

NUMBER	NEUMONIC	DESCRIPTION	
13	INCOM	Total family income for 1979	$56,419
14	PEPLE	Number of people in household as of January 1980	6
15	CARS	Number of cars in household as of January 1980	3
16	GAS	Average bimonthly automobile gas bill as of January 1980	$165
17	GASCA	Average bimonthly automobile gas bill per car as of January 1980	$55
18	GASTR	Average number of trips to the gas station per month as of January 1980	6
19	REPAR	Favorite place to have automotive repairs performed as of January 1980 0 = performs own repairs 1 = repairs done by service station 2 = repairs done by dealer	1
20	FAVGA	Favorite gas station as of January 1980 1 = Paul's Texaco (11th St. & 7th Ave.) 2 = Howie's Gulf (11th St. & Division St.)	2
21	HOSP	Average yearly trips to the hospital by all members of a dwelling unit as of January 1980	6
22	EAT	Average weekly supermarket bill as of January 1980	$128
23	EATPL	Average weekly supermarket bill per person as of January 1980	$21
24	FEAT	Favorite supermarket as of January 1980 1 = Food Fair 2 = Grand Union 3 = A&P	3
25	LSODA	Average weekly purchase of six-packs of diet soda as of January 1980	2
26	HSODA	Average weekly purchase of six-packs of regular soda as of January 1980	0
27	BEER	Average weekly purchase of six-packs of beer as of January 1980	5

The Kilfoyles reside at 406 6th Street, on block 43, in Zone 1, in a house with a monthly mortgage of $502 and an assessed value of $88,264. The house has 10 rooms, an average yearly heating bill of $1,199, an average monthly electric bill of $88, and an average monthly telephone bill of $90. The Kilfoyle family income was $56,419 in 1979. There are six people in the Kilfoyle family. The Kilfoyles own three cars, spend an average of $165 bimonthly on gas ($55 per car), average six trips to the gas station per month, and have most of their car repairs done at Howie's Gulf Station (gas station 2). The Kilfoyle's average about six trips to the hospital per year. The Kilfoyles spend an average of $128 per week on food ($21 per person) in the A&P (supermarket 3). Their many food purchases include a weekly supply of two six-packs of diet soda, no regular soda, and five six-packs of beer.

POPULATION VERSUS SAMPLE

One of the fundamental concepts you will encounter when using statistics is the difference between a POPULATION and a SAMPLE.

> A POPULATION (or Universe) is the total number of items under study. All the dwelling units in Stat City form a population.
> A SAMPLE is a part of the population under investigation selected so that information can be drawn from it about the population. For example, one hundred randomly selected Stat City dwelling units form a sample.

It is important to keep in mind that the dwelling units in one residential zone of Stat City could be considered a population or a sample. Despite the general differentiation between population and sample, it is important to remember that a population is what a client defines it to be, no more, no less.[2] A population may include all dwelling units in a city or it may only include dwelling units in a given area. Your client is responsible to define clearly the population in your problem. For example: If you are working for the Stat City Chamber of Commerce and are asked to do a citywide analysis of dwelling units, the population must be defined as all dwelling units in Stat City. However, if you are working for the Zone 3 Community Planning Board and are asked to do an analysis of Zone 3 dwelling units, all the dwelling units in Zone 3, and only the dwelling units in Zone 3, must be defined as the relevant population.

Let us look at another facet of the same issue. If you assume that the relevant population is all the dwelling units in Stat City and you select all the dwelling units in Zone 4 as your sample, you may very well commit a severe error. The error occurs when the dwelling units in Zone 4 are in some way unique (as the French Quarter is to New Orleans or as Harlem is to New York City). You may then be reporting facts about the dwelling units in Zone 4 and not all the dwelling units in Stat City. Be careful!

What is statistics?

Statistics can be broadly defined as the study of numerical data to better understand the characteristics of a population. Statistics may help in making some rational decision about a population. Statistics can also be thought of as the *art* of extracting a clear picture out of numerical information. I refer to statistics as an art because frequently statisticians must be creative if they are going to be helpful in solving problems. The creative portion of a statistician's job makes his/her work exciting.

The use of statistics in business is pervasive. Some examples of business applications of statistics are:

Opinion polling to predict the outcome of an election.

---

[2] In many instances the statistician and the client are the same person.

Determining the proper dosage level for a drug.

Surveying consumers to establish appeal and nutritional acceptability of a new food product.

Forecasting the population of southern Florida so that an adequate mass transportation system can be built.

Controlling the quality of goods from a production line.

Estimating the population in each state in the United States to disburse federal monies.

It is clear from the above list that statistics is extremely important to business. By the way, another aspect of statistics that makes it exciting is that it can be applied in almost any field of endeavor—from controlling rat populations in Bombay to allocating salespeople's commissions in Detroit. Statistical skills are transferable. A world of opportunity is open to statistician.

Decision makers frequently use two types of information to understand a population, parameters and statistics.

Population parameters versus sample statistics

A PARAMETER is a measure of a population characteristic. For example, the average monthly telephone bill computed from all dwelling units in Stat City or the average monthly electric bill computed from all dwelling units in Zone 2 (if your relevant population is Zone 2) are parameters.

A STATISTIC is an estimate of a population characteristic which is computed from a sample. For example, if 10 dwelling units were randomly selected from Stat City and the average number of rooms computed, this average would be called a statistic. This sample average would be an estimate of the average number of rooms per dwelling unit in Stat City.

There are 1,373 dwelling units in Stat City. To draw a random sample of, say, 10 Stat City dwelling units, the following steps should be followed.

How to select a random sample of Stat City dwelling units

1. Specify the reason(s) you want to draw a sample. For example, you may want to estimate the average[3] number of rooms (variable 9) per dwelling unit in Stat City.
2. Number the dwelling units in Stat City from 0001 through 1373. (This has already been done if you will look at variable 4 in the Stat City Data Base.)
3. Select a page in a table of random numbers. Exhibit A–3, Appendix A, is a random number table.
4. On the selected page in Exhibit A–3 (for example, the first page), select a column of numbers with as many digits in it as are in 1373 (4 digits). For example, begin in the first column of the first page in Exhibit A–3 with 5347.
5. Select the first 10 four-digit numbers in the chosen column on that page which are between 0001 and 1373, inclusive.
    If a number is encountered which is smaller than 0001 (0000) or larger than 1373 (5374 for example), discard the number and continue down the column. If an acceptable number appears more than once, ignore every repetition and continue moving down the column until 10 unique numbers have been selected. Finally, if you get to the bottom of the page before you have obtained all 10 random numbers, go to the top of the page and move down the next four-digit column. Table

---

[3] Arithmetic mean.

1–3 depicts the 10 random numbers drawn, the family names of those dwelling units selected, and the number of rooms in each selected dwelling unit.

Table 1–3

Variable 4 Identification number (selected random numbers)	Variable 1 Name	Variable 9 Number of rooms
933	Althouse	6
481	Comas	7
576	Pomer	10
1318	Shapiro	5
1348	MacMaster	8
966	Slater	5
1215	Lehman	6
573	Reeves	7
997	Singer	5
1144	Halbrooks	6

6. Finally, analyze the information on the variable(s) of interest. The estimated average number of rooms per household is 6.5 rooms, based on the sample of 10 dwelling units.

An important point to remember is that different samples of size 10 may yield different room estimates. We will deal with this frustrating point later on in this book.

The remaining sections of this book will be filled with problems and cases dealing with populations and samples to help you learn how to use statistics in the business world.

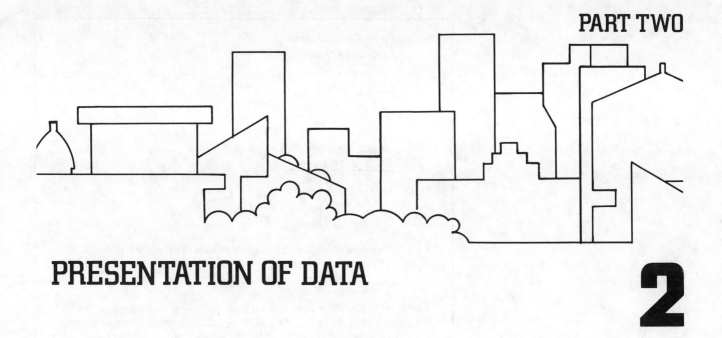

PRESENTATION OF DATA

2

Business people are frequently confronted with huge amounts of data which must be summarized into a workable format so that decisions can be made. In this chapter you will learn how to reduce large amounts of data to manageable proportions for decision-making purposes.

A common method of presenting statistical data is by means of a chart. Charts can be used to summarize large amounts of statistical data in a form which is readily interpreted and easily understood. Charts are widely used in presenting data in newspapers, magazines, and business reports.

The most common types of statistical charts are: bar charts, line charts, pie charts, pictographs, and statistical maps. An example of each type of chart will be illustrated as a needed tool to solve a problem for a Stat City interest group. Each problem in this chapter will present parameters for you to use as input in constructing charts and writing business memos. Don't worry about actually computing the parameters now. We will deal with calculating parameters in later chapters. Before we get involved in calculation, let us get acquainted with writing business memoranda that use statistical data.

INTRODUCTION

Example 2–1 *Stat City Beacon* article on overnight parking

Ms. Donna Nelson, a reporter for the *Stat City Beacon,* is preparing an article on recent problems with overnight parking on public streets within Stat City. She decided that it would be appropriate to describe clearly in the article the number of automobiles in each zone of the city. She chose to present a bar chart which would enable the readers of the article to comprehend readily the preponderance of automobiles in Zones 3 and 4 as compared with Zones 1 and 2. She obtained the information shown in Table 2–1 from the Stat City Department of Traffic as of January 1980.

You have been asked by Ms. Nelson to construct the appropriate bar chart. Type a business memorandum presenting the bar chart to Ms. Nelson (see Exhibit 2–1).

Table 2–1 Automobiles in Stat City by residential zone (January 1980)

Zone	Number of automobiles per zone
1	323
2	365
3	874
4	1,405
Total	2,967

Exhibit 2-1

HOWARD S. GITLOW, PH.D.
STATISTICAL CONSULTANT

MEMORANDUM

TO: Donna Nelson, Reporter
        Stat City Beacon

FROM: Howard Gitlow

DATE: February 21, 1980

RE: Bar chart for article on overnight parking on public
        streets in Stat City as of January 1980.

I have taken the Department of Traffic data you supplied
to me concerning the total number of automobiles per zone in
Stat City and have constructed the desired bar chart per your
request.*

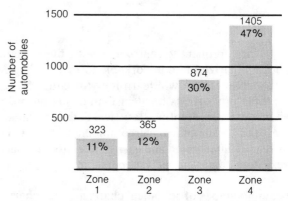

    * Percentages have been rounded so that they
will total 100 percent.
    Source: Department of Traffic, January 1980.

If you have any further questions, please do not hesitate
to call me at 305-999-9999.

The reporter for the *Stat City Beacon* has decided that a pictograph would be understood more widely by her readership. She has asked you once again to type a memorandum presenting her with the pictograph (see Exhibit 2-2).

Exhibit 2-2

HOWARD S. GITLOW, PH.D.
STATISTICAL CONSULTANT

MEMORANDUM

TO: Donna Nelson, Reporter
        Stat City Beacon

FROM: Howard Gitlow

DATE: March 3, 1980

RE: Pictograph for article on overnight parking on
        public streets in Stat City as of January 1980.

   As per your request, I have constructed a pictograph based
upon data supplied by the Department of Traffic concerning
the total number of automobiles per zone in Stat City as of
January 1980. The pictograph appears below.

Exhibit 2–2 (*continued*)

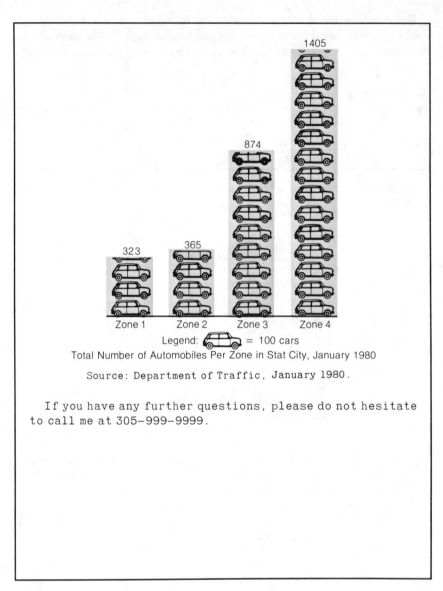

Total Number of Automobiles Per Zone in Stat City, January 1980

Source: Department of Traffic, January 1980.

If you have any further questions, please do not hesitate to call me at 305-999-9999.

Example 2–2 Purchasing books for the Stat City Library

Ms. Elissa Gitlow, chief librarian for the Stat City Library, will increase the number of books purchased by the library if the percentage change in the Stat City population was more than 10 percent between January 1976 and January 1980. The Stat City Chamber of Commerce has supplied Ms. Gitlow with the Stat City population for January 1976 through January 1980. The population figures are shown in Table 2–2.

Ms. Elissa Gitlow has decided that a line chart supplemented by the percentage increase in population for January 1980 over January 1976 would be required to make the correct purchase decision. You have been asked by Ms. Gitlow to construct the line chart and to compute the percentage increase in population. Type a memorandum to Ms. Gitlow reporting the requisite information (see Exhibit 2–3).

Table 2–2

Month	Year	Population
January	1976	5,127
January	1977	5,279
January	1978	5,499
January	1979	5,640
January	1980	5,814

Exhibit 2–3

HOWARD S. GITLOW, PH.D.
STATISTICAL CONSULTANT

MEMORANDUM

TO: Ms. Elissa Gitlow, Chief Librarian
        Stat City Library

FROM: Howard Gitlow

DATE: March 11, 1980

RE: Purchasing books for the Stat City Library

Based upon the data you supplied from the Stat City
Chamber of Commerce I have constructed a line chart to
illustrate the population increase in Stat City from
January 1976 to January 1980.

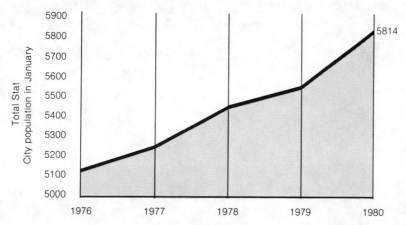

Source: Stat City Chamber of Commerce, Jan. 1976–Jan. 1980.

Exhibit 2-3 (*continued*)

The percentage increase in the population between January 1976 and January 1980 is approximately 13.4 percent [(5,814 − 5,127 = 687) ÷ 5,127 = .13399 = +13.4%]. This increase exceeds your 10.0 percent decision criteria established for the expansion.

If you have any further questions, please do not hesitate to call me at 305−999−9999.

Example 2–3   Department of Public Works population survey

Mr. Joseph Moder, division head of the Stat City Department of Public Works, is preparing a report for the mayor on the need to expand the sewage system in Zone 4. This expansion is required because of the rapid growth in the population of Zone 4. Mr. Moder has decided to enclose a pie chart in his report to illustrate the population inequities existing in the four zones. He obtained the information shown in Table 2–3 from the Stat City Chamber of Commerce.

You have been asked to type a memorandum presenting the pie chart to Mr. Moder.

The degrees of the angles needed to construct the pie chart are computed below.

Zone 1: $504/5,814 = .0867 =$ percentage of population in Zone 1. $.0867\ (360°) = 31.212° =$ number of degrees in the angle describing Zone 1's slice of the pie.

Zone 2: $546/5,814 = .0939 =$ percentage of population in Zone 2. $.0939\ (360°) = 33.804° =$ number of degrees in the angle describing Zone 2's slice of the pie.

Zone 3: $1,411/5,814 = .2427 =$ percentage of population in Zone 3. $.2427\ (360°) = 87.372° =$ number of degrees in the angle describing Zone 3's slice of the pie.

Zone 4: $3,353/5,814 = .5767 =$ percentage of population in Zone 4. $.5767\ (360°) = 207.612° =$ number of degrees in the angle describing Zone 4's slice of the pie.

The pie chart appears in Exhibit 2–4.

Table 2–3   Population of Stat City by zone, January 1980

Zone	Population
1	504
2	546
3	1,411
4	3,353
Total	5,814

Solution

Exhibit 2-4

HOWARD S. GITLOW, PH.D.
STATISTICAL CONSULTANT

MEMORANDUM

TO: Joe Moder, Division Head
        Stat City Department of Public Works

FROM: Howard Gitlow

DATE: March 4, 1980

RE: Pie chart presenting population by zone in Stat City,
        as of January 1980

   I have taken the Chamber of Commerce data concerning the
total population in each zone of Stat City and have
constructed the desired pie chart per your request. The pie
chart appears below.

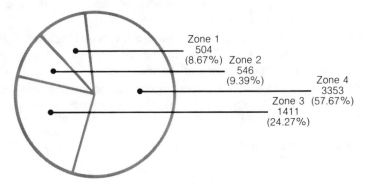

Source: Stat City Chamber of Commerce, January 1980.

   If you have any further questions, please do not hesitate
to call me at 305-999-9999.

Example 2-4 Block analysis of the number of dwelling units in Stat City

Ms. Sharon Vigil, chairperson of the Stat City Real Estate Board, would like a map depicting the density of dwelling units by block. She is interested in obtaining the locations of all blocks with 6 or less dwelling units, 7 to 10 dwelling units, 11 to 29 dwelling units, and 30 or more dwelling units.

You have been hired by Ms. Vigil to construct a statistical map of Stat City reflecting the density of dwelling units per block. (Hint: Use a copy of the Stat City map to construct the statistical map.)

Solution   The density categories and the gradations of gray used to represent those categories are:

Number of dwelling units

1 to 6   7 to 10   11 to 29   30 or more

The map appears as part of the memorandum shown in Exhibit 2–5. It was constructed by counting the number of ID numbers (variable 4) on each block from the Stat City map.

Exhibit 2–5

```
HOWARD S. GITLOW, PH.D.
   STATISTICAL CONSULTANT
   ─────────


                     MEMORANDUM


TO:    Ms. Sharon Vigil, Chairperson
       Stat City Real Estate Board

FROM:  Howard Gitlow

DATE:  March 10, 1980

RE:    Statistical map depicting dwelling unit density by
       block in Stat City, as of January 1980.


   The statistical map requested by your office is herewith
enclosed. If you have any further questions, please call me
at 305-999-9999.
```

Exhibit 2–5 (*continued*) Stat City map depicting dwelling unit density

ADDITIONAL PROBLEMS

2–5 Analysis of electric bills by zone in Stat City The chief financial officer of the Stat City Electric Company, Saul Reisman (ID = 223), is preparing the annual report for stockholders. He has decided to include either a bar or pie chart, or pictograph describing the total of all the average monthly electric bills in each residential zone of Stat City, as of January 1980. He has extracted the following information from the electric company's accounting records:

Zone	Total of average monthly electric bills by zone, as of January 1980
1	$ 9,150
2	$11,189
3	$22,016
4	$34,197
Total	$76,552

Type a memorandum to Mr. Reisman presenting the charts.

2–6 Analysis of telephone bills by zone in Stat City The chairman of the Stat City Telephone Company, Jack Davis (ID = 232), wants you to construct a bar chart for his "Zonal Use" study indicating the total of all average monthly telephone bills for each residential zone as of January 1980. Mr. Davis had his secretary extract the following information from the telephone company's accounting records:

Zone	Total of average monthly telephone bills by zone, as of January 1980
1	$ 4,613
2	$ 5,704
3	$12,346
4	$27,732

Type a memorandum to Mr. Davis presenting the bar chart.

2–7 Construction of an outpatient clinic Ms. Arlene Davis, director of the Park View Hospital, located on Park Street and 10th Avenue, has decided that the hospital will set up an outpatient clinic in any zone which generates more than 2,000 patient visits per year on average. You have been asked by Park View management to construct a pie chart and a pictograph illustrating the average yearly number of hospital visits per zone as of January 1980.

A scrupulous study of the hospital's patient records reveals the following information:

Zone	Average yearly number of hospital visits by zone, as of January 1980
1	414
2	379
3	1,036
4	2,523
Total	4,352

Type a memorandum to Ms. Davis presenting the pie chart and pictograph.

2–8 Promotional campaign for beer in Stat City The Stat City Supermarket Association is considering a special promotional campaign for beer if the rate of growth in the number of six-packs purchased per year was less than 15 percent between 1976 and 1979. The Supermarket Association collected the following data:

Year	Number of six-packs of beer purchased in Stat City
1976	2,720
1977	2,760
1978	2,780
1979	2,800

The association has asked you to prepare a line chart of beer purchases for 1976 through 1979 and to compute the percentage change in beer purchases between 1976 and 1979. Type a memorandum to the Supermarket Association reporting the required information.

2–9 Block analysis of the population in Stat City Mr. Lee Kaplowitz, the mayor of Stat City, is preparing a report for the town council and needs a chart depicting the number of people residing in each block of Stat City. The town council is interested in knowing the location of blocks that have fewer than 10 people, between 10 and 19 people, between 20 and 29 people, between 30 and 39 people, and 40 people or more.

Mr. Kaplowitz obtained a table from the Stat City Chamber of Commerce indicating the population on each block in Stat City. The table is as follows:

Block	Population	Block	Population	Block	Population
1	8	41	24	81	30
2	17	42	28	82	27
3	16	43	28	83	9
4	26	44	24	84	23
5	16	45	18	85	16
6	25	46	23	86	27
7	18	47	4	87	25
8	26	48	10	88	28
9	22	49	24	89	37
10	15	50	22	90	31
11	20	51	21	91	18
12	22	52	12	92	29
13	22	53	13	93	181
14	11	54	25	94	196
15	17	55	33	95	192
16	20	56	32	96	193
17	20	57	20	97	175
18	19	58	15	98	192
19	23	59	35	99	200
20	23	60	17	100	180
21	22	61	15	101	86
22	28	62	30	102	89
23	25	63	23	103	101
24	26	64	32	104	85
25	24	65	18	105	97
26	10	66	36	106	96
27	17	67	15	107	102
28	17	68	34	108	102
29	23	69	34	109	93
30	18	70	175	110	91
31	23	71	162	111	89
32	31	72	172	112	98
33	20	73	12	113	80
34	23	74	25	114	82
35	22	75	11	115	88
36	18	76	22	116	98
37	11	77	30	117	90
38	21	78	26	118	92
39	24	79	27	119	94
40	21	80	24	120	91

You have been asked by the mayor to use this table in preparing a statistical map of Stat City reflecting the block populations according to the previously stated population categories.

FREQUENCY DISTRIBUTIONS

3

In this chapter new techniques will be presented for reducing large amounts of data to manageable proportions for decision-making purposes. There are three tools you will learn to use: frequency distributions, histograms, and frequency polygons. Once you have mastered these tools on Stat City, you will begin to feel the tremendous insight that can be gained about a population by using statistics.

The format of this chapter is to present several Stat City related examples worked out in detail and then to provide you with several more problems to solve for Stat City interest groups. The solutions to most of the examples will have two parts: a population section and a sample section. The population section will show you how to compute parameters. The sample section will show you how to estimate parameters by drawing a random sample and computing statistics. Finally, you will be asked to determine if your statistics are good estimators of the parameters.

Once you have confidence in your ability to sample you will no longer be required to compute parameters. In fact, it is important that you do not develop a dependence on parameters because they are rarely obtainable in real business situations.

Example 3–1   Image study of Howie's Gulf Station

Problem

Mr. Howie, the owner of Howie's Gulf Station, wishes to change the image of his station in the minds of Zone 1 residents. He believes a more favorable, service-conscious image would improve business. Before attempting the image alteration Mr. Howie wants to know how many Zone 1 families think of Howie's Gulf as their favorite station and how many Zone 1 families think of Paul's Texaco as their favorite station. Mr. Howie thinks this information is important in determining whether he should attempt the image change.

You have been retained by Mr. Howie to perform the above study. Conduct the study and prepare a typed memorandum for him (see Exhibit 3–1).

Note to the student: Make two surveys. First, conduct a survey investigating all 130 dwelling units in Zone 1. Second, conduct a survey investigating a random sample of 30 dwelling units in Zone 1. Finally, compare the results of your two surveys. Do you see anything surprising in the comparison?

Solution   **Part 1** The population frequency distribution can be seen in Table 3–1.

Table 3–1

Classes	Frequency	Percent
Howie's Gulf	80	61.5
Paul's Texaco	50	38.5
Total	130	100.0

Part 2 A sample frequency distribution based on 30 randomly selected dwelling units is constructed as shown.

1. Draw 30 random numbers from a random number table using the rules set out in Chapter 1 (see Table 3–2).

Table 3–2

Begin here →

6067	7516	2451	1510	*²⁴*0201	1437
4541	9863	8312	9855	~~0995~~	6025
6987	4802	8975	2847	4413	5997
¹ 0376	8636	9953	4418	2388	8997
8468	5763	3232	1986	7134	4200
9151	4967	3255	8518	2802	8815
*²*1073	4930	1830	2224	2246	*²⁷*1000
5487	1967	5836	2090	3832	0002
4896	4957	6536	7430	6208	3929
9143	7911	*¹³*0368	*¹⁸*0541	2302	5473
9256	2956	4747	6280	7342	*²⁸*0453
4173	*⁸*1219	7744	9241	6354	4211
2525	7811	5417	7824	~~0922~~	8752
9165	*⁹*1156	6603	2852	8370	~~0995~~
³ 0014	8474	6322	5053	5015	6043
5325	7320	8406	5962	6100	3854
2558	1748	5671	4974	7073	3273
⁴ 0117	~~1218~~	*¹⁴*0688	2756	7545	5426
8353	1554	4083	2029	8857	4781
1990	9886	3280	6109	9158	3034
9651	7870	2555	3518	2906	4900
9941	5617	1984	2435	5184	~~0379~~
7769	5785	9321	2734	2890	3105
3224	8379	9952	*¹⁹*0515	2724	4826
*⁵*1287	7275	6646	1378	6433	0005
6389	4191	4548	5546	6651	8248
1625	4327	2654	4129	3509	3217
7555	3020	4181	7498	4022	9122
4177	1844	3468	1389	3884	6900
⁶ 0927	*¹⁰* 0124	8176	0680	*²⁵*1056	~~1008~~
8505	1781	7155	3635	9751	5414
8022	8757	6275	1485	3635	2330
8390	8802	5674	2559	7934	4788
3630	5783	7762	*²⁰* 0223	5328	7731
9154	6388	6053	9633	2080	7269
1441	3381	7823	8767	9647	4445
8246	*¹¹* 0778	*¹⁵* 0993	6687	7212	9968
2730	3984	*¹⁶* 0563	9636	7202	~~0127~~
9196	8276	*¹⁷* 0233	*²¹* 0879	3385	2184
5928	9610	9161	*²²* 0748	3794	9683
⁷ 1042	9600	7122	2135	7868	5596
6552	4103	7957	~~0510~~	5958	*²⁹* 0211
5968	4307	9327	3197	~~0876~~	8480
4445	*¹²* 1018	4356	4653	9302	*³⁰* 0761
8727	8201	5980	7859	6055	1403
9415	9311	4996	2775	8509	7767
2648	7639	9128	*²³* 0341	6875	8957
3707	3454	8829	6863	*²⁶* 1297	5089

2. List the 30 random numbers and the favorite gas station for each dwelling unit to which that random number (identification number—variable 4) belongs (see Table 3–3).

3. Construct a frequency distribution from the sample data and estimate the population frequency distribution (see Table 3–4).

Table 3–4

Class	Sample frequency	Sample percentage	Estimated population frequency
Howie's Gulf	20	66.7	87*
Paul's Texaco	10	33.3	43†
Total	30	100.0	130

* 87 = 130 × .667.
† 43 = 130 × .333.

4. The memorandum appears in Exhibit 3–1.

Exhibit 3–1

```
HOWARD S. GITLOW, PH.D.
STATISTICAL CONSULTANT
_____


              MEMORANDUM


TO:    Mr. Howie, Owner
       Howie's Gulf Station

FROM:  Howard Gitlow

DATE:  March 31, 1980

RE:    Image study of Howie's Gulf Station


   As per your instructions and using commonly accepted
statistical techniques, I have found that approximately 87
Zone 1 dwelling units (66.7 percent of the dwelling units)
consider Howie's Gulf Station their favorite. The remaining
43 Zone 1 dwelling units (33.3 percent of the dwelling units)
prefer Paul's Texaco Station.*
   I direct your attention to the following table for a
summary of my findings.
```

	Percent	Estimated frequency
Howie's Gulf	66.7	87
Paul's Texaco	33.3	43
Total	100.0	130

```
   If you have any questions, please do not hesitate to call
me at 305-999-9999.

   * The above information was obtained through selecting a random
sample of 30 Stat City dwelling units.
```

Table 3–3

Random identification number (variable 4)	Favorite gas station (variable 20)*
37	1
107	2
1	2
11	1
128	1
92	2
104	2
121	1
115	1
12	2
77	2
101	2
36	2
68	2
99	2
56	1
23	1
54	2
51	2
22	1
87	1
74	2
34	2
20	2
105	2
129	2
100	2
45	1
21	2
76	2

* 1 = Paul's Texaco.
2 = Howie's Gulf.

Comments: A comparison of the actual and estimated population frequency distributions are shown in Table 3–5.

Table 3–5

Class	Actual	Estimated	Difference
Howie's Gulf	80	87	−7
Paul's Texaco	50	43	+7
Total	130	130	

It is easy to see that the actual and estimated population frequency distributions are different. This difference is due to sampling error; error introduced by using only 30 Zone 1 dwelling units rather than all 130 Zone 1 dwelling units.

It is important to realize that random samples and statistical procedures yield correct answers only on average, and that a particular sample's statistics may be very different from the parameters the sample was drawn to estimate. On average means that if many random samples were drawn and a statistic was computed from every sample, the average of all the statistics would yield the parameter the statistics were estimating.

Just for practice construct another sample frequency distribution based on a different sample of 30 dwelling units. The second sample frequency distribution is constructed as shown.

1. Draw 30 random numbers from a random number table. Use Table 3–6.

Table 3–6

Begin here →

5347	8111	9803	5122	5952	4023	4057	3935	4321	6925
9734	7032	5811	9196	2624	4464	8328	9739	9282	7757
6602	3827	7452	7111	8489	1395	9889	9231	6578	5964
9977	7572	0317	4311	8308	8198	1453	2616	2489	2055
3017	4897	9215	3841	4243	2663	8390	4472	6921	6911
8187	8333	1498	9993	1321	3017	4796	9379	8669	9885
1983	9063	7186	9505	5553	6090	8410	5534	4847	6379
0933	3343	5386	5276	1880	2582	9619	6651	7831	9701
3115	5829	4082	4133	2109	9388	4919	4487	4718	8142
6761	5251	0303	8169	1710	6498	6083	8531	4781	0807
6194	4879	1160	8304	2225	1183	0434	9554	2036	5593
0481	6489	9634	7906	2699	4396	6348	9357	8075	9658
0576	3960	5614	2551	8615	7865	0218	2971	0433	1567
7326	5687	4079	1394	9628	9018	4711	6680	6184	4468
5490	0997	7658	0264	3579	4453	6442	3544	2831	9900
4258	3633	6006	0404	2967	1634	4859	2554	6317	7522
2726	2740	9752	2333	3645	3369	2367	4588	4151	0475
4984	1144	6668	3605	3200	7860	3692	5996	6819	6258
2931	4046	2707	6923	5142	5851	4992	0390	2659	3306
3046	2785	6779	1683	7427	0579	0290	6349	0078	3509
2870	8408	6553	4425	3386	8253	9839	2638	0283	3683
1318	5065	9487	2825	7854	5528	3359	6196	5172	1421
6079	7663	3015	4029	9947	2833	1536	4248	6031	4277
1348	4691	6468	0741	7784	0190	4779	6579	4423	7723
3491	9450	3937	3418	5750	2251	0406	9451	4461	1048
2810	0481	8517	8649	3569	0348	5731	6317	7190	7118
5923	4502	0117	0884	8192	7149	9540	3404	0485	6591
8743	8275	7109	3683	5358	2598	4600	4284	8168	2145
2904	0130	5534	6573	7871	4364	4624	5320	9486	4871
6203	7188	9450	1526	6143	1036	4205	6825	1438	7943
3885	8004	5997	7336	5287	4767	4102	8229	2643	8737
4066	4332	8737	8641	9584	2559	5413	9418	4230	0736
4058	9008	3772	0866	3725	2031	5331	5098	3290	3209
7823	8655	5027	2043	0024	0230	7102	4993	2324	0086
9824	6747	7145	6954	0116	0332	6701	9254	9797	5272
6997	7855	6543	3262	2831	6181	1459	7972	5569	9134
3984	2307	4081	0371	2189	9635	9680	2459	2620	2600
6288	8727	9989	9996	3437	4255	1167	9960	9801	4886
5613	6492	2945	5296	8662	6242	3016	7618	9531	3926
9080	5602	4899	6456	6746	6018	1297	0384	6258	9385
0966	4467	7476	3335	6730	8054	9765	1134	7877	4501
3475	5040	7663	1276	3222	3454	1810	5351	1452	7212
1215	7332	7419	2666	7808	5363	5230	0000	0570	6353
6938	0773	9445	7642	1612	0930	6741	6858	8793	3884
9335	6456	4376	4504	4493	6997	1696	0827	6775	6029
3887	3554	9956	8540	0491	6254	7840	0101	8618	2207
5831	6029	7239	6966	1247	9305	0205	2980	6364	1279
8356	1022	9947	7472	2207	1023	2157	2032	2131	5712
2806	9115	4056	3370	6451	0706	6437	2633	7965	3114
0513	7555	9316	8092	5587	5410	3480	8315	0453	8136

2. List the 30 random numbers and the favorite gas station for each dwelling unit to which that random number (identification number— variable 4) belongs (Table 3–7).

Table 3–7

Random identification number (variable 4)	Favorite gas station (variable 20)*
93	2
48	2
57	2
96	1
121	1
99	2
114	2
13	2
77	2
102	2
31	1
30	1
116	2
11	1
122	2
26	1
40	1
74	2
88	2
86	1
37	1
127	2
2	1
49	1
124	1
118	2
19	2
34	2
103	1
23	1

* 1 = Paul's Texaco.
  2 = Howie's Gulf.

3. Construct a frequency distribution from the sample data and estimate the population frequency distribution (Table 3–8).

Table 3–8

Class	Sample frequency	Sample percentage	Estimated population frequency
Howie's Gulf	16	53.3	69*
Paul's Texaco	14	46.7	61†
Total	30	100.0	130

* 69 = 130 × .533
† 61 = 130 × .467

Comments: A comparison of the two sample percentages (percent of dwelling units favoring Howie's Gulf) with the population percentage demonstrates that sample statistics differ from sample to sample (66.7 percent versus 53.3 percent), and from the population percentage (61.5 percent). Not all statistics accurately reflect parameters. For example, the first sample percentage (66.7 percent) was reasonably close to the population percentage (61.5 percent). But, the second sample percentage (53.3 percent) reflected a wide variation from the population percentage (61.5 percent).

Later in this course you will learn about techniques for controlling the accuracy of statistics when estimating parameters.

Example 3–2   Distribution of rooms per dwelling unit in Zone 2

Problem

The Stat City Tax Assessor's Office is planning to reassess all dwelling units in Zone 2. The tax office must determine the distribution of rooms per dwelling unit to set the new assessment rates. With the above information the tax office can set assessment guidelines to yield a target total tax.

You have been hired by the tax office to determine the distribution of rooms per dwelling unit in Zone 2. Conduct the necessary survey and write a summary of your findings to the tax office (see Exhibit 3–2). Include a relative frequency polygon in your memorandum.

Note to the student: Conduct the above survey twice. First, conduct a survey investigating all 157 dwelling units in Zone 2. Next, conduct a survey using a random sample of 30 dwelling units. Finally, compare the results of your two surveys. Do you see anything surprising in the comparison?

Part 1   The population frequency distribution can be seen in Table 3–9.

Solution

Table 3–9

Number of rooms	Frequency	Percent
5	7	4.5
6	15	9.6
7	37	23.6
8	49	31.2
9	36	22.9
10	10	6.4
11	3	1.9
Total	157	100.0*

* Percentages do not total 100 percent due to rounding.

Part 2   A sample frequency distribution based on 30 randomly selected dwelling units is constructed as shown.

1.   Draw 30 random numbers from a random number table using the rules set down in Chapter 1 (use Table 3–10).

Table 3–10   Begin here →

5347	8111	9803	1221	5952	4023	4057	3935	4321	6925
9734	7032	5811	9196	*21* 2624	4464	8328	9739	9282	7757
6602	3827	7452	7111	8489	1395	9889	9231	6578	5964
9977	7572	0317	4311	8308	8198	1453	2616	2489	2055
3017	4897	9215	3841	4243	2663	8390	4472	6921	6911
8187	8333	*11* 1498	9993	*22* 1321	3017	4796	9379	8669	9885
1 1983	9063	7186	9505	5553	6090	8410	5534	4847	6379
0933	3343	5386	5276	*23* 1880	2582	9619	6651	7831	9701
3115	5829	4082	4133	*24* 2109	9388	4919	4487	4718	8142
6761	5251	0303	8169	*25* 1710	6498	6083	8531	4781	0807
6194	4879	1160	8304	*26* 2225	1183	0434	9554	2036	5593
0481	6489	9634	7906	*27* 2699	4396	6348	9357	8075	9658
0576	3960	5614	*13* 2551	8615	7865	0218	2971	0433	1567
7326	5687	4079	*14* 1394	9628	9018	4711	6680	6184	4468
5490	0997	7658	0264	3579	4453	6442	3544	2831	9900
4258	3633	6006	0404	2967	1634	4859	2554	6317	7522
2 2726	*8* 2740	9752	*15* 2333	3645	3369	2367	4588	4151	0475
4984	1144	6668	3605	3200	7860	3692	5996	6819	6258
2931	4046	*12* 2707	6923	5142	5851	4992	0390	2659	3306
3046	*9* 2785	6779	*16* 1683	7427	0579	0290	6349	0078	3509
3 2870	8408	6553	4425	3386	8253	9839	2638	0283	3683
4 1318	5065	9487	*17* 2825	7854	5528	3359	6196	5172	1421
6079	7663	3015	4029	9947	2833	1536	4248	6031	4277
5 1348	4691	6468	0741	7784	0190	4779	6579	4423	7723
3491	9450	3937	3418	5750	2251	0406	9451	4461	1048
6 2810	0481	8517	8649	3569	0348	5731	6317	7190	7118
5923	4502	0117	0884	8192	7149	9540	3404	0485	6591
8743	8275	7109	3683	5358	2598	4600	4284	8168	2145
2904	0130	5534	6573	7871	4364	4624	5320	9486	4871
6203	7188	9450	*18* 1526	6143	1036	4205	6825	1438	7943
3885	8004	5997	7336	5287	4767	4102	8229	2643	8737
4066	4332	8737	8641	9584	2559	5413	9418	4230	0736
4058	9008	3772	0866	3725	2031	5331	5098	3290	3209
7823	8655	5027	*19* 2043	0024	0230	7102	4993	2324	0086
9824	6747	7145	6954	0116	0332	6701	9254	9797	5272
6997	7855	6543	3262	*28* 2831	6181	1459	7972	5569	9134
3984	*10* 2307	4081	0371	*29* 2189	9635	9680	2459	2620	2600
6288	8727	9989	9996	3437	4255	1167	9960	9801	4886
5613	6492	2945	5296	8662	6242	3016	7618	9531	3926
9080	5602	4899	6456	6746	6018	1297	0384	6258	9385
0966	4467	7476	3335	6730	8054	9765	1134	7877	4501
3475	5040	7663	1276	3222	3454	1810	5351	1452	7212
1215	7332	7419	*20* 2666	7808	5363	5230	0000	0570	6353
6938	0773	9445	7642	*30* 1612	0930	6741	6858	8793	3884
9335	6456	4376	4504	4493	6997	1696	0827	6775	6029
3887	3554	9956	8540	0491	6254	7840	0101	8618	2207
5831	6029	7239	6966	1247	9305	0205	2980	6364	1279
8356	1022	9947	7472	2207	1023	2157	2032	2131	5712
7 2806	9115	4056	3370	6451	0706	6437	2633	7965	3114
0573	7555	9316	8092	5587	5410	3480	8315	0453	8136

Table 3–11

Random identification number (*variable 4*)	Number of rooms (*variable 9*)
198	6
272	9
287	9
131	6
134	7
281	9
280	8
274	7
278	10
230	8
149	8
270	5
255	7
139	9
233	8
168	7
282	9
152	8
204	8
266	7
262	5
132	7
188	8
210	8
171	10
222	8
269	7
283	8
218	7
161	8

2. List the 30 random numbers and the number of rooms for each dwelling unit to which that random (identification number—variable 4) number belongs (Table 3–11).

3. Construct a frequency distribution from the sample data and estimate the population frequency distribution (Table 3–12).

Table 3–12

Number of rooms	Sample frequency	Sample percentage	Estimated population frequency
5	2	6.7	10
6	2	6.7	10
7	8	26.7	42
8	11	36.7	58
9	5	16.7	26
10	2	6.7	10
Total	30	100.0*	157†

* Percentages do not add up to 100 due to rounding.
† Estimated frequencies do not add up to 157 due to rounding.

4. The memorandum appears in Exhibit 3–2.

Exhibit 3–2

HOWARD S. GITLOW, PH.D.
STATISTICAL CONSULTANT
———

MEMORANDUM

TO: Stat City Tax Assessor's Office

FROM: Howard Gitlow

DATE: September 22, 1980

RE: Distribution of rooms per dwelling unit in Zone 2 of
        Stat City, as of January 1980

   As per your instructions and using commonly accepted statistical techniques, I have estimated the distribution of rooms per dwelling unit in Zone 2 of Stat City. I direct your attention to the chart below for a summary of my findings.

Exhibit 3–2 (*continued*)

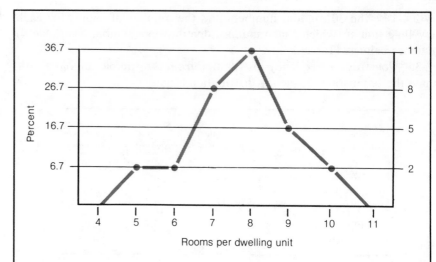

The above information was obtained through selecting a
sample of 30 Zone 2 dwelling units.

If you have any further questions, please do not hesitate
to call me at 305-999-9999.

Comments: A comparison of the actual and estimated population frequency distributions is shown in Table 3–13.

Table 3–13

Number of rooms	Actual frequency	Estimated population frequency	Difference = actual − estimated
5	7	10	−3
6	15	10	+5
7	37	42	−5
8	49	58	−9
9	36	26	+10
10	10	10	0
11	3	0	+3
Total	157	157*	

* Estimated frequencies do not add up to 157 due to rounding.

It is apparent that the actual and estimated population frequency distributions are different. This difference is once again due to sampling error. However, you should notice that eight rooms is the most commonly occurring class in both distributions and the number of rooms trails off on either side of eight rooms in both distributions. Consequently, the final picture we get from the sample is not as inaccurate as we had originally thought. Both the actual and the estimated frequency distributions tell us that the typical number of rooms per dwelling unit in Zone 2 is approximately eight.

Remember, random samples and statistical procedures yield correct answers only on average, and that a particular sample's statistics may be very different from the parameters the sample was drawn to estimate.

Just for practice construct another sample frequency distribution based on a different sample of 30 dwelling units. The second sample frequency distribution is constructed as shown.

1. Draw 30 random numbers from Table 3–14.

Table 3–14

Begin here →

2668	7422	4354	4569	9446	8212	3737	2396	6892	3766
6067	7516	2451	1510	0201	1437	6518	1063	6442	6674
4541	9863	8312	9855	0995	6025	4207	4093	9799	9308
6987	4802	8975	2847	4413	5997	9106	2876	8596	7717
0376	8636	9953	4418	2388	8997	1196	5158	1803	5623
8468	5763	3232	1986	7134	4200	9699	8437	2799	2145
9151	4967	3255	8518	2802	8815	6289	9549	2942	3813
1073	4930	1830	2224	2246	1000	9315	6698	4491	3046
5487	1967	5836	2090	3832	0002	9844	3742	2289	3763
4896	4957	6536	7430	6208	3929	1030	2317	7421	3227
9143	7911	0368	0541	2302	5473	9155	0625	1870	1890
9256	2956	4747	6280	7342	0453	8639	1216	5964	9772
4173	1219	7744	9241	6354	4211	8497	1245	3313	4846
2525	7811	5417	7824	0922	8752	3537	9069	5417	0856
9165	1156	6603	2852	8370	0995	7661	8811	7835	5087
0014	8474	6322	5053	5015	6043	0482	4957	8904	1616
5325	7320	8406	5962	6100	3854	0575	0617	8019	2646
2558	1748	5671	4974	7073	3273	6036	1410	5257	3939
0117	1218	0688	2756	7545	5426	3856	8905	9691	8890
8353	1554	4083	2029	8857	4781	9654	7946	7866	2535
1990	9886	3280	6109	9158	3034	8490	6404	6775	8763
9651	7870	2555	3518	2906	4900	2984	6894	5050	4586
9941	5617	1984	2435	5184	0379	7212	5795	0836	4319
7769	5785	9321	2234	2890	3105	6581	2163	4938	7540
3224	8379	9952	0515	2724	4826	6215	6246	9704	1651
1287	7275	6646	1378	6433	0005	7332	0392	1319	1946
6389	4191	4548	5546	6651	8248	7469	0786	0972	7649
1625	4327	2654	4129	3509	3217	7062	6640	0105	4422
7555	3020	4181	7498	4022	9122	6423	7301	8310	9204
4177	1844	3468	1389	3884	6900	1036	8412	0881	6678
0927	0124	8176	0680	1056	1008	1748	0547	8227	0690
8505	1781	7155	3635	9751	5414	5113	8316	2737	6860
8022	8757	6275	1485	3635	2330	7045	2206	6381	2986
8390	8802	5674	2559	7934	4788	7791	5202	8430	0289
3630	5783	7762	0223	5328	7731	4010	3845	9221	5427
9154	6388	6053	9633	2080	7269	0894	0287	7489	2259
1441	3381	7823	8767	9647	4445	2509	2929	5067	0779
8246	0778	0993	6687	7212	9968	8432	1453	0841	4595
2730	3984	0563	9636	7202	0127	9283	4009	3177	4182
9196	8276	0233	0879	3385	2184	1739	5375	5807	4849
5928	9610	9161	0748	3794	9683	1544	1209	3669	5831
1042	9600	7122	2135	7868	5596	3551	9480	2342	0449
6552	4103	7957	0510	5958	0211	3344	5678	1840	3627
5968	4307	9327	3197	0876	8480	5066	1852	8323	5060
4445	1018	4356	4653	9302	0761	1291	6093	5340	1840
8727	8201	5980	7859	6055	1403	1209	9547	4273	0857
9415	9311	4996	2775	8509	7767	6930	6632	7781	2279
2648	7639	9128	0341	6875	8957	6646	9783	6668	0317
3707	3454	8829	6863	1297	5089	1002	2722	0578	7753
8383	8957	5595	9395	3036	4767	8300	3505	0710	6307

Table 3–15

Random identification number (variable 4)	Number of rooms (variable 9)
266	7
252	9
255	7
199	9
162	8
144	7
273	6
264	10
196	8
174	9
155	8
184	9
178	9
245	6
183	7
198	6
265	7
151	8
284	8
222	8
209	9
285	7
275	9
202	11
243	8
137	8
138	8
148	9
213	7
277	9

2. List the 30 random numbers and the number of rooms for each dwelling unit to which that random number (identification number—variable 4) belongs (Table 3–15).

3. Construct a frequency distribution from the sample data and estimate the population frequency distribution (Table 3–16).

Table 3–16

Number of rooms	Sample frequencies	Sample percent	Estimated population frequencies
6	3	10.00	16
7	7	23.34	37
8	9	30.00	47
9	9	30.00	47
10	1	3.33	5
11	1	3.33	5
Total	30	100.00	157

Comments: A comparison of the actual and both estimated population frequency distributions is shown in Table 3–17.

Table 3–17

Number of rooms	Actual frequencies	First estimated population frequencies	Second estimated population frequencies
5	7	10	0
6	15	10	16
7	37	42	37
8	49	58	47
9	36	26	47
10	10	10	5
11	3	0	5
Total	157	157*	157

* Estimated frequencies do not add up to 157 due to rounding.

Once again, a comparison of the second sample's estimated population frequencies and the actual population frequencies show a difference in the breakdown of the number of rooms per dwelling unit in Zone 2. Hence, the second sample's estimated population frequency distribution is also not a perfect estimator of the actual population frequency distribution, a consequence of sampling error. However, once again, both the estimated and actual frequency distributions center around eight rooms per dwelling unit. Consequently, both samples yield a reasonably accurate picture of the typical number of rooms per dwelling unit in Zone 2.

Remember, it is important that you do *not* develop a dependence on parameters because they are obtainable rarely in real business situations.

Example 3–3   Analysis of electric bills in Stat City

Problem   Mr. Saul Reisman, the chief financial officer of the Stat City Electric Company, needs to know the distribution of average monthly electric bills for Stat City families. In particular, Mr. Reisman needs to know the number of Stat City dwelling units with average monthly electric bills in the following categories: $10 to less than $40, $40 to less than $70, $70 to less than $100, and $100 to less than $130. If any category contains more than 500 Stat City dwelling units, the electric company must change the capacity of their generators.

Conduct a survey and write a summary of your findings to Mr. Reisman (see Exhibit 3–4). Use a histogram and frequency polygon in your report.

Note to student: Draw a random sample of 30 Stat City dwelling units to perform your appointed task. In this problem you are *not* being asked to compute parameters; consequently, you must rely on statistics for decision-making purposes. From now on we will only work with statistics and will draw inferences about the unknown parameters. Reliance on statistics, and nonreliance on parameters, is a big step into the world of statistics.

It will be assumed from now on that you know how to use a random number table. Consequently, in most problems you will not be required to actually select the random identification numbers used in obtaining your sample. Instead, the random numbers to be used in drawing your random samples will be listed in the problems. The only reason for doing this is to save time and effort that could more wisely be invested elsewhere.

Table 3–18 lists the random sample of 30 dwelling units' identification numbers (variable 4) to be used in solving Mr. Reisman's problem. Remember, the numbers are just random identification numbers. You must still draw the sample of 30 average monthly electric bills (variable 11) from Stat City.

Table 3–18

4,	37,	163,	164,	191,
208,	391,	395,	408,	485,
511,	540,	541,	632,	780,
805,	815,	923,	954,	968,
973,	1038,	1060,	1063,	1153,
1188,	1206,	1209,	1281,	1291,

A sample frequency distribution based on the 30 randomly selected dwelling units is constructed as shown.

Solution

1. List the 30 random identification numbers and the average monthly electric bills for each dwelling unit to which those random identification numbers belong (Table 3–19).

Table 3–19

Random identification number (variable 4)	Average monthly electric bill (variable 11)
4	$78
37	83
163	66
164	70
191	41
208	75
391	73
395	97
408	48
485	64
511	47
540	72
541	80
632	24
780	40
805	37
815	31
923	46
954	54
968	48
973	43
1038	38
1060	54
1063	34
1158	29
1188	38
1206	56
1209	34
1281	37
1291	35

2. Construct a sample frequency distribution and estimate the population frequency distribution, population histogram, and population frequency polygon (Table 3–20 and Exhibit 3–3).

Table 3–20

Class	Sample frequency	Sample percentage	Estimated population frequency
$ 10 to less than $ 40	10	33.3	.333 × 1,373 = 458
40 to less than 70	12	40.0	.400 × 1,373 = 549
70 to less than 100	8	26.7	.267 × 1,373 = 366
100 to less than 130	0	0.0	0
Total	30	100.0	1,372

Exhibit 3–3

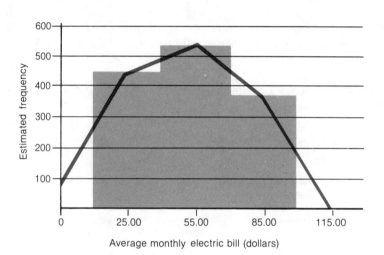

3. The memorandum appears in Exhibit 3–4.

Exhibit 3–4

HOWARD S. GITLOW, PH.D.
STATISTICAL CONSULTANT

MEMORANDUM

TO: Mr. Saul Reisman, Chief Financial Officer
        Stat City Electric Company

FROM: Howard Gitlow

DATE: April 15, 1980

RE: Average monthly electric bills in Stat City, as of
        January 1980

As per your instructions and using commonly accepted statistical techniques, I have estimated the distribution of average monthly electric bills in Stat City. I direct your attention to the charts below for a summary of my findings. I am enclosing two pictorial representations of the same information so that you may select the one which best meets your needs.

Exhibit 3–4 (*continued*)

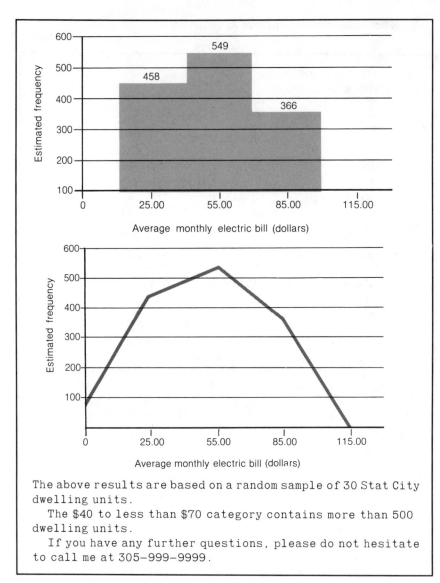

The above results are based on a random sample of 30 Stat City dwelling units.

The $40 to less than $70 category contains more than 500 dwelling units.

If you have any further questions, please do not hesitate to call me at 305-999-9999.

Comments: This is the first time you do not have parameters (population frequency distribution) to compare with your statistics (sample frequency distribution). Consequently, you do not know if your random sample of 30 dwelling units is giving you a true picture of Stat City average monthly electric bills. WELCOME TO THE WORLD OF DECISION MAKING!

ADDITIONAL PROBLEMS

3–4 Image study of the Grand Union Mr. Hector Chavez (ID = 251), manager of the Stat City Grand Union, is interested in conducting an image study of consumers' perceptions of Stat City supermarkets so he can prepare his annual marketing plan. Basically, Mr. Chavez wants to know the number of Stat City dwelling units that favor the Grand Union. Also, Mr. Chavez wants to know the number of Stat City dwelling units that favor the Food Fair and the A&P.

You have been hired as a marketing research consultant by Mr. Chavez to perform the above study. Type a memorandum summarizing your results for Mr. Chavez.

In order that everyone in your class will arrive at the same results, use the following list of randomly selected dwelling unit identification numbers (variable 4).

634, 933, 1008, 47, 178, 579, 1039, 278, 420, 1007, 222,
484, 1363, 620, 709, 1299, 1279, 573, 628, 552, 819, 783,
433, 809, 939, 990, 972, 535, 1353, 575.

Remember, the above numbers are just random identification numbers. You must still draw the sample of 30 favorite supermarket designations (variable 24) from the Stat City Data Base.

3–5 Heating bill survey Mr. Saul Reisman, chief financial officer of the Stat City Electric Company, needs to know the distribution of average yearly heating bills for Stat City residents. In particular, Mr. Reisman needs to know the percentage of dwelling units with average yearly heating bills in the following categories: under $500, $500–$899, and $900 and over. If any category contains more than 45 percent of the dwelling units in Stat City, the electric company must change their master plan for providing service to Stat City residents.

Conduct a survey and type a summary of your findings to Mr. Reisman.

In order that everyone in your class will arrive at the same results, use the following randomly selected list of dwelling unit identification numbers (variable 4):

279, 1131, 985, 222, 127, 212, 643, 1224, 881, 154,
254, 1330, 287, 441, 1317, 543, 70, 920, 1119, 957,
859, 1180, 1138, 536, 808, 1367, 280, 489, 1178, 996.

Remember, the above numbers are just random identification numbers. You must still draw the sample of 30 average yearly heating bills (variable 10) from the Stat City data base.

3–6 Widening 12th Street between 9th and 13th Avenues The Stat City Chamber of Commerce is contemplating widening 12th Street between 9th and 13th Avenues in answer to residents' complaints concerning congestion caused by on-street parking.

You have been retained by the Stat City Chamber of Commerce to determine how many cars are owned per family in the apartment buildings bordering the proposed construction. If more than 20 percent of the dwelling units along the proposed construction area own two or more cars, then the construction should be performed. Conduct the necessary survey and write a summary of your findings to the Stat City Chamber of Commerce.

Note to the student This problem uses addresses to define the relevant population.

In order that everyone in your class will come up with the same sample frequency distribution, use the random number table (shown in margin) to draw 20 random identification numbers (variable 4).

0918	7703	6832	1273
2827	0907	8327	4883
2184	3554	1263	3477
2202	7955	1969	2525
0904	2077	0908	0916
0731	7112	0742	2051
0922	0901	2025	0931
2260	5671	2178	6878
3669	0751	3427	6549
5980	6998	7180	6498
5647	0905	5851	9540
0756	3487	8320	0313
7448	3399	0917	9258
1979	2248	9423	8105
1194	2785	6052	2414
0921	6063	6570	7788
6927	2320	6103	8426
8365	0740	0737	4108
7861	5110	6682	4043
1498	7415	7821	3898

3–7 Distribution of telephone bills in Stat City Mr. Jack Davis, chairman of the Stat City Telephone Company, needs to know the distribution of average monthly telephone bills for Stat City families. In particular, Mr. Davis needs to know the number of Stat City dwelling units with average monthly telephone bills in the following categories: under $30, $30 to less than $60, $60 to less than $90, $90 to less than $120, and $120 to less than $150. If any category contains more than 500 of the dwelling units in Stat City, the telephone company must change the capacity of their switching stations.

Conduct a survey and type a summary of your findings to Mr. Davis. Use a histogram and a frequency polygon in your report.

Please use the following random sample of 30 dwelling units' identification numbers (variable 4).

72,	86,	180,	195,	271,
285,	322,	340,	426,	432,
468,	520,	611,	616,	620,
759,	853,	855,	941,	1061,
1078,	1104,	1180,	1184,	1217,
1222,	1240,	1246,	1264,	1331.

Remember, the above numbers are just random identification numbers. You must still draw the sample of 30 average monthly telephone bills (variable 12) from the Stat City data base.

COMPUTATIONAL EXERCISE   **3–8** Mr. Kaplowitz, the mayor of Stat City, is preparing a report describing several characteristics of Stat City families. The characteristics to be included in the report are: housing cost (rent or mortgage) as of January 1980 (HCOST), average monthly telephone bill as of January 1980 (PHONE), total 1979 family income (INCOM), average bimonthly automobile gasoline bill as of January 1980 (GAS), and average weekly supermarket bill as of January 1980 (EAT). The mayor has hired you to conduct a survey of 50 Stat City dwelling units and to construct frequency distributions, histograms, and frequency polygons for each of the aforementioned variables.

The mayor has specified the following class limits for each variable:

HCOST ($0 to less than $250)
($250 to less than $500)
($500 to less than $750)
($750 to less than $1,000)
($1,000 to less than $1,250)
($1,250 to less than $1,500)

PHONE ($0 to less than $25)
($25 to less than $50)
($50 to less than $75)
($75 to less than $100)
($100 to less than $125)
($125 to less than $150)

INCOM ($0 to less than $25,000)
($25,000 to less than $50,000)
($50,000 to less than $75,000)
($75,000 to less than $100,000)
($100,000 to less than $125,000)
($125,000 to less than $150,000)
($150,000 to less than $175,000)

GAS ($0 to less than $50)
($50 to less than $100)
($100 to less than $150)
($150 to less than $200)
($200 to less than $250)

EAT ($0 to less than $50)
($50 to less than $100)
($100 to less than $150)
($150 to less than $200)
($200 to less than $250)

Use the accompanying simple random sample of 50 Stat City dwelling units.

Random sample for Exercise 3–8

CASE-N	ID	HCOST	PHONE	INCOM	PEPLE	GAS	GASTR	REPAR	EAT	LSODA	HSODA
1	6.	340.	24.	45454.	2.	133.	3.	1.	50.	0.	0.
2	47.	610.	20.	51671.	4.	162.	5.	2.	80.	0.	0.
3	53.	433.	25.	43451.	1.	65.	6.	1.	27.	2.	0.
4	68.	305.	21.	41836.	1.	65.	4.	1.	27.	1.	0.
5	88.	359.	19.	40283.	2.	99.	2.	2.	41.	1.	0.
6	92.	629.	19.	52548.	5.	190.	8.	1.	97.	0.	1.
7	164.	827.	25.	68773.	3.	175.	3.	1.	71.	0.	1.
8	167.	975.	47.	91910.	5.	211.	5.	2.	112.	0.	5.
9	291.	505.	68.	21086.	4.	176.	8.	2.	80.	2.	0.
10	459.	424.	90.	42806.	6.	120.	6.	0.	124.	0.	0.
11	469.	370.	16.	36907.	5.	197.	6.	1.	99.	0.	0.
12	477.	379.	20.	29803.	6.	222.	11.	1.	111.	0.	0.
13	490.	433.	16.	27809.	4.	163.	6.	1.	84.	0.	0.
14	498.	424.	19.	41587.	5.	231.	4.	0.	117.	0.	0.
15	528.	315.	34.	41557.	5.	199.	9.	1.	102.	2.	0.
16	538.	589.	15.	42033.	7.	165.	5.	2.	170.	2.	0.
17	539.	279.	17.	49245.	3.	142.	6.	0.	77.	2.	0.
18	588.	396.	21.	44824.	4.	160.	7.	1.	88.	0.	1.
19	597.	346.	67.	48075.	4.	183.	2.	1.	84.	0.	4.
20	599.	366.	19.	34509.	4.	173.	6.	2.	90.	2.	0.
21	616.	484.	48.	27071.	6.	217.	5.	2.	138.	0.	2.
22	657.	450.	20.	28604.	4.	214.	2.	0.	94.	0.	0.
23	684.	667.	20.	12539.	5.	65.	4.	1.	117.	0.	4.
24	705.	425.	80.	16726.	4.	65.	6.	1.	74.	0.	0.
25	724.	518.	20.	17943.	2.	120.	6.	1.	53.	0.	0.
26	764.	143.	81.	8102.	2.	120.	3.	0.	55.	2.	0.
27	788.	286.	63.	7548.	7.	120.	3.	1.	99.	2.	0.
28	795.	333.	73.	9558.	5.	65.	4.	1.	107.	2.	0.
29	797.	278.	19.	13705.	4.	65.	3.	1.	94.	2.	0.
30	816.	603.	16.	25718.	4.	179.	8.	1.	97.	0.	5.
31	835.	483.	12.	13661.	3.	120.	7.	0.	69.	0.	1.
32	854.	395.	23.	12097.	7.	65.	5.	1.	150.	0.	0.
33	873.	485.	25.	11509.	6.	120.	7.	1.	97.	2.	0.
34	897.	340.	26.	11969.	7.	65.	4.	1.	138.	0.	0.
35	962.	365.	10.	7852.	6.	120.	3.	0.	139.	2.	0.
36	970.	337.	14.	21258.	6.	223.	12.	1.	126.	0.	0.
37	996.	491.	10.	18157.	4.	65.	4.	1.	102.	2.	0.
38	1073.	359.	71.	7306.	6.	120.	7.	0.	138.	0.	0.
39	1086.	222.	23.	20539.	5.	191.	5.	1.	88.	0.	0.
40	1092.	525.	13.	26576.	4.	149.	6.	2.	78.	0.	0.
41	1134.	468.	17.	17395.	8.	65.	4.	1.	163.	2.	0.
42	1136.	424.	105.	23923.	8.	120.	5.	2.	162.	0.	5.
43	1192.	405.	16.	5786.	7.	65.	2.	1.	124.	0.	0.
44	1215.	538.	20.	21094.	5.	207.	7.	0.	123.	2.	0.
45	1244.	352.	23.	16095.	3.	65.	2.	1.	78.	2.	1.
46	1279.	313.	36.	5555.	2.	120.	6.	1.	45.	0.	0.
47	1297.	583.	15.	18922.	5.	65.	6.	0.	109.	1.	0.
48	1299.	305.	76.	29004.	4.	160.	7.	2.	85.	0.	6.
49	1320.	450.	10.	19193.	7.	120.	4.	1.	142.	0.	0.
50	1348.	611.	19.	19695.	4.	65.	3.	0.	72.	2.	0.

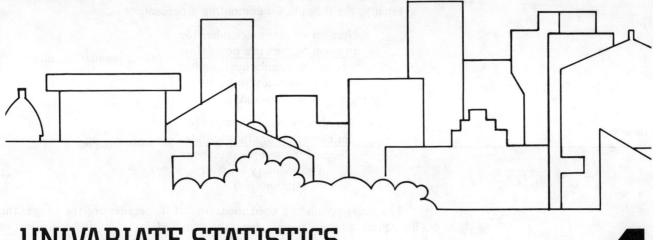

UNIVARIATE STATISTICS

4

In the last chapter you learned how to summarize large amounts of data using frequency distributions, histograms, and frequency polygons. In this chapter you will work with more frugal statistical techniques for summarizing data. These statistical techniques fall into three distinct categories: measures of location, measures of dispersion, and measures of shape.

There are many measures of location that can be used to summarize data; for example, the arithmetic mean (simple average), the geometric mean, the harmonic mean, the weighted mean, the mode, the median, percentages and quantiles. In this chapter you will only work with the arithmetic mean, the mode, the median (for ungrouped data), and percentages.

The formula that most textbooks use for the arithmetic mean (of ungrouped data) is:

$$\mu = \sum_{i=1}^{N} X_i / N \qquad \text{for a population, and}$$

$$\bar{x} = \sum_{i=1}^{n} x_i / n \qquad \text{for a sample.}$$

The formula for the mode of ungrouped data is too complex to mention here; consequently, consider the mode the most frequently occurring number in a population or a sample. Also, the formula for the median of ungrouped data is too complex to mention here, hence, consider the median the middlemost data point in a population or sample in which the numbers have been ranked from lowest to highest. One final point about the median is that its position in a ranked sequence of numbers is:

$$\frac{N + 1}{2} \qquad \text{for a population, and}$$

$$\frac{n + 1}{2} \qquad \text{for a sample.}$$

Be careful, the position of the median is not necessarily the median.

Finally, the formula for computing a percentage is:

$$\pi = \frac{\text{Number of times a number (or an event) occurs in a population}}{\text{Size of the population (number of elementary units in the population)}} \quad \text{for a population, and}$$

$$\bar{p} = \frac{\text{Number of times a number (or an event) occurs in a sample}}{\text{Size of the sample (number of elementary units in the sample)}} \quad \text{for a sample.}$$

The most commonly used measures of dispersion are the range, the interquartile range, the variance, the standard deviation, the coefficient of variation, sums of squares, and the average deviation. In this chapter you will work with the range and the standard deviation of ungrouped data. The formula that most textbooks use for the range (of ungrouped data) is:

$R = \text{MAXIMUM} - \text{MINIMUM}$ for a population, and
$r = \text{maximum} - \text{minimum}$ for a sample (lower case letters indicate sample values).

The formulas that most textbooks use for the standard deviation (of ungrouped data) are shown in Table 4–1.

This book will cover measures of shape, such as skewness and kurtosis, in the "Computational Exercises" section.

The structure of this chapter will be very much like Chapter 3. I will present several Stat City related examples worked out in detail and several additional problems to solve for Stat City interest groups.

Example 4–1 Revision of the report on the distribution of rooms per dwelling unit in Zone 2 of Stat City

The Stat City Tax Assessor's Office wants you to rewrite your report and include the mean, mode, median, minimum, maximum, range, and standard deviation of the number of rooms per dwelling unit in Zone 2 (see Exhibit 4–1).

Note to the student: Please use the same sample that you used in constructing your original report to the Stat City Tax Assessor's Office (Table 3–11). This random sample is repeated in Table 4–2 for your convenience.

Table 4–1

Population formulas

$$\sigma = \sqrt{\frac{\sum_{i=1}^{N}(X_i - \mu)^2}{N}}$$

or

$$\sigma = \sqrt{\frac{\sum_{i=1}^{N}(X_i^2) - \frac{(\sum_{i=1}^{N}X_i)^2}{N}}{N}}$$

Sample formulas

$$s = \sqrt{\frac{\sum_{i=1}^{n}(x_i - \bar{x})^2}{n - 1}}$$

or

$$s = \sqrt{\frac{\sum_{i=1}^{n}(x_i^2) - \frac{[\sum_{i=1}^{n}(x_i)]^2}{n}}{n - 1}}$$

Table 4–2

Random identification number (variable 4)	Number of rooms (variable 9)
198	6
272	9
287	9
131	6
134	7
281	9
280	8
274	7
278	10
230	8
149	8
270	5
255	7
139	9
233	8
168	7
282	9
152	8
204	8
266	7
262	5
132	7
188	8
210	8
171	10
222	8
269	7
283	8
218	7
161	8

The sample statistics are shown below.   **Solution**

Mean[1]	7.700
Mode[2]	8.000
Median[3]	8.000
Minimum	5.000
Maximum	10.000
Range[4]	5.000
Standard deviation[5]	1.236

[1] Mean $= x = \dfrac{\Sigma x}{n} = \dfrac{231}{30} = 7.7$.

[2] Mode $= 8$.

Rooms per dwelling unit (variable 9)	Frequency
5	2
6	2
7	8
8	11*
9	5
10	2
Total	30

* There are more dwelling units with 8 rooms than any other number of rooms.

[3] Median position $= \dfrac{30 + 1}{2} = 15.5$.

5, 5, 6, 6, 7, 7, 7, 7, 7, 7, 7, 7, 8, 8, ⟨8, 8⟩, 8, 8, 8, 8, 8, 8, 9, 9, 9, 9, 9, 10, 10

Median $= \dfrac{8 + 8}{2} = \dfrac{16}{2} = 8$.

[4] Range $= 10 - 5 = 5$.

[5] $s = 1.236$. (See Table 4–3).

Table 4–3

Random identification number	Number of rooms (x)	$(x - \bar{x}) = (x - 7.7)$	$(x - 7.7)^2$
198	6	−1.7	2.89
272	9	1.3	1.69
287	9	1.3	1.69
131	6	−1.7	2.89
134	7	−.7	.49
281	9	1.3	1.69
280	8	.3	.09
274	7	−.7	.49
278	10	2.3	5.29
230	8	.3	.09
149	8	.3	.09
270	5	−2.7	7.29
255	7	−.7	.49
139	9	1.3	1.69
233	8	.3	.09
168	7	−.7	.49
282	9	1.3	1.69
152	8	.3	.09
204	8	.3	.09
266	7	−.7	.49
262	5	−2.7	7.29
132	7	−.7	.49
188	8	.3	.09
210	8	.3	.09
171	10	2.3	5.29
222	8	.3	.09
269	7	−.7	.49
283	8	.3	.09
218	7	−.7	.49
161	8	.3	.09

$\Sigma(x - \bar{x})^2 = 44.3$

$$s = \sqrt{\frac{\Sigma(x - \bar{x})^2}{n - 1}} = \sqrt{\frac{44.3}{30 - 1}} = \sqrt{1.5276} = 1.236$$

The memorandum appears in Exhibit 4–1.

Exhibit 4–1

HOWARD S. GITLOW, PH.D.
STATISTICAL CONSULTANT
_____

MEMORANDUM

TO: Stat City Tax Assessor's Office

FROM: Howard Gitlow

DATE: October 18, 1980

RE: Revision of the report on the distribution of rooms
        per dwelling unit in Zone 2 of Stat City, as of January
        1980.

   As per your request and using commonly accepted
statistical techniques, I have computed several summary
measures concerning the number of rooms per dwelling unit in
Zone 2.[1] My research indicates that the typical Zone 2
dwelling unit has approximately eight rooms.[2] Further, the
number of rooms per dwelling unit in Zone 2 ranges between 5
and 10,[3] indicating some degree of variability in dwelling
unit size.[4]
   If you have any further questions, please do not hesitate
to call me at 305–999–9999.

Exhibit 4–1 (*continued*)

Footnotes

[1] The summary statistics reported are based on a random sample of 30 Zone 2 dwelling units.

[2] Typical refers to the fact that the average number of rooms per dwelling unit is 7.7, and the mode and median for the number of rooms per dwelling unit in Zone 2 are both 8.

[3] Dwelling units range in size from two bedrooms, one bathroom, one kitchen, one living room up to four bedrooms, three bathrooms, one kitchen, one den, one living room.

[4] The degree of variability in the number of rooms per dwelling unit in Zone 2 of Stat City is also evidenced by the fact that the standard deviation of the number of rooms per dwelling unit in Zone 2 is 1.236 rooms. This number indicates that approximately 95 percent of Zone 2 dwelling units have between 5.28 and 10.12 rooms. The above statement is true if the sample mean and standard deviation accurately reflect the true, but unknown, population mean and standard deviation and the distribution of rooms in Zone 2 is bell-shaped. I believe that this assumption is tenable; consequently, the standard deviation can be interpreted as stated above.

Please do not be confused by the discrepancy between 95 percent of the dwelling units having between 5.28 and 10.12 rooms and that the number of rooms per dwelling unit in Zone 2 ranges between 5 and 10. The discrepancy can be easily explained by realizing that all the statistics presented are only sample estimates. If the true, but unknown, minimum and maximum number of rooms per dwelling unit in Zone 2, and the true, but unknown, standard deviation of the number of rooms per dwelling unit in Zone 2 were available, this discrepancy would no longer exist. Briefly expressed, the interval created by subtracting the minimum from the maximum number of rooms per dwelling unit would be larger than the interval created in which 95 percent of Zone 2 dwelling units would fall.

Example 4–2   Construction of an emergency clinic in Zone 1

Ms. Arlene Davis, director of the Stat City Hospital, is considering the advisability of locating a new emergency clinic in Zone 1 of Stat City. Ms. Davis decided that if the average number of trips to the hospital per year in Zone 1 is more than three, the emergency clinic should be constructed.

Problem

She has asked you to provide her with appropriate statistics. Conduct the required survey and type a memorandum to Ms. Davis (see Exhibit 4–2). Make sure you include information concerning the mean, mode, minimum, maximum, range, and standard deviation for the relevant variable.

Please use the following list of 19 randomly selected identification numbers (ID) in solving the problem:

8, 17, 29, 36, 37, 44, 55, 57, 65, 67, 69, 74, 80, 85, 90, 98, 110, 122, and 130.

Solution   The hospital statistics are shown below.

Mean[1] 3.316
Mode[2] .... 1 and 2
Median[3] 3.000
Minimum 0.000
Maximum 8.000
Range[4] 8.000
Standard deviation[5] 2.237

[1] Mean $= \bar{x} = \dfrac{\Sigma x}{n} = \dfrac{63}{19} = 3.316$.

[2] Mode* = 1 and 2.

Yearly trips to hospital	*Frequency*	*Percent*
0	1	5.3
1	4*	21.1
2	4*	21.1
3	1	5.3
4	3	15.8
5	2	10.5
6	3	15.8
8	1	5.3
Total	19	100.0*

* Percentages do not total 100 percent due to rounding.

[3] Median position $= \dfrac{19 + 1}{2} = \dfrac{20}{2} = 10$.

0, 1, 1, 1, 1, 2, 2, 2, 2, ③ 4, 4, 4, 5, 5, 6, 6, 6, 8

Median = 3.

[4] Range = 8 − 0 = 8.

[5] $s = 2.237$ (See Table 4–4).

Table 4–4

Random identification number (variable 4)	*Average yearly trips to hospital per dwelling unit (variable 21)* x	$(x - \bar{x})$ $(x - 3.316)$	$(x - \bar{x})^2$ $(x - 3.316)^2$
8	5	1.684	2.8359
17	4	.684	.4679
29	2	1.316	1.7319
36	4	.684	.4679
37	2	1.316	1.7319
44	2	1.316	1.7319
55	0	−3.316	10.9959
57	5	1.684	2.8359
65	2	1.316	1.7319
67	3	−.316	.0999
69	8	4.684	21.9399
74	1	−2.316	5.3639
80	6	2.684	7.2039
85	1	−2.316	5.3639
90	1	−2.316	5.3639
98	1	−2.316	5.3639
110	6	2.684	7.2039
122	6	2.684	7.2039
130	4	.684	.4679

$\Sigma(x - \bar{x})^2 = 90.106$

$$s = \sqrt{\dfrac{\Sigma(x - \bar{x})^2}{n - 1}}$$

$$s = \sqrt{\dfrac{90.106}{19 - 1}} = \sqrt{5.0059} = 2.237$$

The memorandum appears in Exhibit 4–2.

Exhibit 4–2

HOWARD S. GITLOW, PH.D.
STATISTICAL CONSULTANT

MEMORANDUM

TO: Ms. Arlene Davis, Director
        Stat City Hospital

FROM: Howard Gitlow

DATE: April 30, 1980

RE: Construction of an emergency clinic in Zone 1.

   As per your request and using commonly accepted
statistical techniques, I have computed several summary
measures concerning the average number of trips to the
hospital per year for Zone 1 families (dwelling units)[1] My
research indicates that the most common number of trips to
the hospital per year for a Zone 1 family is 1 or 2.[2] However,
the average number of yearly trips to the hospital per family
is 3.316.[3] Further, the number of trips to the hospital per
year ranges between 0 and 8, indicating some degree of
variability in yearly trips to the hospital.[4]
   If you have any further questions, please do not hesitate
to call me at 305–999–9999.

Exhibit **4–2** (*continued*)

Footnotes

¹ The summary statistics reported are based on a random sample of 19 Zone 1 dwelling units.

² The distribution of yearly trips to the hospital is bimodal, with one and two trips being the modes.

³ The mean yearly number of trips to the hospital per year is 3.316. The mean is greater than the modes indicating that the distribution of yearly trips to the hospital is skewed to the right; that is, most families make a few trips while a few families make many trips.

⁴ The variability is evidenced by the fact that the standard deviation of average yearly trips to the hospital is 2.237 trips. This number indicates that at least 75 percent of Zone 1 families make between 0 and 7.79 trips to the hospital per year. The above statement is true if the sample mean and standard deviation accurately reflect the true, but unknown, population mean and standard deviation. The 75 percent was derived by invoking Chebyshev's inequality.

Example 4–3   Construction of an automotive parts outlet in Stat City

Sears, Roebuck & Company is considering building an automobile parts outlet in Stat City. Sears management decided that if more than 30 percent of the families living in Stat City perform their own automotive repairs, they will construct the automobile parts outlet. Sears's management has asked you to conduct a survey and type a memorandum presenting them with your findings (see Exhibit 4–3).

Please use the following list of 30 randomly selected identification numbers in solving the problem:

794, 616, 1326, 131, 55, 1306, 516, 493, 734, 94,
859, 223, 1283, 378, 57, 1339, 517, 65, 541, 451,
1337, 1104, 93, 1311, 797, 238, 1261, 533, 383, 770.

Remember, the above numbers are just random identification numbers. You must still draw the sample of 30 favorite places for car repairs (variable 19) from the Stat City data base.

The "favorite place for car repairs" variable has to be redefined as **Solution** follows: 1 = does own repair (original 0) and 0 = other (original 1 and 2). The "favorite place for car repairs" statistics are shown below:

$$\bar{p}^1 = \frac{6}{30} = .2 = \text{Percentage which does own repair}$$

Mode[2] = 0

Standard deviation[3] = .407
(See Table 4–5.)

[1] $p = 6/30 = 0.20$
[2] Mode = 0

Favorite place for car repair	Frequency
Do own repairs (1)	6
Other (0)	24
Total	30

[3] $s = .407$, where variable 19 is redefined as 1 – does own repairs and 0 = other. (See Table 4–5).

Table 4–5

Random identification number (variable 4)	Favorite place for car repairs (variable 19)	$(x - \bar{x})$ $(x - .2)$	$(x - \bar{x})^2$ $(x - .2)^2$
794	0 (1)	−.2	.04
616	0 (2)	−.2	.04
1326	0 (2)	−.2	.04
131	0 (2)	−.2	.04
55	0 (2)	−.2	.04
1306	1 (0)	.8	.64
516	0 (2)	−.2	.04
493	0 (1)	−.2	.04
743	0 (1)	−.2	.04
94	0 (1)	−.2	.04
859	1 (0)	.8	.64
223	0 (2)	−.2	.04
1283	1 (0)	.8	.64
378	0 (1)	−.2	.04
57	0 (2)	−.2	.04
1339	1 (0)	.8	.64
517	0 (1)	−.2	.04
65	0 (1)	−.2	.04
541	0 (1)	−.2	.04
451	0 (1)	−.2	.04
1337	0 (1)	−.2	.04
1104	0 (2)	−.2	.04
93	0 (1)	−.2	.04
1311	1 (0)	.8	.64
797	0 (1)	−.2	.04
238	0 (1)	−.2	.04
1261	0 (1)	−.2	.04
533	1 (0)	.8	.64
383	0 (1)	−.2	.04
770	0 (1)	−.2	.04

$$\Sigma(x - \bar{x})^2 = 4.8$$

$$s = \sqrt{\frac{\Sigma(x - \bar{x})^2}{n - 1}} = \sqrt{\frac{4.8}{29}} = .407$$

or

$$s = \sqrt{\frac{n\bar{p}(1 - \bar{p})}{n - 1}} = \sqrt{\frac{30(.2)(.8)}{29}} = .407$$

The memorandum appears in Exhibit 4–3.

Exhibit 4–3

HOWARD S. GITLOW, PH.D.
STATISTICAL CONSULTANT

MEMORANDUM

TO: Sears, Roebuck and Company

FROM: Howard Gitlow

DATE: May 1, 1980

RE: Construction of an automotive parts outlet in Stat
       City.

   I have estimated that 20 percent of Stat City families
perform their own automotive repairs using commonly
accepted statistical techniques. Based on your decision
criteria (that more than 30 percent of the families living in
Stat City must perform their own repairs to open the outlet)
the automotive parts outlet should not be opened.
   If I can be of any further assistance, please do not
hesitate to call me at 305–999–9999.

ADDITIONAL PROBLEMS

4–4 Federal funding for Stat City Mr. Lee Kaplowitz, the mayor of
Stat City, is attempting to obtain federal funding for several projects. To
apply for the funds, the mayor must prepare a statistical summary concerning the number of persons per dwelling unit in Stat City as of January
1980.

Mr. Kaplowitz has asked you to conduct a survey and type a
memorandum indicating the mean, mode, minimum, maximum, range,
and standard deviation of family size for Stat City dwelling units.

Please use the following random sample of 10 dwelling units' identification numbers (variable 4).

694,   286,   917,   457,   319,
817, 1105,   363,   233,   268.

Remember, the above numbers are just random identification numbers.
You must still draw the sample of 10 family sizes (variable 14) from the
Stat City data base.

4–5 Housing cost survey for the Stat City Real Estate Board Ms. Sharon Vigil (ID = 231), chairperson of the Stat City Real Estate Board, wants to obtain some information concerning the housing costs (mortgages and rents) of Stat City families. She needs this information so the board can revise the prospectus they give to potential Stat City residents.

You have been hired by Ms. Vigil to conduct a survey of Stat City housing costs. Use the following randomly selected identification numbers (variable 4) to draw your sample:

> 25, 61, 306, 310, 337, 370, 407, 488, 597, 609, 633,
> 669, 673, 720, 734, 768, 795, 849, 902, 916, 923,
> 1014, 1015, 1039, 1061, 1106, 1150, 1196, 1216, 1321.

Type a memorandum to Ms. Vigil indicating the mean, median, minimum, maximum, range, and standard deviation of housing costs.

4–6 Diet soda survey Mr. Marc Cooper, manager of the A&P, would like to determine the average weekly purchase of six-packs of diet soda per dwelling unit in Stat City. This information will be helpful to Mr. Cooper in establishing an A&P inventory policy.

You have been hired by Mr. Cooper to conduct a survey to supply him with the required information. Use the following random sample.

Random ID number (variable 4)	Average weekly purchase of six-packs of diet soda (variable 25)	Random ID number (variable 4)	Average weekly purchase of six-packs of diet soda (variable 25)
5	2	519	2
58	1	560	2
68	1	563	0
150	2	586	2
198	1	604	0
201	1	624	1
210	0	662	0
213	0	684	0
216	0	727	0
217	0	741	0
226	2	776	2
278	1	850	0
279	2	947	2
320	0	962	2
340	2	988	0
379	0	996	0
388	2	1089	0
397	2	1144	1
425	2	1145	1
465	2	1152	2
467	2	1223	2
492	0	1236	0
499	0	1255	2
504	0	1333	0
511	0	1338	0

Type a memorandum to Mr. Cooper indicating the mean, mode, minimum, maximum, range, percentage breakdown, and standard deviation of average weekly purchases of six-packs of diet soda in Stat City, as of January 1980.

COMPUTATIONAL EXERCISE   4–7 Mr. Paul Lund, chairman of the Stat City Chamber of Commerce, is preparing a demographic profile of Stat City families. The demographic variables that Mr. Lund has chosen to include in the profile are: total 1979 family incomes (INCOM), assessed value of a home as of January 1980 (ASST), average total yearly heating bill as of January 1980 (HEAT), average monthly electric bill as of January 1980 (ELEC), average monthly telephone bill as of January 1980 (PHONE), number of cars in the household as of January 1980 (CARS), average bimonthly automobile gasoline bill as of January 1980 (GAS), and average number of trips to the gas station per month as of January 1980 (GASTR). Mr. Lund has provided you with enough funds to survey 10 randomly selected Stat City families. Further, he wants you to compute the mean, standard deviation, variance, minimum, maximum, range, skewness, and kurtosis for each of the above variables.

Use the following simple random sample of 10 Stat City families.

CASE-N	ID	ASST	HEAT	ELEC	PHONE	INCOM	CARS	GAS	GASTR
1	258.	92419.	1136.	73.	24.	116203.	2.	127.	3.
2	295.	23773.	538.	42.	65.	25328.	1.	99.	6.
3	386.	39558.	792.	72.	23.	32667.	2.	160.	6.
4	415.	0.	413.	37.	22.	26972.	1.	65.	6.
5	416.	0.	457.	37.	17.	41549.	3.	174.	7.
6	736.	0.	392.	38.	14.	18404.	2.	120.	12.
7	858.	0.	256.	21.	21.	17052.	2.	120.	5.
8	1016.	0.	512.	46.	19.	24800.	3.	185.	5.
9	1149.	0.	651.	69.	27.	19670.	2.	120.	5.
10	1242.	0.	471.	47.	22.	24743.	3.	168.	8.

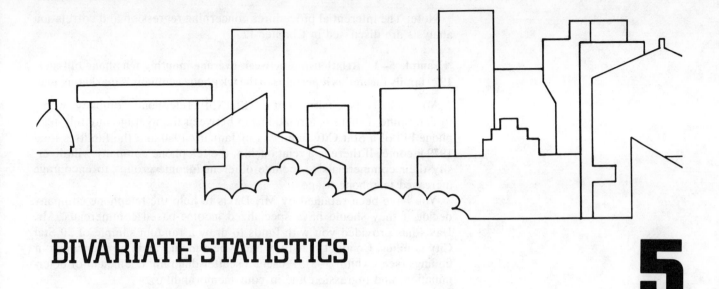

BIVARIATE STATISTICS

5

This chapter focuses on understanding the relationship between two variables. The notion of relationship will be discussed in much more detail in Chapters 12 ("Regression and Correlation Analysis") and 14 ("Some Nonparametric Statistics").

There are essentially two types of relationship measurement problems: (1) problems in which the variables under investigation are categorical (race, religion, region of the county, and so on); and there is no basis for estimating the value of one variable from knowledge of the other variable, and (2) problems in which the variables under investigation are numerical (age, height, price, number of units purchased, and so forth) where there is a basis for estimating the value of one variable (called the dependent, criterion, or response variable) from knowledge of the other variable (called the independent or predictor variable).

There are two types of numerical relationship problems: deterministic (functional) and statistical. A deterministic (functional) relationship exists between two variables if the value of the dependent variable (Y) is exactly and uniquely defined by specifying the value of the independent variable (X). For example, if the rental fee for a chain saw (Y, in dollars) is $8 per day ($X$) plus a fixed charge of $10, then the deterministic relationship is as follows:

$$Y = \$10 + \$8 \ (X)$$

A statistical relationship exists between two variables if the value of the dependent variable (Y) is not exactly or uniquely defined by specifying the value of the independent variable (X). For example, if the weekly dollar sales volume in a store (Y) is related to the store's weekly advertising budget (X), then a statistical relationship can be formed! Please note that this is a statistical relationship because there can be several weekly sales volumes for each weekly advertising budget; that is, the sales volume is not uniquely defined for an advertising budget.

The analysis of categorical relationship is called contingency table analysis or cross-tabulation analysis. The analysis of statistical-numeric relationships is called regression and correlation analysis.

Several Stat City examples will be presented and worked out in detail to aid in your understanding bivariate statistics. All problems in this chapter will rely on sample statistics for their solution, not on population parameters.

INTRODUCTION

Note: The inferential procedures concerning regression and correlation analysis are discussed in Chapter 12.

Example 5–1 Relationship between average monthly telephone bill and 1979 family income as it pertains to the telephone company's marketing plan

Mr. Jack Davis, chairman of the Stat City Telephone Company, wants to determine if there is a relationship between the average monthly telephone bill of a Stat City family as of January 1980 and the family's total 1979 income. If there is a relationship, the telephone company will diversify their commercials to include different income groups to encourage increased telephone usage.

You have been retained by Mr. Davis to help the telephone company decide if they should have specialized income-based commercials. Mr. Davis has provided you with funds to draw a random sample of 30 Stat City families. Conduct the study and type a memorandum reporting your findings (see Exhibit 5–3). Present a scatter diagram, coefficient of determination, and regression line in your memorandum.

Please use the following randomly selected identification numbers (variable 4) to draw your sample:

72, 86, 180, 195, 271, 285, 322, 340, 426, 432, 468,
520, 611, 616, 620, 759, 807, 853, 855, 941, 1061,
1078, 1104, 1180, 1184, 1217, 1222, 1240, 1246, 1331.

Solution   The random sample is shown in Table 5–1.

Table 5–1

ID	PHONE (Y)	INCOM (X)	ID	PHONE (Y)	INCOM (X)
72.	39.	29910.	759.	77.	12717.
86.	17.	42818.	807.	78.	24995.
180.	12.	65255.	853.	41.	15113.
195.	16.	60714.	855.	38.	14460.
271.	14.	116346.	941.	33.	9726.
285.	21.	50449.	1061.	26.	1399.
322.	19.	13452.	1078.	91.	6894.
340.	15.	18485.	1104.	20.	24259.
426.	87.	26316.	1180.	19.	11156.
432.	93.	32862.	1184.	13.	6549.
468.	21.	47583.	1217.	27.	20175.
520.	19.	39430.	1222.	15.	20390.
611.	14.	18591.	1240.	76.	25241.
616.	48.	27071.	1246.	58.	24855.
620.	21.	24156.	1331.	73.	19738.

Some of the following quantities should be helpful in computing the desired statistics; see Exhibit 5–1 for details.

$$\Sigma x = 851,105 \qquad \Sigma(xy) = 27,547,299$$
$$\Sigma(x^2) = 39,120,966,527 \qquad (\Sigma x)^2 = 724,379,721,025$$
$$\Sigma y = 1,141 \qquad (\Sigma y)^2 = 1,301,881$$
$$\Sigma(y^2) = 64,931 \qquad n = 30$$

$$\Sigma(x - \bar{x})^2 = 14,974,975,825.5$$
$$\Sigma(y - \bar{y})^2 = 21,534.965$$
$$\Sigma[(x - \bar{x})(y - \bar{y})] = -4,823,065.8$$

Exhibit 5-1

SYSTEM		PUNCHING INSTRUCTIONS		PAGE 1 OF 4
PROGRAM *Problem 5-1*		GRAPHIC		*
PROGRAMMER	DATE	PUNCH	CARD FORM #	

COBOL STATEMENT — $(x-\bar{x})^2$

	X	X²	$(X-\bar{X})$	$(X-\bar{X})^2$	y	y²	$(y-\bar{y})$	$(y-\bar{y})^2$
	29910	894608100	1539.833	2371085.7	39	1521	0.967	0.934
	42818	1833381124	14447.833	208739878.4	17	289	-21.03	442.401
	65255	4258215025	36884.833	1360490905	12	144	-26.03	677.1734
	60714	3686189796	32343.833	1046123533	16	256	-22.03	485.1468
#	116346	13536391716	87975.833	7739747192	14	196	-24.03	577.601
	50449	2545101601	22078.833	487474866.6	21	441	-17.03	290.134
	13452	180956304	-14918.167	222551706.6	19	361	-19.03	362.268
	18485	341695225	-9885.167	97716526.6	15	225	-23.03	530.534
	26316	692531856	-2054.167	4219602.1	87	7569	48.97	2397.735
	32862	1079911044	4491.833	20176563.7	93	8649	54.97	3021.335
	47583	2264141889	19212.833	369132951.9	21	441	-17.03	290.134
	39430	1554724900	11059.833	122319906	19	361	-19.03	362.268
	18591	345625281	-9779.167	95632107.2	14	196	-24.03	577.601
	27071	732839041	-1299.167	1687834.9	48	2304	9.97	99.334
	24156	583512336	-4214.167	17759203.5	21	441	-17.03	290.134
	12717	161722089	-15653.167	245021637.1	77	5929	38.97	1518.401
	24995	624750025	-3375.167	11391752.3	78	6084	39.97	1597.335
	15113	228402769	-13257.167	175752476.9	41	1681	2.97	8.801
	14460	209091600	-13910.167	193492746	38	1444	-0.03	0.001
	9726	94595076	-18644.167	347604963.1	33	1089	-5.03	25.334
	1399	1957201	-26971.167	727443849.3	26	676	-12.03	144.801
	6894	47527236	-21476.167	461235749	91	8281	52.97	2805.468
	24259	588499081	-4111.167	16901694.1	20	400	-18.03	325.201
	11156	124456336	-17214.167	296327545.5	19	361	-19.03	362.268

SYSTEM		PUNCHING INSTRUCTIONS		PAGE 2 OF 4
PROGRAM *Problem 5-1*		GRAPHIC		*
PROGRAMMER	DATE	PUNCH	CARD FORM #	

COBOL STATEMENT — $(x-\bar{x})^2$

	X	X²	$(X-\bar{X})$	$(X-\bar{X})^2$	y	y²	$(y-\bar{y})$	$(y-\bar{y})^2$
	6549	42889401	-21821.167	476163329.2	13	169	-25.03	626.668
	20175	407030625	-8195.167	67160762.2	27	729	-11.03	121.1734
	20390	415752100	-7980.167	63683065.4	15	225	-23.03	530.534
	25241	637108081	-3129.167	9791686.1	76	5776	37.97	1441.468
	24855	617771025	-3515.167	12356399	58	3364	19.97	398.668
	19738	389588644	-8632.167	74514307.1	73	5329	34.97	1222.668
2	851105	39120966527	-0.01	14974975825.5	1141	64931	0.00	21534.965

Exhibit 5–1 (*continued*)

SYSTEM		PUNCHING INSTRUCTIONS		PAGE 3 OF 4
PROGRAM *Problem 5–1*		GRAPHIC		
PROGRAMMER	DATE	PUNCH	CARD FORM #	*

SEQUENCE (Page) (Serial)	CONT.	A	B $(x)(y)$	$(x-\bar{x})(y-\bar{y})$	COBOL STATEMENT	IDENTIFICATION
			1166490	−1489.0185		
			7217906	−303837.93		
			783060	−960112.2		
			971424	−712534.64		
			1628844	−2114059.3		
			1059429	−376002.53		
			2555588	283892.72		
			277275	227655.4		
			2289492	−100592.56		
			3056166	246916.06		
			999243	−327194.55		
			7419170	−210468.62		
			260274	234993.38		
			1299408	−12952.695		
			507276	71767.264		
			979209	−610003.92		
			1949610	−134905.42		
			6119633	−39373.786		
			549480	417.305		
			320958	93780.16		
			36374	324463.14		
			627354	−1137592.6		
			485180	74124.341		
			211964	327585.6		

SYSTEM		PUNCHING INSTRUCTIONS		PAGE 4 OF 4
PROGRAM *Problem 5–1*		GRAPHIC		
PROGRAMMER	DATE	PUNCH	CARD FORM #	*

SEQUENCE (Page) (Serial)	CONT.	A	B $(x)(y)$	$(x-\bar{x})(y-\bar{y})$	COBOL STATEMENT	IDENTIFICATION
			85137	546183.81		
			5414725	90392.692		
			305850	183783.25		
			1918316	−118814.47		
			1441590	−70197.885		
			1440874	−301866.88		
2			27547299	−4823065.8		

Exhibit 5–2 lists various statistics of interest.

Exhibit 5–2

Mean income ($\bar{x}$)	\$28,370.17
Standard deviation of income (s_x)	\$22,723.96
Mean telephone bill ($\bar{y}$)	\$ 38.03
Standard deviation of telephone bill (s_y)	\$ 27.25
Correlation (r)	−.26858
Coefficient of determination (r^2)	.07213
Y-intercept (b_0 or a or $\hat{\beta}_0$)	\$ 47.17
Slope (b_1 or b or $\hat{\beta}_1$)	−0.00032

The memorandum appears in Exhibit 5–3.

Exhibit 5–3

HOWARD S. GITLOW, PH.D.
STATISTICAL CONSULTANT

MEMORANDUM

TO: Mr. Jack Davis, Chairman
Stat City Telephone Company

FROM: Howard Gitlow

DATE: May 13, 1980

RE: Relationship between average monthly telephone
bill and 1979 family income as it pertains to the
telephone company's marketing plan (advertising
copy).

As per your instructions and using commonly accepted
statistical techniques, I have investigated the
relationship between "average monthly telephone bill as of
January 1980 (PHONE)" and "family income in 1979 (INCOM)". A
plot of PHONE versus INCOM is shown below:[1]

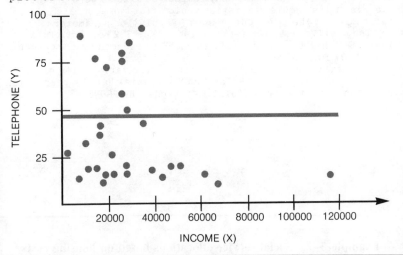

Exhibit 5–3 (*continued*)

Only 7.2 percent of the variation in PHONE is explained by INCOM.[2] The weak relationship between PHONE and INCOM is expressed in the following equation:[3]

$$PHONE = \$47.17 - .00032 \, (INCOM)$$

Consequently, one standard commercial stressing increased telephone usage or multiple commercials stressing increased telephone usage by variables other than income should be employed according to the telephone company's marketing strategy.

If you have any further questions, please do not hesitate to call me at 305-999-9999.

<u>Footnotes</u>

[1] The statistics reported are based on a random sample of 30 Stat City dwelling units and are representative of parameters only on average. Simply stated, if many samples of size 30 were drawn, the average of the sample statistics would afford an accurate reflection of the true, but unknown, population parameters.

[2] The coefficient of determination (r^2) is 7.2 percent. This indicates that the variability in PHONE is reduced by 7.2 percent when INCOM is considered.

[3] The small proportionate reduction in the variability of PHONE (7.2 percent) when considering INCOM suggests that the regression relation may not be very useful in forecasting PHONE.

Table 5–2

ID	HCOST (Y)	INCOM (X)
72.	618.	29910.
86.	880.	42818.
180.	541.	65255.
195.	728.	60714.
271.	738.	116346.
285.	749.	50449.
322.	225.	13452.
340.	193.	18485.
426.	352.	26316.
432.	571.	32862.
468.	441.	47583.
520.	572.	39430.
611.	287.	18591.
616.	484.	27071.
620.	337.	24156.
759.	622.	12717.
807.	540.	24995.
853.	320.	15113.
855.	422.	14460.
941.	569.	9726.
1061.	430.	1399.
1078.	283.	6894.
1104.	468.	24259.
1180.	608.	11156.
1184.	424.	6549.
1217.	267.	20175.
1222.	365.	20390.
1240.	498.	25241.
1246.	432.	24855.
1331.	461.	19738.

Example 5–2 Social service allocations based on housing costs

Mr. Paul Lund, chairman of the Stat City Chamber of Commerce, would like to determine if there is a relationship between a Stat City family's housing cost as of January 1980 (HCOST) and the family's total 1979 income (INCOM). The Chamber of Commerce needs this information to decide if they can use family income as a proxy measure for housing cost when deciding how to disburse social services. It is easy for the Chamber of Commerce to obtain family income data, but difficult for the Chamber of Commerce to obtain housing cost data.

You have been retained by Mr. Lund to ascertain if the Chamber of Commerce can use INCOM as a proxy measure for HCOST. Mr. Lund has provided you with funds to draw a random sample of 30 Stat City families. Conduct the study and type a memorandum reporting findings (see Exhibit 5–6). Present a scatter diagram, coefficient of the determination, and regression line in your memorandum.

Please use the following randomly selected identification numbers (variable 4) to draw your sample.

72, 86, 180, 195, 271, 285, 322, 340, 426, 432, 468, 520, 611, 616, 620, 759, 807, 853, 855, 941, 1061, 1078, 1104, 1180, 1184, 1217, 1222, 1240, 1246, 1331.

Solution The random sample is shown in Table 5–2.

Some of the following quantities may be helpful in computing your statistics; see Exhibit 5-4 for details.

$$\Sigma x = 851,105 \qquad \Sigma(xy) = 470,635,698$$
$$\Sigma(x^2) = 39,120,966,527 \qquad \Sigma(x)^2 = 724,379,720,000$$
$$\Sigma y = 14,425 \qquad (\Sigma y)^2 = 208,080,625$$
$$\Sigma(y^2) = 7,742,157 \qquad n = 30$$
$$\Sigma(x - \bar{x})^2 = 14,974,975,825.5$$
$$\Sigma(y - \bar{y})^2 = 806,136.08$$
$$\Sigma[(x - \bar{x})(y - \bar{y})] = 61,396,044$$

Exhibit 5-4

COBOL Coding Form

SYSTEM — PUNCHING INSTRUCTIONS — PAGE 1 OF 4
PROGRAM *Problem 5-2* — GRAPHIC — CARD FORM #
PROGRAMMER — DATE — PUNCH

X	X^2	$(x-\bar{x})$	$(x-\bar{x})^2$	y	y^2	$(y-\bar{y})$
29910	894608100	1539.833	2371085.7	618	381924	137.167
42818	1833381124	14447.833	208739878.4	880	774400	399.167
65255	4258215025	36884.833	1360490905	541	292681	60.167
60714	3686189796	32343.833	1046123533	728	529984	247.167
116346	13536391716	87975.833	7739747192	738	544644	257.167
50449	2545101601	22078.833	487474866.6	749	561001	268.167
13452	180956304	-14918.167	222551706.6	225	50625	-255.833
18485	341695225	-9885.167	97716526.6	193	37249	-287.833
26316	692531856	-2054.167	4219602.1	352	123904	-128.833
32862	1079911044	4491.833	20176563.7	571	326041	90.167
47583	2264141889	19212.833	369132951.7	441	194481	-39.833
39430	1554724900	11059.833	122319906	572	327184	91.167
18591	345625281	-9779.167	95632107.2	287	82369	-193.833
27071	732839041	-1299.167	1687834.9	484	234256	3.167
24156	583512336	-4214.167	17759203.5	337	113569	-143.833
12717	161722089	-15653.167	245021637.1	622	386884	141.167
24995	624750025	-3375.167	11391752.3	540	291600	59.167
15113	228402769	-13257.167	175752476.9	320	102400	-160.833
14460	209091600	-13910.167	193492746	422	178084	-58.833
9726	94595076	-18644.167	347604963.1	569	323761	88.167
1399	1957201	-26971.167	727443849.3	430	184900	-50.833
6894	47527236	-21476.167	461225749	283	80089	-197.833
24259	588499081	-4111.167	16901694.1	468	219024	-12.833
11156	124456336	-17214.167	296327545.5	608	369664	137.167

COBOL Coding Form

SYSTEM — PUNCHING INSTRUCTIONS — PAGE 2 OF 4
PROGRAM *Problem 5-2* — GRAPHIC — CARD FORM #
PROGRAMMER — DATE — PUNCH

Serial	X	X^2	$(x-\bar{x})$	$(x-\bar{x})^2$	y	y^2	$(y-\bar{y})$
	6549	42889401	-21821.167	476163329.2	424	179776	-56.833
	20175	407030625	-8195.167	67160762.2	267	71289	-213.833
	20390	415752100	-7980.167	63683065.4	365	133225	-115.833
	25241	637108081	-3129.167	9791686.1	498	248004	17.167
	24855	617771025	-3515.167	12356399	432	186624	-48.833
	19738	389588644	-8632.167	745143307.1	461	212521	-19.833
Σ	851105	39120966527	-0.01	14974975825.5	14425	7742157	0.0001

Exhibit 5–4 (*continued*)

COBOL Coding Form

SYSTEM		PUNCHING INSTRUCTIONS		PAGE 3 OF 4
PROGRAM *Problem 5-2*		GRAPHIC		*
PROGRAMMER	DATE	PUNCH	CARD FORM #	

COBOL STATEMENT

$(y-\bar{y})^2$	$(x)(y)$	$(x-\bar{x})(y-\bar{y})$
18814.70	18484380	211214.27
159334.03	37679840	5767098.2
3620.03	35302955	2219249.7
61091.36	44199792	7994328.2
66134.70	85863348	2262448.1
71913.36	37786301	5920814.4
65450.69	3026700	3816559.4
82848.03	3567605	2845277.3
6598.03	9263232	264644.5
8130.03	18764202	405015.11
1586.69	20984103	-765304.78
8311.36	22553960	1008291.8
3757.36	5335617	1895525.3
10.03	13102364	-4114.4619
20688.03	8140572	606136.28
9928.03	7909974	-2209710.6
3500.70	13497300	-199698.51
25867.36	4836160	2132189.9
3461.36	6102120	818376.86
7773.36	5534094	-1643800.3
2584.03	601570	1371025.3
39138.03	1951002	4248694.5
164.69	11353212	527258.606
6171.36	6782849	-2189074

COBOL Coding Form

SYSTEM		PUNCHING INSTRUCTIONS		PAGE 4 OF 4
PROGRAM *Problem 5-2*		GRAPHIC		*
PROGRAMMER	DATE	PUNCH	CARD FORM #	

COBOL STATEMENT

$(y-\bar{y})^2$	$(x)(y)$	$(x-\bar{x})(y-\bar{y})$
3230.03	2776776	1240162.4
457214.69	5386725	1752397.1
13417.36	7442350	924366.68
294.69	12570018	-53718.41
2384.69	10737360	171656.15
393.36	9099218	171201.77
806136.08	470635698	61396044

Exhibit 5–5 lists the various statistics of interest.

Exhibit 5–5

Correlation (r)	0.55880
Coefficient of determination (r^2)	0.31225
y-intercept (b_0 or a or $\hat{\beta}_0$)	$364.51822
Slope (b_1 or b or $\hat{\beta}_1$)	0.00410

The memorandum appears in Exhibit 5–6.

Exhibit 5–6

HOWARD S. GITLOW, PH.D.
STATISTICAL CONSULTANT

MEMORANDUM

TO: Mr. Paul Lund, Chairman
        Stat City Chamber of Commerce

FROM: Howard Gitlow

DATE: May 14, 1980

RE: Social Service allocations based on housing costs

   I have investigated the relationship between "housing
cost as of January 1980 (HCOST)" and "total family income in
1979 (INCOM)" using commonly accepted statistical
techniques. A plot of HCOST versus INCOM is shown below:[1]

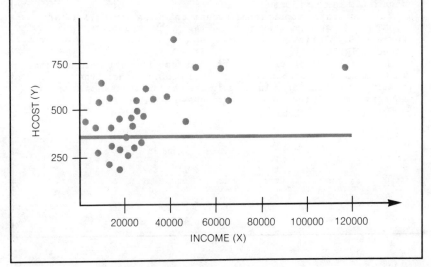

Exhibit 5–6 (*continued*)

Approximately 31.2 percent of the variation in HCOST is explained by INCOM.[2] This moderate relationship between HCOST and INCOM is expressed in the following equation:[3]

$$HCOST = \$364.51822 + .0041(INCOM)$$

Consequently, INCOM can be used as a proxy measure for HCOST only for gross approximations.

If you have any further questions, please do not hesitate to call me at 305–999–9999.

Footnotes

[1] The statistics reported are based on a random sample of 30 Stat City dwelling units and are representative of parameters only on average. In other words, if many samples of size 30 were drawn, the average of the sample statistics would afford an accurate reflection of the true, but unknown, population parameters. Any given sample's statistics may not accurately reflect the population parameters.

[2] The coefficient of determination (r^2) is 31.2 percent. This indicates that the variability in HCOST is reduced by 31.2 percent when INCOM is considered.

[3] The moderate proportional reduction in the variability of HCOST (31.2 percent) when considering INCOM suggests that the regression relation may provide a useful proxy measure of HCOST based on INCOM.

The small coefficient in front of income does not indicate that INCOM is not an important predictor of HCOST; rather, the coefficient transforms incomes in the tens of thousands into housing costs in the hundreds and consequently is small.

Example 5–3   Revision of Howie's Gulf Station marketing plan

The owner of Howie's Gulf Station, Mr. Howie, wants to know if there is a relationship between "a family's favorite gas station as of January 1980 (FAVGA)" and "a family's favorite place for car repairs as of January 1980 (REPAR)." If such a relationship does exist, Mr. Howie will revise his marketing plan to cater to the types of customers he is currently not receiving.

You have been retained by Mr. Howie to determine if a meaningful relationship exists between FAVGA and REPAR. Mr. Howie has provided you with funds to draw a random sample of 30 Stat City dwelling units. Conduct the survey and type a memorandum to Mr. Howie reporting your findings (see Exhibit 5–7). Use a cross-tabulation (contingency) table in your memorandum.

Please use the following randomly selected identification numbers (variable 4) to draw your sample:

8, 12, 22, 103, 142, 229, 258, 260, 321, 352, 363, 446, 452, 517, 687, 693, 729, 745, 828, 932, 967, 1013, 1075, 1100, 1112, 1138, 1250, 1297, 1305, 1341.

The random sample is shown in Table 5–3.

Table 5–3

ID	REPAR	FAVGA	ID	REPAR	FAVGA
8.	2.	2.	693.	1.	2.
12.	1.	2.	729.	0.	1.
22.	2.	1.	745.	2.	2.
103.	2.	1.	828.	2.	1.
142.	2.	2.	932.	1.	2.
229.	1.	2.	967.	1.	2.
258.	2.	2.	1013.	1.	2.
260.	2.	1.	1075.	1.	2.
321.	2.	2.	1100.	1.	2.
352.	1.	1.	1112.	1.	2.
363.	1.	1.	1138.	0.	2.
446.	2.	1.	1250.	0.	2.
452.	0.	2.	1297.	0.	1.
517.	1.	1.	1305.	2.	2.
687.	0.	1.	1341.	2.	1.

Set up a contingency table using the above data (Table 5–4).

Table 5–4

	Performs own repair (0)	Repairs done by service station (1)	Repairs done by dealer (2)	Total
Paul's Texaco Station (1)	3 (10%)*	3 (10%)	6 (20%)	12 (40%)
Howie's Gulf Station (2)	3 (10%)	9 (30%)	6 (20%)	18 (60%)
Total	6 (20%)	12 (40%)	12 (40%)	30 (100%)

* All percents are computed by dividing the cell count by 30.

Use the sample cell percentages to estimate the population cell frequencies (Table 5–5).

Table 5–5

	Performs own repairs (0)	Repairs done by service station (1)	Repairs done by dealer (2)	Total
Paul's Texaco Station	137*	137	275	549
Howie's Gulf Station	137	412	275	824
Total	274	549	550	1,373

* All estimated cell counts are computed as follows:
  Estimated cell count = 1,373 × sample cell percent.

The memorandum appears in Exhibit 5–7.

Exhibit 5–7

HOWARD S. GITLOW, PH.D.
STATISTICAL CONSULTANT

MEMORANDUM

TO: Mr. Howie, Owner
        Howie's Gulf Station

FROM: Howard Gitlow

DATE: May 15, 1980

RE: Revision of marketing plan based on customer's
        favorite place of car repair.

As per your request and using commonly accepted
statistical techniques, I have investigated the
relationship between "a family's favorite gas station as of
January 1980 (FAVGA)" and "a family's favorite place for car
repairs as of January 1980 (REPAR)." The following chart
shows the estimated breakdown of Stat City residents by
FAVGA and REPAR.[1]

	Performs own repairs (0)	Repairs performed by service station (1)	Repairs done by dealers (2)	Total
Paul's Texaco Station	137	137	275	549
Howie's Gulf Station	137	412	275	824
Total	274	549	550	1,373

Exhibit 5–7 (*continued*)

As you can see, there is little relationship between FAVGA and REPAR. Howie's and Paul's split the "performs own repairs" and "repairs done by dealer" customers. However, Howie's Gulf Station does have a significant inroad to the "repairs done by service station" customers. Consequently, according to your marketing strategy, Howie's Gulf Station should maintain the current marketing plan for the "repairs performed by service station" customers and create a new marketing plan to attract the "performs own repairs" and "repairs done by dealer" customer.

If you have any further questions, please do not hesitate to call me at 305-999-9999.

Footnotes

[1] The statistics reported are based on a random sample of 30 Stat City dwelling units and are representative of parameters only on average. In other words, if many samples of size 30 were drawn, the average of the sample statistics would afford an accurate reflection of the true, but unknown, population parameters. Any given sample's statistics may not accurately reflect the population parameters.

ADDITIONAL PROBLEMS

5–4 Forecasting the number of cars in a household from the size of a household   Ms. Sharon Lowe, division head of the Stat City Department of Traffic, would like to know if it is possible to forecast the number of cars in a household as of January 1980 (CARS) from the numbers of people in a household as of January 1980 (PEPLE). Further, she wants to know the estimated number of cars per household for households ranging in size from 1 to 10 people.

You have been retained by Ms. Lowe to provide this information. Conduct the required study and type a memorandum reporting your findings. Present a scatter diagram, a coefficient of determination, a regression line, and a chart indicating the estimated number of cars for households of varying sizes.

Please use the following randomly selected identification numbers (variable 4) to draw your sample:

58, 127, 584, 615, 767, 806, 1121, 1160, 1164, 1366.

Bivariate statistics

5-5 Forecasting the assessed value of a dwelling unit from the number of rooms in a dwelling unit The Stat City Tax Assessor's Office wants to determine if the number of rooms in a dwelling unit as of January 1980 can be used to forecast the assessed value of a dwelling unit as of January 1980. If the number of rooms in a dwelling unit can be used to forecast the assessed value of the dwelling unit, then the tax assessor's office only needs to determine the number of rooms in a dwelling unit.

You have been retained by the Stat City Tax Assessor's Office to furnish this information. Conduct the required study and type a memorandum reporting your findings. Present a scattergram, a coefficient of determination, a regression line, and a chart indicating the proposed tax assessments for dwelling units containing different numbers of rooms.

Please use the following randomly selected identification numbers (variable 4) to draw your sample:

170, 358, 391, 610, 735, 785, 936, 1119, 1151, 1350.

5-6 Relationship between family size and number of automobiles in a household Ms. Marsha Lubitz, manager of the Stat City Chapter of the American Automobile Association (AAA), to better serve the needs of her constituents, needs to know if larger families own more cars than smaller families. Specifically, she wants to determine if "family size as of January 1980 (PEPLE)" and "the number of automobiles owned by a household as of January 1980 (CARS)" are positively related. Ms. Lubitz is interested in the following classifications for CARS and PEPLE.

ID	PEPLE	CARS
67.	5.	4.
127.	2.	1.
132.	4.	3.
152.	3.	3.
245.	1.	1.
336.	5.	4.
363.	5.	3.
396.	1.	1.
453.	3.	2.
459.	6.	2.
520.	3.	2.
521.	4.	2.
752.	4.	1.
803.	3.	2.
826.	6.	2.
837.	3.	2.
860.	3.	2.
929.	3.	2.
977.	1.	1.
1064.	5.	2.
1069.	3.	2.
1097.	2.	2.
1126.	4.	2.
1139.	5.	2.
1214.	2.	2.
1237.	3.	2.
1272.	5.	1.
1278.	4.	2.
1318.	6.	2.
1334.	7.	2.

Cars	People
One car	Single person
Two cars	Married couple—no children
Three or more cars	Married couple—one or two children
	Married couple—three or more children

Ms. Lubitz knows that single-parent families are rare in Stat City. Consequently, it is safe to assume that if three or more persons live in a household, two are adults and the remainder are children.

You have been asked by the Stat City Chapter of the AAA to ascertain if CARS is positively related to PEPLE. The AAA has provided you with funds to draw a random sample of 30 Stat City dwelling units. Conduct the survey and type a memorandum to Ms. Lubitz reporting your findings. Use a cross-tabulation (contingency) table in your memo.

Please use the randomly selected sample of Stat City dwelling units shown in the margin.

5-7 Territorial shopping behavior study The Stat City Supermarket Association would like to find out if Stat City families are "territorial" in their shopping behavior. "Territorial shoppers" are consumers who shop as close to their neighborhood as possible. This would mean that: (1) most of Zone 1 residents would shop at the A&P (supermarket 3), (2) most of Zone 2 residents would shop at either the Food Fair (supermarket 1) or the A&P (supermarket 3), (3) most of Zone 3 residents would shop at either the Food Fair (supermarket 1) or the Grand Union (supermarket 2), and (4) most of Zone 4 residents would shop at the Grand Union (supermarket 2).

The Supermarket Association is providing funds for a survey of "territorial shopping" to determine if there is a relationship between "residential housing zone (ZONE)" and "a family's favorite supermarket as of

January 1980 (FEAT)." The association has provided you with enough funds to draw a random sample of 30 Stat City dwelling units. Conduct the necessary survey and type a memorandum to the Supermarket Association reporting your findings. Use a contingency table in your memorandum.

Please use the randomly selected sample of Stat City dwelling units shown in the margin.

ID	ZONE	FEAT
51.	1.	2.
69.	1.	3.
106.	1.	3.
119.	1.	3.
188.	2.	1.
189.	2.	1.
218.	2.	3.
259.	2.	3.
340.	3.	1.
435.	3.	2.
536.	3.	2.
590.	3.	2.
605.	3.	1.
651.	4.	2.
661.	4.	2.
699.	4.	2.
828.	4.	2.
909.	4.	2.
926.	4.	2.
966.	4.	2.
1002.	4.	2.
1012.	4.	2.
1029.	4.	2.
1056.	4.	2.
1093.	4.	2.
1142.	4.	2.
1177.	4.	2.
1187.	4.	2.
1327.	4.	2.
1370.	4.	2.

COMPUTATIONAL EXERCISES

5–8 A survey of 50 Stat City families was conducted to collect information on the following variables: favorite gas station as of January 1980 (FAVGA), average number of trips to the gas station per month as of January 1980 (GASTR), favorite supermarket as of January 1980 (FEAT), type of dwelling unit as of January 1980 (DWELL), residential housing zone as of January 1980 (ZONE); and average weekly purchase of six-packs of diet soda (LSODA), regular soda (HSODA), and beer (BEER) as of January 1980. Construct the following contingency tables from the above variables:

FAVGA by GASTR	DWELL by ZONE
FAVGA by ZONE	FEAT by DWELL
FAVGA by DWELL	LSODA by BEER
GASTR by ZONE	LSODA by HSODA
GASTR by DWELL	HSODA by BEER

Use the accompanying random sample for your computations.

CASE-N	ID	ZONE	DWELL	GASTR	FAVGA	FEAT	LSODA	HSODA	BEER
1	6.	1.	1.	3.	1.	3.	0.	0.	1.
2	42.	1.	1.	4.	1.	2.	2.	0.	0.
3	51.	1.	1.	5.	2.	2.	0.	6.	0.
4	79.	1.	1.	9.	2.	1.	2.	0.	7.
5	80.	1.	1.	6.	2.	3.	2.	0.	1.
6	84.	1.	1.	5.	1.	3.	0.	0.	0.
7	155.	2.	1.	5.	2.	1.	2.	0.	2.
8	159.	2.	1.	4.	1.	2.	0.	6.	2.
9	196.	2.	1.	12.	1.	3.	0.	1.	3.
10	205.	2.	1.	3.	2.	3.	1.	0.	1.
11	401.	3.	0.	3.	2.	2.	2.	0.	10.
12	429.	3.	0.	7.	2.	2.	2.	0.	0.
13	498.	3.	0.	4.	1.	1.	0.	0.	2.
14	520.	3.	0.	3.	2.	1.	2.	0.	5.
15	553.	3.	1.	5.	1.	1.	0.	1.	0.
16	580.	3.	1.	6.	2.	1.	1.	0.	1.
17	586.	3.	1.	5.	2.	1.	2.	0.	0.
18	615.	3.	1.	5.	2.	2.	0.	0.	0.
19	616.	3.	1.	5.	2.	1.	0.	2.	4.
20	622.	3.	1.	2.	1.	1.	2.	0.	2.
21	636.	4.	0.	8.	1.	2.	2.	0.	8.
22	687.	4.	0.	5.	1.	2.	0.	0.	1.
23	728.	4.	0.	10.	2.	2.	0.	2.	0.
24	749.	4.	0.	5.	1.	2.	0.	0.	4.
25	852.	4.	0.	3.	1.	2.	0.	0.	0.
26	862.	4.	0.	7.	1.	2.	0.	0.	0.
27	875.	4.	0.	6.	2.	3.	0.	1.	7.
28	943.	4.	0.	11.	2.	1.	0.	0.	0.
29	967.	4.	0.	3.	2.	3.	0.	2.	0.
30	998.	4.	0.	7.	1.	2.	0.	0.	0.
31	1002.	4.	0.	8.	1.	2.	0.	0.	0.
32	1041.	4.	0.	10.	1.	2.	2.	0.	0.
33	1063.	4.	0.	10.	2.	2.	0.	0.	0.
34	1136.	4.	0.	5.	2.	2.	0.	5.	1.
35	1139.	4.	0.	4.	2.	2.	0.	1.	4.
36	1144.	4.	0.	8.	1.	2.	1.	0.	9.
37	1164.	4.	0.	5.	2.	2.	2.	0.	2.
38	1169.	4.	0.	6.	1.	2.	0.	0.	9.
39	1172.	4.	0.	3.	1.	2.	0.	0.	7.
40	1189.	4.	0.	6.	2.	2.	0.	1.	1.
41	1219.	4.	0.	5.	1.	2.	0.	5.	1.
42	1233.	4.	0.	4.	2.	2.	1.	0.	3.
43	1237.	4.	0.	5.	1.	3.	1.	0.	6.
44	1239.	4.	0.	7.	1.	2.	0.	2.	0.
45	1259.	4.	0.	7.	1.	2.	0.	5.	5.
46	1303.	4.	0.	7.	1.	2.	1.	0.	0.
47	1335.	4.	0.	5.	2.	2.	0.	1.	3.
48	1345.	4.	0.	2.	2.	2.	0.	6.	4.
49	1350.	4.	0.	4.	2.	2.	1.	0.	6.
50	1365.	4.	0.	6.	2.	2.	0.	6.	7.

5–9 A survey of 10 Stat City families was conducted to collect information of the following variables: HEAT, ELEC, GAS, and GASTR. Perform the following simple linear regression analyses:

$$\text{HEAT} = f(\text{ELEC})$$
$$\text{GAS} = f(\text{GASTR})$$
$$\text{HEAT} = f(\text{GAS})$$

Compute the slope, intercept, and coefficient of determination for each regression line.

Use the following random sample for your computations.

CASE-N	ID	HEAT	ELEC	GAS	GASTR
1	152.	1174.	74.	172.	5.
2	215.	1053.	58.	163.	8.
3	274.	959.	66.	144.	9.
4	313.	1079.	92.	120.	6.
5	384.	1198.	111.	165.	4.
6	394.	489.	44.	105.	7.
7	497.	538.	42.	153.	3.
9	1018.	616.	52.	65.	5.
10	1156.	484.	50.	224.	8.

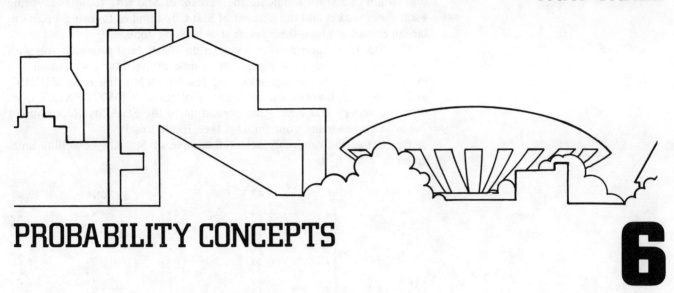

PART THREE

PROBABILITY CONCEPTS

6

Part Two of this book was devoted to investigating various characteristics of Stat City; for example, the distribution of average monthly telephone bills or the possibility of a relationship between residential zone and a family's favorite supermarket. This chapter will introduce probability concepts to determine the chances of various phenomena occurring. For example, you will learn how to determine the likelihood of a randomly selected family in Zone 4 earning over $30,000 a year or the chances of a family with children going to the hospital at least once during a year.

The basic probability concepts you learn in this chapter will increase your capabilities of making sound inferences about a population based only on a sample. The topics discussed in this chapter are:

Marginal, joint, and conditional probabilities.
Addition and multiplication rules.
Independent and dependent events.

Probabilities indicate the relative frequency of the occurrence of events. For example, the probability of randomly selecting a Stat City family that lives in Zone 1 is 0.0947 (9.47% = 130/1,373); there are 130 dwelling units in Zone 1 and 1,373 dwelling units in Stat City. Probabilities give decision makers a handle on the likelihood that a particular event will occur. If the probability of an event's occurring is small, then the event is a rare event and the chances of it occurring are not likely. On the other hand, if the probability of an event's occurring is large, then the event is a common event and the chances of it occurring are likely. It is important to realize that the definitions of rare and common are subjective judgments on the part of the decision maker, not the statistician.

Example 6–1 Reinvestigation of the territorial shopping behavior study

The Stat City Supermarket Association would like to reinvestigate the "territorial shopping behavior" of Stat City residents. You will recall that "territorial shoppers" are consumers who shop as close to their neighborhoods as possible. This would mean that (1) most of Zone 1 residents would shop at the A&P (supermarket 3), (2) most of Zone 2 residents would shop at either the Food Fair (supermarket 1) or the A&P (supermarket 3), (3) most of Zone 3 residents would shop either at Food Fair (supermarket 1) or the Grand Union (supermarket 2), and (4) most of Zone

4 residents would shop at the Grand Union (supermarket 2). The association would like you to indicate the percent of Stat City families favoring each supermarket and the percent of Stat City families favoring a particular supermarket given their residential housing zone.

The Stat City Supermarket Association insists (and provides you with the funds) that you draw a random sample of 100 Stat City families to investigate the relationship between "residential housing zone (ZONE)" and a "families favorite supermarket as of January 1980 (FEAT)." Conduct the survey and type a memorandum to the Stat City Supermarket Association reporting your findings (see Exhibit 6-1).

Please use the randomly selected sample of Stat City dwelling units given in Table 6-1.

Table 6-1	ID	ZONE	FEAT	ID	ZONE	FEAT
	6.	1.	3.	734.	4.	2.
	8.	1.	3.	740.	4.	2.
	21.	1.	3.	747.	4.	2.
	76.	1.	3.	765.	4.	2.
	79.	1.	1.	785.	4.	2.
	85.	1.	3.	787.	4.	2.
	92.	1.	3.	802.	4.	2.
	112.	1.	3.	808.	4.	2.
	123.	1.	3.	825.	4.	2.
	140.	2.	2.	858.	4.	2.
	150.	2.	3.	861.	4.	2.
	153.	2.	1.	867.	4.	2.
	161.	2.	1.	881.	4.	2.
	228.	2.	3.	885.	4.	2.
	233.	2.	3.	896.	4.	2.
	243.	2.	1.	902.	4.	2.
	246.	2.	1.	913.	4.	2.
	253.	2.	1.	914.	4.	2.
	289.	3.	1.	916.	4.	2.
	291.	3.	2.	921.	4.	2.
	315.	3.	3.	929.	4.	2.
	322.	3.	2.	931.	4.	2.
	327.	3.	2.	951.	4.	2.
	337.	3.	1.	953.	4.	2.
	388.	3.	1.	956.	4.	2.
	390.	3.	1.	994.	4.	2.
	414.	3.	2.	996.	4.	2.
	429.	3.	1.	1005.	4.	2.
	440.	3.	2.	1007.	4.	2.
	460.	3.	2.	1020.	4.	2.
	468.	3.	2.	1030.	4.	2.
	482.	3.	2.	1037.	4.	2.
	492.	3.	1.	1046.	4.	2.
	517.	3.	2.	1094.	4.	2.
	545.	3.	2.	1122.	4.	2.
	561.	3.	2.	1141.	4.	1.
	585.	3.	1.	1156.	4.	2.
	586.	3.	1.	1159.	4.	2.
	622.	3.	1.	1171.	4.	2.
	629.	4.	2.	1186.	4.	2.
	630.	4.	2.	1227.	4.	2.
	636.	4.	2.	1239.	4.	2.
	638.	4.	2.	1243.	4.	2.
	668.	4.	2.	1245.	4.	2.
	675.	4.	1.	1263.	4.	2.
	678.	4.	2.	1277.	4.	2.
	692.	4.	2.	1282.	4.	2.
	699.	4.	2.	1295.	4.	2.
	721.	4.	2.	1297.	4.	2.
	730.	4.	1.	1365.	4.	2.

The memorandum appears in Exhibit 6–1.

Exhibit 6–1

HOWARD S. GITLOW, PH.D.
STATISTICAL CONSULTANT

MEMORANDUM

TO: Stat City Supermarket Association

FROM: Howard Gitlow

DATE: May 27, 1980

RE: Reinvestigation of the territorial shopping
        behavior study.

In accordance with the association's request I have
reinvestigated the "territorial shopping behavior" of Stat
City residents using commonly accepted statistical
techniques. The following table depicts the estimated
breakdown of Stat City families into housing zone and
favorite supermarket as of January 1980.[1]

Favorite supermarket	Zone*				Total
	1	2	3	4	
Food Fair	14 (1%)	69 (5%)	124 (9%)	41 (3%)	248 (18%)
Grand Union	0 (0%)	14 (1%)	151 (11%)	796 (58%)	961 (70%)
A&P	110 (8%)	40 (3%)	14 (1%)	0 (0%)	164 (12%)
Total	124 (9%)	123 (9%)	289 (21%)	837 (61%)	1,373 (100%)

* Numbers in this table are rounded to sum to 1,373.

The above table indicates that 18 percent of Stat City
families favor the Food Fair, 70 percent favor the Grand

Exhibit 6–1 (*continued*)

Union, and 12 percent favor the A&P. Approximately 11
percent of Zone 1 families frequent the Food Fair
(14/124 = 11.1 percent), while the majority of Zone 1
residents (88.9 percent) frequent the A&P (110/124).
Approximately 56 percent of Zone 2 families prefer to shop at
the Food Fair (69/123 = 55.6 percent), 11 percent shop at the
Grand Union (14/123 = 11.1 percent), while 33 percent choose
the A&P (40/123 = 32.5 percent). Approximately 43 percent of
Zone 3 families market at Food Fair (124/289 = 42.9
percent), 52 percent at Grand Union (151/289 = 52.4 percent),
and only 5 percent favor the A&P (14/289 = 4.8 percent).
Approximately 5 percent of Zone 4 families prefer the Food
Fair (41/837 = 4.9 percent), but the Grand Union is the
overwhelming favorite of the rest (796/837 = 95.1 percent).

All of the above statistics indicate that Zone 1 families
primarily shop in the A&P, Zone 2 families mainly shop in the
Food Fair or the A&P, Zone 3 families generally shop in the
Grand Union or the Food Fair, and Zone 4 families almost
exclusively shop at the Grand Union. Consequently, our
second analysis revalidated the existence of Stat City
"territorial shopping" behavior.

I hope that this memorandum satisfies your informational
needs. If you have any further questions, please do not
hesitate to call me at 305–999–9999.

Footnotes

[1] The statistics reported are based on a random sample of 100 Stat
City dwelling units and are representative of parameters only on
average. In other words, if many samples of size 100 were drawn, the
average of the sample statistics would afford an accurate
reflection of the true, but unknown, population parameters. Any
given sample's statistics may not accurately reflect the
population's parameters.

Example 6–2 Relationship between alcohol consumption and income

Dr. Marsha Cox, chief psychiatrist for the Stat City Hospital, believes
that poorer families (families earning under $20,000 per year) consume
more alcohol than wealthier families (families earning $20,000 or more per
year). She believes that understanding the relationship between alcohol
consumption and income will enhance her staff's ability to detect families
with drinking problems.

If Dr. Cox's belief is correct, then the Stat City Hospital psychological
counselors should be alerted for possible alcohol-related problems when
working with poorer families. Before issuing a directive to her staff, she
would like to sample Stat City families with reference to this problem.
Basically, she needs comparative statistics depicting the relationship between the consumption of alcohol of poor and wealthy families.

She has decided that the most "practical" measure of alcohol consumption is if a family purchases at least one six-pack of beer per week.

You have been asked by Dr. Cox to perform a study which will shed
light upon her "income-alcohol" premise. She has provided you with
funds to draw a random sample of 50 Stat City families. Conduct the
survey and type a memorandum reporting your findings to Dr. Cox (see
Exhibit 6–2).

Use the randomly selected sample of Stat City dwelling units shown in Table 6–2.

The memorandum appears in Exhibit 6–2.

Exhibit 6–2

HOWARD S. GITLOW, PH.D.
STATISTICAL CONSULTANT

MEMORANDUM

To: Dr. Marsha Cox, Chief Psychiatrist
      Stat City Hospital

FROM: Howard Gitlow

DATE: July 16, 1980

RE: Income–alcohol premise

In order to conduct your requested investigation I used the following definitions for income status and alcohol consumption.

1. Income status =
   - 0 if income was less than $20,000 in 1979 (poor family)
   - 1 if income was $20,000 or more in 1979 (wealthy family).

2. Alcohol consumption =
   - 0 if a family's average weekly purchase of beer as of January 1980 was zero six-packs (nondrinking family).
   - 1 if a family's average weekly purchase of beer as of January 1980 was one or more six-packs (drinking family).

The following table which was constructed using commonly accepted statistical techniques, depicts the estimated breakdown of Stat City families by alcohol consumption and income status.[1]

Table 6–2

ID	INCOM	BEER
5.	40603.	1.
82.	17046.	4.
94.	36165.	3.
122.	62205.	3.
123.	56479.	0.
134.	63330.	0.
140.	23576.	4.
211.	96618.	5.
292.	48054.	0.
316.	28784.	0.
320.	31665.	4.
328.	22232.	4.
408.	33300.	0.
439.	25971.	5.
487.	23527.	9.
501.	26669.	0.
568.	51522.	3.
581.	41974.	0.
637.	16917.	0.
655.	23675.	0.
689.	14369.	7.
712.	13527.	0.
718.	19359.	3.
723.	13377.	3.
743.	16473.	0.
767.	7809.	0.
778.	6476.	0.
793.	18811.	10.
851.	12642.	0.
862.	17522.	0.
884.	11445.	0.
897.	11969.	3.
912.	16601.	3.
931.	8987.	5.
942.	3938.	1.
967.	26820.	0.
972.	21765.	0.
992.	33790.	0.
1018.	14985.	5.
1024.	12822.	5.
1032.	18651.	0.
1078.	6894.	0.
1090.	24361.	2.
1127.	18341.	1.
1155.	14018.	0.
1242.	24743.	2.
1247.	21735.	1.
1299.	29004.	0.
1358.	9207.	0.
1368.	8603.	4.

Exhibit 6–2 (*continued*)

Alcohol consumption (average weekly purchase of beer as of January 1980)	Income status		
	Under $20,000 in 1979 0	$20,000 or more in 1979 1	Total
0 six-packs	357 (26%)	330 (24%)	687 (50%)
1 or more six-packs	329* (24%)	357 (26%)	686 (50%)
Total	686 (50%)	687 (50%)	1,373

* This number is rounded so cells will add to 1,373.

From the above table we note that:

1. The chances of randomly selecting a beer purchasing family out of all poor Stat City families are about 48 in 100 (48 = 329/686) times.
2. The chances of randomly selecting a beer purchasing family out of all wealthy Stat City families are about 52 in 100 (.52 = 357/687) times.

Consequently the "income–alcohol premise" does not hold up upon statistical investigation. However, I would take note that 50 percent of Stat City families do drink so that you may want to issue a directive to your staff telling them to be on guard for alcohol–related problems with all families they counsel.

If you have any further questions, please do not hesitate to call me at 305–999–9999.

Footnotes

[1] The statistics reported are based on a random sample of 50 Stat City dwelling units and are representative of parameters only on average. In other words, if many samples of size 50 were drawn, the average of the sample statistics would afford an accurate reflection of the true, but unknown, population parameters. Any given sample's statistics may not accurately reflect the population's parameters.

Example 6–3   Soda purchase survey

Mr. Marc Cooper, manager of the A&P, would like to determine:

a. The percentage of Stat City families that did not purchase any soda, as of January 1980.
b. The percentage of Stat City families that purchased either diet or regular soda (or both), as of January 1980.
c. The percentage of Stat City families that purchased only diet soda (no regular soda), as of January 1980.
d. The percentage of Stat City families that purchased only regular soda (no diet soda), as of January 1980.
e. If diet soda purchases were related to regular soda purchases, as of January 1980.

You have been asked by Mr. Cooper to conduct a study to answer the above questions. Mr. Cooper has provided you with sufficient funds to draw a simple random sample of 120 Stat City families. Conduct the survey and type a memorandum reporting your findings to Mr. Cooper (see Exhibit 6–3).

Please use the randomly selected sample of 120 Stat City dwelling units given in Table 6–3.

Table 6-3

CASE-N	ID	LSODA	HSODA	CASE-N	ID	LSODA	HSODA	CASE-N	ID	LSODA	HSODA
1	3.	1.	0.	41	469.	0.	0.	81	917.	0.	0.
2	12.	2.	0.	42	477.	0.	0.	82	922.	0.	0.
3	23.	0.	0.	43	494.	0.	1.	83	924.	1.	0.
4	45.	0.	1.	44	496.	2.	0.	84	943.	0.	0.
5	65.	1.	0.	45	504.	0.	1.	85	945.	0.	0.
6	75.	0.	0.	46	529.	0.	1.	86	946.	0.	1.
7	93.	2.	0.	47	531.	0.	0.	87	949.	2.	0.
8	103.	0.	0.	48	533.	0.	0.	88	967.	0.	2.
9	123.	0.	0.	49	546.	0.	0.	89	975.	1.	0.
10	128.	0.	1.	50	571.	0.	5.	90	978.	0.	5.
11	132.	0.	1.	51	588.	0.	1.	91	982.	0.	0.
12	138.	0.	0.	52	608.	1.	0.	92	985.	0.	1.
13	145.	0.	1.	53	611.	1.	0.	93	997.	0.	0.
14	154.	0.	6.	54	613.	0.	0.	94	998.	0.	0.
15	161.	1.	0.	55	627.	0.	6.	95	1007.	2.	0.
16	162.	1.	0.	56	650.	0.	0.	96	1033.	0.	0.
17	183.	0.	0.	57	654.	0.	0.	97	1043.	0.	3.
18	190.	0.	5.	58	666.	2.	0.	98	1077.	0.	1.
19	194.	0.	0.	59	676.	0.	0.	99	1108.	2.	0.
20	209.	2.	0.	60	690.	2.	0.	100	1121.	0.	0.
21	222.	2.	0.	61	695.	0.	3.	101	1126.	0.	1.
22	247.	2.	0.	62	700.	0.	4.	102	1130.	2.	0.
23	249.	2.	0.	63	713.	2.	0.	103	1131.	2.	0.
24	260.	0.	1.	64	720.	0.	0.	104	1133.	0.	0.
25	261.	0.	5.	65	725.	0.	2.	105	1171.	1.	0.
26	266.	0.	0.	66	735.	0.	2.	106	1204.	0.	1.
27	279.	2.	0.	67	737.	2.	0.	107	1206.	2.	3.
28	284.	1.	0.	68	744.	0.	0.	108	1207.	0.	0.
29	327.	0.	1.	69	747.	0.	0.	109	1211.	0.	1.
30	329.	0.	0.	70	776.	2.	0.	110	1231.	0.	4.
31	339.	0.	0.	71	784.	0.	4.	111	1238.	2.	0.
32	346.	0.	1.	72	801.	0.	0.	112	1246.	0.	0.
33	356.	2.	0.	73	818.	0.	1.	113	1251.	2.	0.
34	367.	0.	3.	74	823.	0.	0.	114	1274.	0.	0.
35	382.	0.	0.	75	826.	0.	0.	115	1276.	0.	1.
36	383.	0.	4.	76	832.	0.	2.	116	1291.	2.	0.
37	400.	0.	0.	77	836.	0.	3.	117	1317.	1.	0.
38	423.	0.	1.	78	858.	2.	0.	118	1356.	2.	0.
39	424.	0.	0.	79	865.	0.	1.	119	1359.	0.	4.
40	445.	0.	0.	80	868.	2.	0.	120	1367.	0.	1.

Solution   First, Mr. Cooper's questions must be restated statistically.

	Mr. Cooper's question	*Statistical restatement of Mr. Cooper's question*
a.	Determine the percentage of Stat City families that did not purchase any soda, as of January 1980.	Determine the probability of randomly selecting a Stat City family that did not purchase diet soda and regular soda, as of January 1980. P (LSODA = 0 ∩ HSODA = 0)
b.	Determine the percentage of Stat City families that purchased either diet or regular soda (or both), as of January 1980.	Determine the probability of randomly selecting a Stat City family that purchased diet soda, or regular soda, or both, as of January 1980. P (LSODA > 0 ∪ HSODA > 0)
c.	Determine the percentage of Stat City families that purchased only diet soda (no regular soda), as of January 1980.	Determine the probability of randomly selecting a Stat City family that purchased only diet soda (no regular soda), as of January 1980. P (LSODA > 0 ∩ HSODA = 0)
d.	Determine the percentage of Stat City families that purchased only regular soda (no diet soda), as of January 1980.	Determine the probability of randomly selecting a Stat City family that purchased only regular soda (no diet soda), as of January 1980. P (LSODA = 0 ∩ HSODA > 0)
e.	Determine if diet soda purchases were related to regular soda purchases, as of January 1980.	Determine if diet soda purchases were statistically independent of regular soda purchases, as of January 1980. P (LSODA = 0 ∩ HSODA = 0) $\overset{?}{=}$ P (LSODA = 0) × P (HSODA = 0)

Second, a contingency table must be constructed from the sample data (see Table 6–4).

Table 6–4

Purchases of six-packs of regular soda, as of January 1980 (HSODA)	*Purchases of six-packs of diet soda, as of January 1980 (LSODA)*		
	0	*1 or more*	*Total*
0	42 (0.35)	36 (0.30)	78 (0.65)
1 or more	42 (0.35)	0 (0.00)	42 (0.35)
Total	84 (0.70)	36 (0.30)	120 (100)

Third, the relevant probabilities must be computed.

$$P\,(\text{LSODA} = 0 \cap \text{HSODA} = 0) = \frac{42}{120} = 0.35$$

Hence, it will be inferred that 35 percent of Stat City families did not purchase any soda, as of January 1980.

$$
\begin{aligned}
P\,(\text{LSODA} > 0 \cup \text{HSODA} > 0) &= P\,(\text{LSODA} > 0) + P\,(\text{HSODA} > 0) \\
&\quad - P\,(\text{HSODA} > 0 \cap \text{LSODA} > 0) \\
&= \frac{36}{120} + \frac{42}{120} - \frac{0}{120} \\
&= 0.30 + 0.35 - 0.00 = .65
\end{aligned}
$$

Consequently, it will be inferred that 65 percent of Stat City families purchased either diet or regular soda (or both), as of January 1980.

$$P\,(\text{LSODA} > 0 \cap \text{HSODA} = 0) = \frac{36}{120} = 0.30$$

Hence, it will be inferred that 30 percent of Stat City families purchased only diet soda (no regular soda), as of January 1980.

$$P\,(\text{LSODA} = 0 \cap \text{HSODA} > 0) = \frac{42}{120} = 0.35$$

Consequently, it will be assumed that 35 percent of Stat City families purchased only regular soda (no diet soda), as of January 1980.

$$
\begin{aligned}
P\,(\text{LSODA} = 0 \cap \text{HSODA} = 0) &\overset{?}{=} P\,(\text{LSODA} = 0) \times P\,(\text{HSODA} = 0) \\
0.35 &\neq (0.70 \times 0.65 = 0.455)
\end{aligned}
$$

Since the above joint probability [P(LSODA = 0 $\cap$ HSODA = 0)] is not equal to the product of the above marginal probabilities [P(LSODA = 0) $\times$ P(HSODA = 0)], the purchase of diet soda was statistically dependent upon the purchase of regular soda (and vice versa), as of January 1980.

The memorandum appears in Exhibit 6–3.

The memorandum appears in Exhibit 6–3.

Exhibit 6–3

HOWARD S. GITLOW, PH.D.
STATISTICAL CONSULTANT
_____

MEMORANDUM

TO: Mr. Marc Cooper, Manager
       A&P

FROM: Howard Gitlow

DATE: February 11, 1981

RE: Soda purchase survey, as of January 1980

   In accordance with our consulting contract, I have
investigated the soda purchasing behavior of Stat City
families. All statistics presented in this memorandum are
based on commonly accepted statistical practices.
   The following table presents the estimated breakdown of
Stat City families into nonusers and users of diet and
regular soda, as of January 1980.[1]

Purchases of six-packs of regular soda, as of January, 1980 (HSODA)	Purchases of six-packs of diet soda, as of January 1980 (LSODA)		
	Nonusers (0 six-packs)	Users (1 or more six-packs)	Total
Nonusers (0 six-packs)	481 (35%)	411* (30%)	892 (65%)
Users (1 or more six-packs)	481 (35%)	0 (0%)	481 (35%)
Total	962 (70%)	411 (30%)	1,373 (100%)

* Rounded down to force sum to 1,373.

Exhibit 6-3 (*continued*)

The above table indicates that as of January 1980:

a. 35 percent of Stat City families did not purchase any soda.
b. 65 percent of Stat City families purchased either diet or regular soda (or both).
c. 30 percent of Stat City families purchased only diet soda (no regular soda).
d. 35 percent of Stat City families purchased only regular soda (no diet soda).
e. Diet soda purchases were related to regular soda purchases.

I hope that the above statistics satisfy your informational needs. If you have any questions, please do not hesitate to call me at 305-999-9999.

Footnotes

[1] The statistics reported are based on a random sample of 120 Stat City dwelling units and are representative of parameters only on average. In other words, if many samples of size 120 were drawn, the average of the sample statistics would afford an accurate reflection of the true, but unknown, population parameters. Any given sample's statistics may not accurately reflect the population's parameters.

ADDITIONAL PROBLEMS

6-4 Distribution of dwelling types by residential housing zone Ms. Sharon Vigil, chairperson of the Stat City Real Estate Board, would like information concerning the distribution of apartments and houses in each of the four residential zones in Stat City. This information is important to real estate agents when showing prospective Stat City residents the composition of dwelling units by residential zone.

You have been asked by Ms. Vigil to review the Stat City Census of Housing to determine the distribution of apartments and houses by residential zone.

An investigation of the said census (conducted in January 1980) revealed the following data:

Dwelling type	Zone				
	1	*2*	*3*	*4*	*Total*
Apartments (0)	0	0	126	748	874
Houses (1)	130	157	212	0	499
Total	130	157	338	748	1,373

Type a memorandum to Ms. Vigil reporting your findings.

6–5 Crowding and health status in Stat City   The federal government sent a message to all U.S. cities indicating funds would be available to alleviate crowded housing if it was found that crowding adversely affects health. Consequently, Mr. Lee Kaplowitz, the mayor of Stat City, would like to determine if crowded living space affects health in his city.

The federal government has defined a living space to be crowded if the number of persons per room in a dwelling unit is one or more, and not crowded if the number of persons per room in a dwelling unit is less than one. Further, the federal government has defined the health status of a family to be the average number of visits per year a family makes to the hospital. The health categories are no visits, one visit, or two or more visits per year to a hospital.

You have been asked by Mr. Kaplowitz to conduct a study to investigate the above problem. Mr. Kaplowitz has provided you with sufficient funds to sample 60 Stat City dwelling units. Conduct the study and type a memorandum reporting your findings to Mr. Kaplowitz.

Please assume that you obtained the following table from your random sample of 60 Stat City dwelling units.

Health status	Living space		
	Not crowded (less than one person per room)	*Crowded (one or more persons per room)*	*Total*
No visits per year	4	0	4
One visit per year on average	14	1	15
Two or more visits per year on average	29	12	41
Total	47	13	60

6–6 Beer and soda survey   The Stat City Supermarket Association needs to determine, as of January 1980:

a. The percentage of Stat City families that did not purchase soda and beer.
b. The percentage of Stat City families that purchased either soda or beer (or both).
c. The percentage of Stat City families that purchased only soda (no beer).
d. The percentage of Stat City families that purchased only beer (no soda).
e. If beer purchases were related to soda purchases.

You have been asked by the Stat City Supermarket Association to conduct a survey to answer the above questions. The association has provided you with funds to draw a simple random sample of 90 Stat City families.

Random sample for Problem 6–6

CASE-N	ID	LSODA	HSODA	BEER	SODA	CASE-N	ID	LSODA	HSODA	BEER	SODA
1	4.	0.	6.	0.	6.	46	688.	0.	0.	0.	0.
2	10.	0.	4.	3.	4.	47	691.	1.	0.	0.	1.
3	21.	0.	0.	0.	0.	48	708.	1.	0.	4.	1.
4	27.	0.	0.	0.	0.	49	715.	0.	2.	0.	2.
5	48.	0.	0.	0.	0.	50	770.	0.	0.	4.	0.
6	87.	0.	3.	2.	3.	51	774.	0.	6.	6.	6.
7	93.	2.	0.	2.	2.	52	784.	0.	4.	3.	4.
8	95.	2.	0.	5.	2.	53	800.	2.	0.	0.	2.
9	106.	0.	5.	1.	5.	54	813.	2.	0.	5.	2.
10	114.	2.	0.	2.	2.	55	822.	0.	0.	0.	0.
11	127.	2.	0.	4.	2.	56	829.	2.	0.	2.	2.
12	142.	1.	0.	3.	1.	57	877.	0.	3.	0.	3.
13	181.	0.	0.	7.	0.	58	883.	0.	0.	0.	0.
14	212.	2.	0.	1.	2.	59	891.	0.	2.	4.	2.
15	239.	0.	3.	3.	3.	60	892.	0.	0.	0.	0.
16	254.	0.	1.	1.	1.	61	897.	0.	0.	0.	0.
17	257.	0.	0.	0.	0.	62	908.	0.	0.	0.	0.
18	268.	0.	5.	6.	5.	63	914.	0.	5.	5.	5.
19	273.	0.	0.	0.	0.	64	943.	0.	0.	0.	0.
20	276.	0.	4.	7.	4.	65	960.	2.	0.	0.	2.
21	293.	0.	0.	0.	0.	66	993.	0.	0.	0.	0.
22	305.	0.	4.	0.	4.	67	1004.	0.	4.	10.	4.
23	351.	0.	1.	0.	1.	68	1064.	0.	6.	0.	6.
24	364.	0.	4.	11.	4.	69	1066.	0.	0.	0.	0.
25	380.	2.	0.	3.	2.	70	1108.	2.	0.	4.	2.
26	392.	2.	0.	6.	2.	71	1117.	0.	0.	9.	0.
27	427.	0.	0.	0.	0.	72	1138.	0.	0.	10.	0.
28	450.	0.	4.	3.	4.	73	1139.	0.	1.	4.	1.
29	451.	2.	0.	1.	2.	74	1150.	0.	0.	0.	0.
30	497.	2.	0.	3.	2.	75	1159.	0.	0.	0.	0.
31	519.	2.	0.	1.	2.	76	1186.	0.	0.	0.	0.
32	526.	2.	0.	0.	2.	77	1199.	0.	0.	11.	0.
33	529.	0.	1.	5.	1.	78	1232.	0.	0.	0.	0.
34	546.	0.	0.	0.	0.	79	1237.	1.	0.	6.	1.
35	549.	1.	0.	4.	1.	80	1256.	2.	0.	7.	2.
36	557.	0.	0.	0.	0.	81	1266.	0.	0.	0.	0.
37	561.	0.	0.	3.	0.	82	1267.	0.	0.	0.	0.
38	562.	0.	0.	0.	0.	83	1289.	0.	1.	6.	1.
39	568.	0.	3.	3.	3.	84	1291.	2.	0.	0.	2.
40	605.	1.	0.	0.	1.	85	1297.	1.	0.	0.	1.
41	625.	1.	0.	0.	1.	86	1304.	0.	3.	9.	3.
42	646.	1.	0.	11.	1.	87	1330.	0.	5.	0.	5.
43	664.	0.	0.	0.	0.	88	1337.	2.	0.	3.	2.
44	683.	0.	0.	1.	0.	89	1342.	0.	0.	0.	0.
45	687.	0.	0.	1.	0.	90	1372.	0.	1.	4.	1.

Conduct the survey and type a memorandum reporting your findings to the association.

Please use the accompanying randomly selected sample of 90 Stat City dwelling units.

COMPUTATIONAL EXERCISES

6–7 Using the random sample and contingency tables from the "Computational Exercises" in Chapter 5, compute the following probabilities:

$P(\text{FAVGA} = 1 \cup \text{GASTR} \geq 4)$, $\qquad$ $P(\text{DWELL} = 1 \cap \text{ZONE} \leq 2)$,

$P(\text{FAVGA} = 2 \cup \text{ZONE} = 1)$, $\qquad$ $P(\text{FEAT} = 1 \cap \text{DWELL} = 1)$,

$P(\text{FAVGA} = 1 \cup \text{ZONE} \geq 3)$, $\qquad$ $P(\text{LSODA} = 0 \cup \text{BEER} = 0)$,

$P(\text{GASTR} \geq 3 \cap \text{ZONE} = 2)$, $\qquad$ $P(\text{LSODA} = 0 \cap \text{HSODA} = 0)$, and

$P(\text{GASTR} \leq 2 \cap \text{DWELL} = 0)$, $\qquad$ $P(\text{HSODA} = 0 \cup \text{BEER} \geq 1)$.

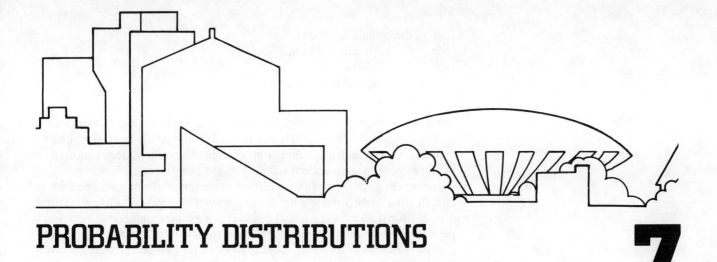

PROBABILITY DISTRIBUTIONS

7

Chapter 3 dealt with constructing and using frequency distributions to describe the characteristics of a population. Chapter 6 presented applications of the basic probability concepts. In this chapter you will use probability concepts to describe frequency distributions for certain Stat City characteristics (variables).

The probability distributions presented in this chapter will help you understand more about Stat City characteristics while using less information than before; however, you will have to rely on certain critical assumptions. If the assumptions you make are reasonable, then you will make appropriate decisions. This chapter will help you become accustomed to knowing when and where these assumptions should be invoked. The topics discussed in this chapter are: the binomial distribution, the normal distribution, the normal approximation to the binomial distribution, and the Poisson distribution.

The binomial distribution is a tool used to determine the probability of x successes occurring in n trials. The assumptions required to use the binomial distribution are: (1) each trial has only two outcomes, (2) the probability of success remains constant for all n trials, and (3) each trial is independent of all other trials. The formula used in most textbooks for computing binomial probabilities is:

$$P(x \text{ successes in } n \text{ trials}|\pi, n) = \frac{n!}{x!(n-x)!} \pi^x (1-\pi)^{n-x}$$

where

n = the number of trials
x = the number of successes
π = the probability of success

Binomial probabilities for various combinations of n, x, and π can be seen in Table B–1 in Appendix B.

The normal distribution is a model used to determine the probability of a random data point falling between two values in the domain of a bell-shaped variable. The formula used in most textbooks for the normal distribution is:

$$f(x) = \frac{1}{\sqrt{2\pi}\sigma} e^{\left[-\frac{1}{2}\left(\frac{x-\mu}{\sigma}\right)^2\right]}$$

where

μ = the population mean
σ = the population standard deviation
π = 3.141592654
x = the value of the variable
e = 2.71828

Standard normal probabilities can be seen in Table B–3 in Appendix B.

The Poisson distribution is a model used to determine the probability of x events occurring in a specified unit of time. The assumptions required to use the Poisson distribution are: (1) there is only one of two possible outcomes in each unit of time, (2) the occurrence of the outcome in one unit of time in no way affects the occurrence of the outcome in another unit of time (independent trials), and (3) the average number of outcomes per specified time unit remains constant for the time period under examination. The formula used in most textbooks for computing Poisson probabilities is:

$$P(x) = \frac{e^{-\mu t}(\mu t)^x}{x!}$$

where

μ = the average number of occurrences per unit of time
t = the number of time units under study
e = 2.71828
x = the number of occurrences
$P(x)$ = the probability of x and occurrences in t units of time

Poisson probabilities for various combinations of μt and x can be seen in Table B–2 in Appendix B.

Example 7–1 Representativeness of the sample to be used in estimating the average number of rooms per dwelling unit in Stat City

Mr. Paul Lund, chairman of the Stat City Chamber of Commerce, selected a sample of 20 Stat City dwelling units to determine the average number of rooms per dwelling unit as of January 1980. Once the sample was drawn, it was determined that of the 20 dwelling units surveyed all were apartments. Mr. Lund knows that 36.3 percent of Stat City dwelling units are houses. Consequently, he is concerned that the sample may not adequately reflect the actual number of rooms per dwelling unit.

Mr. Lund would like to know if the composition of dwelling unit types in his sample could be reasonably expected to occur by chance. You have been asked by Mr. Lund to resolve his dilemma. Type a memorandum to Mr. Lund reporting your findings (see Exhibit 7–1).

Solution   To determine the chances of obtaining zero houses in a sample of 20 dwelling units (if 36.3 percent of all dwelling units are houses), necessitates the use of the binomial distribution. Also, to use the binomial distribution model one must assume the following:

	Assumption	*Assumption met*
1.	Each trial has only two outcomes (house or apartment)	Yes
2.	π remains constant over all n trials (the percent of houses in Stat City remains constant over the duration of the sampling process)	"Yes
3.	Each trial is independent (selecting one dwelling unit in no way affects the selection of another dwelling unit)	Yes

Consequently:

$$n = 20$$
$$x = 0$$
$$\pi = 0.363$$
$$P(x|n,\pi) = \frac{20!}{(20 - 0)!0!}(0.363^0)(0.637^{20-0})$$
$$= 0.000\ 12$$

Approximately 12 chances in 100,000

The memorandum appears in Exhibit 7-1.

Exhibit 7-1

HOWARD S. GITLOW, PH.D.
STATISTICAL CONSULTANT

MEMORANDUM

TO: Mr. Paul Lund, Chairman
        Stat City Chamber of Commerce

FROM: Howard Gitlow

DATE: June 10, 1980

RE: Representativeness of the sample to be used in
        estimating the average number of rooms per dwelling
        unit in Stat City, as of January 1980.

As requested and implementing commonly accepted
statistical techniques, I have determined that the chances
of obtaining zero houses in a sample of 20 Stat City dwelling
units (provided that 36.3 percent[1] of all Stat City dwelling
units are houses) is approximately 12 chances in 100,000.[2]
This result is so unlikely that if type of dwelling unit is
felt to seriously affect the statistic of concern (average
number of rooms per dwelling unit in Stat City as of January
1980), then another sample should be selected and attention
should be given to ensuring proper representation of houses
and apartments in the sample.

    If you have any further questions, please do not hesitate
to call me at 305-999-9999.

Exhibit 7–1 (*continued*)

_____
<u>Footnotes</u>

[1] The Stat City Census revealed that 499 out of 1,373 dwelling units (36.3 percent), as of January 1980, were houses.
[2] To determine the chances of obtaining zero houses in a sample of 20 dwelling units, if 36.3 percent of all dwelling units are houses, requires using the binomial distribution.

Example 7–2   Composition of the Stat City Zoning Board

The Stat City Zoning Board is comprised of 10 supposedly representative head-of-households. However, Mr. Pinero, a Zone 4 resident realized that Zone 4 had no representation on the zoning board despite the fact that the Stat City census recorded 54.5 percent of all Stat City dwelling units are in Zone 4.

You have been retained by Mr. Pinero to determine the probability of Zone 4 families being excluded from the Stat City Zoning Board. Type a memorandum reporting your results to Mr. Pinero (see Exhibit 7–2).

Solution   To ascertain the likelihood of obtaining zero Zone 4 families in a committee of 10 Stat City families, if 54.5 percent of all Stat City families live in Zone 4, requires using the binomial distribution. To use the binomial distribution model requires that the following assumptions have been met:

Assumption	Assumption met
1. Each trial has only two outcomes (family lives in Zone 4 or not)	Yes
2. π remains constant over all n trials (the percent of Zone 4 families in Stat City remains constant over the duration of the zoning board selection process)	Yes
3. Each trial is independent (selecting one family to be on the board in no way affect the selection of other families)	Yes

Consequently:

$$n = 10$$
$$x = 0$$
$$\pi = 0.545$$
$$P(x|n,\pi) = \frac{10!}{(10 - 0)!0!}(0.545)^0(0.455)^{10-0}$$
$$= 0.00038$$
$$\approx \text{less than 4 times in 10,000}$$

Exhibit 7–2

HOWARD S. GITLOW, PH.D.
STATISTICAL CONSULTANT

MEMORANDUM

TO: Mr. Pinero
       902 12th Street
       Zone 4
       Stat City

FROM: Howard Gitlow

DATE: June 12, 1980

RE: Composition of the Stat City Zoning Board

As per your request and using commonly accepted statistical techniques, I have determined that the chances of obtaining "zero Zone 4 head-of-households" on the Stat City Zoning Board are less than 4 times in 10,000.[1] In other words, if 10,000 such zoning boards were formed, one would expect "zero Zone 4 head-of-households" to occur only in 4 of these 10,000 zoning boards. The chances of obtaining "zero Zone 4 head-of-households" is so slight that it is beyond serious discussion to assume that the Stat City Zoning Board was formed by a chance happening.

If you have any further questions, please do not hesitate to call me at 305-999-9999.

Probability distributions

Exhibit 7–2 (*continued*)

_____

<u>Footnotes</u>

[1] This analysis required information from the Stat City Census
indicating that 54.5 percent of Stat City families live in Zone 4.
Further, to determine the chances of obtaining zero Zone 4 families
in a committee of 10 Stat City families, if 54.5 percent of all
Stat City families live in Zone 4, requires using the binomial
distribution.

Example 7–3   Distribution of weekly supermarket bills, per person, in Stat City

The Stat City Supermarket Association would like to know the percentage of families that have weekly supermarket bills per person, as of January 1980, in the categories as shown in Table 7–1. The Supermarket

Table 7–1

Category of purchase	Definition
Small	Under $19.00 per week, per person
Medium	Between $19.00 and $24.99 per week, per person
Large	$25.00 or more per week, per person.

Association obtained the mean weekly supermarket bill per person ($21.86) and the standard deviation of weekly supermarket bills per person ($3.22), as of January 1980, from the Stat City Census. Unfortunately, they were unable to get any more detailed information on the distribution of weekly supermarket bills per person in Stat City.

You have been hired by the Supermarket Association to ascertain the desired information. Type a memorandum reporting your results to the Stat City Supermarket Association (see Exhibit 7–3).

Solution

To compute the percentage of Stat City families in the small, medium, and large categories of weekly supermarket bills per person, you must make an assumption about the shape of the frequency distribution. It seems reasonable to assume that the distribution of weekly supermarket bills per person is normal. The normal assumption is based on the idea that most families consume $21.86 of supermarket items per person, per week, and that the percentage of families deviating from $21.86 trails off in both directions in a symmetric, bell-shaped pattern.

The percent of small purchasers is 18.67 percent.

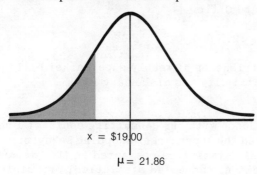

The percent of medium purchasers is 64.98 percent.

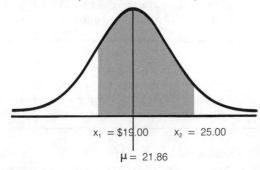

$$Z_1 = \frac{x_1 - \mu}{\sigma} \qquad\qquad Z_2 = \frac{x_2 - \mu}{\sigma}$$

$$= \frac{19.00 - 21.86}{3.22} \qquad = \frac{25.00 - 21.86}{3.22}$$

$$= \frac{-2.86}{3.22} = -0.89 \qquad = \frac{3.14}{3.22} = +0.98$$

$$P(-0.89 \le Z_1 \le 0) = 0.3133 \qquad P(0 \le Z_2 \le 0.98) = 0.3365$$

$$P(-0.89 \le Z \le 0.98) = P(\$19.00 \le X \le \$25.00) = 0.6498$$

The percent of large purchasers is 16.35 percent.

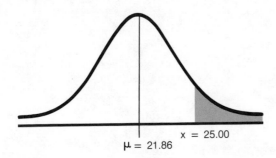

The memorandum appears in Exhibit 7–3.

Exhibit 7–3

HOWARD S. GITLOW, PH.D.
STATISTICAL CONSULTANT

MEMORANDUM

TO: Stat City Supermarket Association

FROM: Howard Gitlow

DATE: June 12, 1980

RE: Distribution of weekly supermarket bills per person
        in Stat City, as of January 1980

In accordance with our consulting contract I have investigated the supermarket shopping behavior of Stat City families. All statistics presented in this memorandum are based on commonly accepted statistical principles.

The following table presents the percent of Stat City families who have weekly supermarket bills per person, as of January 1980, in the following categories:[1]

Category of purchase	Definition of category	Percent
Small	Under $19.00 per week per person.	18.67
Medium	Between $19.00 and $24.99 per week per person.	64.98
Large	$25.00 or more per week per person.	16.35

I hope the above table satisfies your informational needs. If you have any further questions, please do not hesitate to call me at 305–999–9999.

Exhibit 7-3 (*continued*)

Footnotes

¹ I had to make an assumption about the distribution of weekly
supermarket bills per person, as of January 1980, to determine the
percentage of Stat City families in the small, medium, and large
categories of weekly supermarket bills per person. It seems
reasonable to assume that weekly supermarket bill per person is
normally distributed. The normal assumption is based on the idea
that most families consume $21.86 of supermarket items per person,
per week, and that the percent of families deviating from $21.86
trials off in both directions in a symmetric, bell-shaped pattern.

Example 7-4   HUD "home improvement" funding for Stat City

Mr. Lee Kaplowitz, the mayor of Stat City, would like to determine if
his city is eligible for HUD (Housing and Urban Development) funding to
improve the standard of living for homeowners. The HUD guideline for
funding eligibility is that at least 15 percent of the houses in a city must
have assessed values under $20,000. Consequently, Mr. Kaplowitz needs
to determine the percent of Stat City houses with assessed values under
$20,000, as of January 1980.

Mr. Kaplowitz obtained the average assessed value for houses
($67,980.65) and the standard deviation of assessed value for houses
($25,711.26), as of January 1980, from the Stat City Census. Unfortu-
nately, he was unable to get any more detailed information on the distribution of assessed values of houses.

You have been retained by the mayor to determine the required information. Type a memorandum reporting your findings to Mr. Kaplowitz
(see Exhibit 7-4).

First, the relevant population must be defined. The population consists
of the 499 houses in Stat City. Second, an assumption about the shape of
the frequency distribution of assessed values, as of January 1980, must be

Solution

made. It seems reasonable to assume that assessed housing values are normally distributed. This assumption will be made because no other distributional assumption is more viable than the normal assumption. If the normal assumption is false, then the results reported to Mr. Kaplowitz will be wrong. The degree of error in the memorandum is directly related to how far the distribution of assessed values is from the normal distribution.

The percent of houses in Stat City that have assessed values under $20,000 is 3.07 percent.

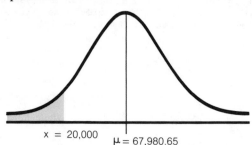

x = 20,000 μ = 67,980.65

Consequently, Stat City is not eligible for HUD home improvement funding.

The memorandum appears in Exhibit 7–4.

Exhibit 7–4

HOWARD S. GITLOW, PH.D.
STATISTICAL CONSULTANT

MEMORANDUM

TO: Mr. Lee Kaplowitz, Mayor
        Stat City

FROM: Howard Gitlow

DATE: June 16, 1980

RE: HUD "home improvement" funding for Stat City

   As per your request and using commonly accepted
statistical techniques, I have determined that the
percentage of Stat City houses with assessed values under
$20,000, as of January 1980, is 3.07 percent.[1] I hope the
above statistic meets your informational needs. If you have
any further questions, please do not hesitate to call me at
305-999-9999.

_____

                                        Footnotes

   [1] The computation of 3.07 percent was based on the assumption that
the distribution of assessed values for Stat City houses, as of
January 1980, is normal. The reported 3.07 percent would change if
the distribution of assessed values was not normal. However, the
distribution of assessed values would have to depart rather
severely from the normal distribution before 15 percent or more of
Stat City houses would have assessed values under $20,000.

Example 7–5 Composition of the Stat City Planning Board

The Stat City Planning Board consists of 25 members who are purportedly representative "head-of-households." A Zone 4 resident (Mr. Pinero again) realized that only three Zone 4 families were on the planning board. It is known from the Stat City Census that 54.5 percent of all Stat City dwelling units are in Zone 4.

You have been retained by Mr. Pinero to determine the likelihood of acquiring three or less Zone 4 "head-of-households" on the planning board by chance. Type a memorandum reporting your results to Mr. Pinero (see Exhibit 7–5).

Solution

To determine the chances of obtaining three or fewer Zone 4 "head-of-households" on the 25 member Stat City Planning Board, if 54.5 percent of all Stat City "head-of-households" live in Zone 4, requires using the binomial distribution. To use the binomial distribution model requires that the following assumptions have been met:

	Assumption	Assumption met
1.	Each trial has only two outcomes (Zone 4 head-of-household or not).....	Yes
2.	π remains constant over all n trials (the percent of Zone 4 head-of-households remains constant each time a head-of-household is selected onto the board)	Yes
3.	Each trial is independent (selecting one head-of-household in no way affects the selection of any other head-of-household...................	Yes

Consequently:

$$n = 25$$
$$x \leq 3$$
$$\pi = 0.545$$

Unfortunately, $P(x \leq 3 | n = 25$ and $\pi = 0.545)$ is cumbersome to compute. Consequently, this problem requires that the normal approximation to the binomial distribution be used because n is large and computations would be extremely tedious and time consuming. However, before the normal approximation can be used we must check to make sure that π is close to 0.50. Whenever $\pi = 0.50$, the binomial distribution is symmetric and resembles the normal distribution. The further π strays from 0.50 (either higher or lower), the less symmetrical (normal looking) the binomial distribution will be. As long as π stays between 0.10 and 0.90 most statisticians agree that the normal approximation is acceptable.

There are two commonly accepted guidelines that can be used in determining if the normal approximation can be used: $n\bar{p} > 5$ and $n(1 - \bar{p}) > 5$ if you are dealing with a sample, and $N\pi > 5$ and $N(1 - \pi) > 5$ if you are dealing with a population.

In the problem at hand $n = 25$ and $\pi = 0.545$. Consequently,

$$n\pi = 25(0.545) = 13.625 > 5$$
$$n(1 - \pi) = 25(0.455) = 11.375 > 5$$

Hence, the normal approximation to the binomial distribution is appropriate.

The following graph and supportive formulas are used to compute the likelihood of finding three or fewer Zone 4 head-of-households on the 25-member planning board, given that 54.5 percent of Stat City head-of-households live in Zone 4.

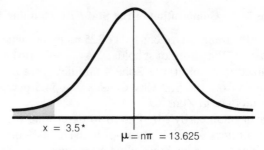

$$x = 3.5^*$$ $$\mu = n\pi = 13.625$$

* 3 + 0.5 is correction for continuous approximation.

The memorandum appears in Exhibit 7–5.

Exhibit 7–5

HOWARD S. GITLOW, PH.D.
STATISTICAL CONSULTANT

MEMORANDUM

TO: Mr. Pinero
       902 12th Street
       Zone 4
       Stat City

FROM: Howard Gitlow

DATE: June 24, 1980

RE: Composition of the Stat City Planning Board

   As per your request and using commonly accepted
statistical techniques, I have determined that the
likelihood of finding three or fewer Zone 4
"head–of–households" on the 25 member Stat City Planning
Board is less than 26 times in 1 million.[1] This result is so
unlikely that it precludes serious discussion of the board
having been randomly selected. In other words, it is not
reasonable to assume that the planning board is a fair
representation of Stat City with respect to zonal
representation.
   If you have any further questions, please do not hesitate
to call me at 305–999–9999.

Exhibit 7–5 (*continued*)

Footnotes

[1] To determine the chances of obtaining three or fewer Zone 4 head–of–households being on a 25–member planning board, given 54.5 percent of all Stat City head–of–households reside in Zone 4, requires using the normal approximation to the binomial distribution. (Note: The Stat City Census revealed that 54.5 percent of all dwelling units in Stat City are in Zone 4.)

Example 7–6 Possibility of false advertising claims by Paul's Texaco Station

Mr. Paul Sugrue, the owner of Paul's Texaco Station, recently formed a consumer panel consisting of 100 Stat City "head-of-households." Mr. Sugrue asked the panel to name their favorite gas station. Fifty-two of the 100 panelists said they prefer Paul's Texaco Station. Consequently, Mr. Sugrue claimed in a newspaper advertisement that his station was rated number one by a consumer panel.

Mr. Howie, the owner of Howie's Gulf Station was extremely upset at this "false" advertising. From the Stat City Census, Mr. Howie knows that only 40.4 percent of all Stat City residents regularly choose Paul's Texaco Station. Mr. Howie can accept that Mr. Sugrue was unaware of the Census figures and had conducted a study of his own to establish the 52 percent (52/100) statistic. However, Mr. Howie is worried that Mr. Sugrue selected his consumer panel unfairly; for example, Mr. Sugrue may have selected the first 100 "head-of-households" to come to his station on a given day.

Consequently, Mr. Howie has hired you to ascertain the likelihood of finding 52 or more "head-of-households" that favor Paul's Texaco Station, out of 100 "head-of-households," given that 40.4 percent of all "head-of-households" favor Paul's Texaco Station. Type a memorandum to Mr. Howie reporting your findings (see Exhibit 7–6).

Solution   To determine the likelihood of finding 52 or more "head-of-households" that favor Paul's Texaco Station, out of 100 "head-of-households," given that 40.4 percent of all "head-of-households" favor Paul's Texaco Station, requires using the normal approximation to the binomial distributions. The normal approximation is necessary because computations would be too cumbersome otherwise. Interest in this problem centers on determining the chances of finding 52 or more "head-of-households" because Mr. Howie wants to know the likelihood that 52, or, worse yet, even more "head-of-households" favor Paul's Texaco Station. If the chances of getting 52 or more "head-of-households" favoring Paul's Texaco Station are small, then certainly the chances of getting exactly 52 "head-of-households" favoring Paul's Texaco Station are even smaller. Hence, using 52 or more is a very conservative method for estimating the representativeness of the sample.

Before using the normal approximation to the binomial distribution we must be certain that the three assumptions of the binomial distribution have been met and that $n\pi$ and $n(1 - \pi)$ are both greater than five. In this problem the three binomial assumptions seem reasonable, $n\pi = 100(0.404) = 40.4$, and $n(1 - \pi) = 100(0.596) = 59.6$; hence, we can procede.

The likelihood of randomly finding 52 or more "head-of-households" that favor Paul's Texaco Station, out of 100 "head-of-households," given that 40.4 percent of all Stat City "head-of-households" favor Paul's Texaco Station, is computed as follows:

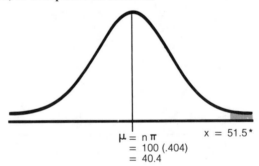

$$\mu = n\pi$$
$$= 100 \, (.404)$$
$$= 40.4$$

$$x = 51.5*$$

The memorandum appears in Exhibit 7–6.

Exhibit 7–6

HOWARD S. GITLOW, PH.D.
STATISTICAL CONSULTANT
———

MEMORANDUM

TO: Mr. Howie, Owner
       Howie's Gulf Station

FROM: Howard Gitlow

DATE: June 30, 1980

RE: Possibility of false advertising claims by Paul's
       Texaco Station

   Using commonly accepted statistical techniques, I have
determined that there is little likelihood of randomly
finding 52 or more "head–of–households" that favor Paul's
Texaco Station out of 100 "head–of–households." In fact, it
is less than 12 times in 1,000.' This result is so unlikely
that it precludes serious discussion of the consumer panel
having been randomly chosen. In other words, it is not
reasonable to assume that the consumer panel is a fair and
representative sample of Stat City "head–of–households,"
in respect to favorite gas station.
   If you have any further questions, please do not hesitate
to call me at 305–999–9999.

Exhibit 7–6 (*continued*)

```
                              Footnotes
    ¹ To determine the likelihood of finding 52 or more
"head-of-households" that favor Paul's Texaco Station out of 100,
given that 40.4 percent of all "head-of-households" favor Paul's
Texaco Station, requires using the normal approximation to the
binomial distribution. (Note: The Stat City Census revealed that
40.4 percent of all Stat City "head-of-households" think of
Paul's Texaco Station as their favorite gas station.)
```

Example 7–7   Staffing the Stat City Chamber of Commerce switchboard

The Stat City Chamber of Commerce receives an average of five telephone calls per minute. An average switchboard operator can handle three telephone calls per minute. The Chamber of Commerce is trying to determine the number of switchboard operators they need so that no more than 1 percent of all incoming telephone calls go unanswered. The chamber can hire part-time employees.

You have been hired by Mr. Paul Lund, chairman of the Chamber of Commerce, to determine how many switchboard operators the Chamber must hire. Type a memorandum to Mr. Lund reporting your findings (see Exhibit 7–7).

Solution   To compute the number of switchboard operators required to answer at least 99 percent of all incoming telephone calls to the Stat City Chamber of Commerce, requires using the Poisson distribution. Use of the Poisson distribution is predicated upon the following assumptions:

	Assumption	Assumption met
1.	Number of events occurring in one time interval is independent of what happened in previous time intervals.............................	Yes
2.	The average number of occurrences per unit of time remain constant for the time period under examination................................	Yes
3.	It is extremely rare for more than one event to occur in a small interval of time...	Yes

Consequently, $\mu t = 5$, and

$$\left\{ P(x > X) = 1 - \sum_{i=0}^{X} \frac{e^{-(\mu t)}(\mu t)^i}{i!} \right\} \leq 0.01$$

	i	$P(x = i)$	$P(x \leq i)$	$P(x > i)$
	0	.006738	.006738	
	1	.033690	.040428	
	2	.084224	.124652	
	3	.140374	.265026	
	4	.175467	.440493	
	5	.175467	.615960	
	6	.146223	.762183	
	7	.104445	.866628	
	8	.065278	.931906	
	9	.036266	.968172	
$X = i \rightarrow$	10	.018133	.986305	.013695
	11	.008242	.994570	.005430 $= P(x > 10) = P(x \geq 11)$

As you can see, the chamber switchboard receives 11 or more telephone calls per minute approximately 0.50 percent of the time. Hence, the chamber must have the capability of answering at least 10 calls per minute if they are to miss only 1 percent or less of incoming telephone calls. To answer 10 calls per minute, given an average operator can answer three calls per minute, the chamber must hire three and one-third operators (10 calls per minute divided by 3 calls per minute equals 3⅓ operators).

The memorandum appears in Exhibit 7–7.

Exhibit 7–7

HOWARD S. GITLOW, PH.D.
STATISTICAL CONSULTANT

MEMORANDUM

TO: Mr. Paul Lund, Chairman
        Stat City Chamber of Commerce

FROM: Howard Gitlow

DATE: February 23, 1981

RE: Staffing the Stat City Chamber of Commerce
        switchboard

The Stat City Chamber of Commerce must employ three and
one–third switchboard operators if you are to have the
capability of answering at least 99 percent of all incoming
telephone calls.[1] The above calculations presume that an
average of five telephone calls come into the chamber per
minute and that a typical switchboard operator can handle
three calls per minute.

The above computations are based on commonly accepted
statistical techniques. If you have any further questions.
please do not hesitate to call me at 305–999–9999.

Footnote

[1] The computations presented in this memorandum are based
upon the Poisson distribution and consequently are subject
to its assumptions.

ADDITIONAL PROBLEMS

7–8 Composition of the Stat City task force on problem drinking Mr. Lee Kaplowitz, the mayor of Stat City, convened a five-person task force to discuss problem drinking in Stat City. Mr. Upton (identification number 115), the local tavern owner claimed that the task force had not been randomly chosen but had been personally selected to come up with anti-alcohol recommendations. Mr. Upton claimed that not one person on the task force came from a family that purchased one or more six-packs of beer on a weekly basis. This determination was made by canvasing each member of the task force. Mr. Upton decided to use beer drinking as an extremely conservative measure of attitude toward alcohol; that is, nonbeer drinkers have negative attitudes toward alcohol and beer drinkers have positive attitudes.

Mr. Upton checked the Stat City Census and found that 46.76 percent of Stat City families (642) purchase one or more six-packs of beer weekly. You have been hired by Mr. Upton to determine the chances of randomly selecting a five-person task force in which not one member comes from a family that purchase beer on a weekly basis. Type a memorandum reporting your findings to Mr. Upton.

7–9 Federal request for information on housing costs in Stat City The mayor of Stat City, Mr. Lee Kaplowitz, must determine if more than 30 percent of Stat City families spend less than $200 per month on housing (includes rent or mortgage), as of January 1980. This information is required by the federal Department of Health and Human Services (HHS).

Mr. Kaplowitz obtained the average housing cost ($462.80) and the standard deviation of housing costs ($171), as of January 1980, from the Stat City Census. Unfortunately, he was unable to obtain more detailed information on the distribution of housing costs in Stat City.

You have been retained by Mr. Kaplowitz to determine the required information. Type a memorandum reporting your findings to the mayor. (Assume housing costs are normally distributed.)

7–10 False advertising claims by the A&P Mr. Marc Cooper, the manager of the A&P, recently formed a consumer panel consisting of 50 Stat City families (each family is represented by the person most responsible for the family's supermarket shopping). He asked the panel which supermarket in Stat City they prefer. Thirty-one of the 50 family representatives said they prefer the A&P. Consequently, Mr. Cooper claimed in a TV commercial that the A&P is Stat City's favorite supermarket.

The management of the Food Fair and the Grand Union (especially the Grand Union) were extremely upset at this "false" claim. Grand Union management knew from the Stat City Census that only 19.5 percent of all Stat City families prefer the A&P. Food Fair and Grand Union management can accept that Mr. Cooper had not known about the Census figures and had conducted a study of his own to arrive at the 62 percent (31/50) statistic. However, Food Fair and Grand Union management are concerned that Mr. Cooper unfairly selected his consumer panel; for example, he may have selected the first 50 people who came into the A&P on a given day.

Consequently, Food Fair and Grand Union management have hired you to determine the chances of finding 31 or more families that favor the A&P, out of 50 families, given that 19.5 percent of all families favor the A&P. Report your findings to Food Fair and Grand Union management.

7–11 Staffing the maternity ward in Park View Hospital Ms. Arlene Davis, director of the Park View Hospital, knows from experience that approximately three women enter the hospital per day to give birth. Further, she also knows that the staff obstetrician, Dr. Deborah Jones, can deliver a maximum of five babies per day.

Ms. Davis has hired you to determine the average number of days per year that Dr. Jones will be unable to handle the inflow of expectant mothers. Type a memorandum reporting your findings to Ms. Davis.

7–12 Aiding library patrons Ms. Elissa Gitlow, chief librarian (the only librarian) of the Stat City Library, knows that an average of five people per hour enter the library seeking aid from a librarian. She also knows that she can help an average of seven people per hour.

She has retained you to determine the percentage of the time that she will not be able to help patrons seeking the advices of a librarian. Type a memorandum reporting your findings to Ms. Gitlow.

COMPUTATIONAL EXERCISES

7–13 Assume that HEAT, ELEC, PHONE, GAS, and EAT are all normally distributed random variables. The Stat City Census revealed the following parameters for the aforementioned random variables:

	Mean (μ)	Standard deviation (σ)
HEAT	$648.89	$271.57
ELEC	$ 55.76	$ 19.73
PHONE	$ 36.70	$ 26.50
GAS..................	$134.37	$ 44.88
EAT	$ 90.19	$ 33.14

Compute the following quantities:

$P(\text{HEAT} > \$700) =$
$P(\text{HEAT} < \$600) =$
$P(190 \leq \text{HEAT} \leq \$600) =$
$P(\text{ELEC} > \$40) =$
$P(\$60 \leq \text{ELEC} \leq \$70) =$
$P(\$25 \leq \text{PHONE} \leq \$35) =$
$P(\text{PHONE} \leq \$10) =$
$P(\text{GAS} < \$100 \text{ OR GAS} > \$200) =$
$P(\text{EAT} < \$70 \text{ OR EAT} > \$100) =$

7–14 The Stat City Census revealed the following parameters:

$P(\text{HSODA} = 0) = .673,$
$P(\text{LSODA} = 0) = .689,$
$P(\text{BEER} > 3) = .238,$
$P(\text{FEAT} = 2) = .606,$
$P(\text{FAVGA} = 1) = .404,$
$P(\text{PEPLE} = 1) = .046,$
$P(\text{CARS} < 2) = .731,$
$P(3 \leq \text{ROOMS} \leq 8) = .829.$

Compute the following binomial probabilities:

1. The probability of randomly selecting 3 families that do not purchase any six-packs of regular soda weekly out of 10 randomly selected families.
2. The probability of randomly selecting more than 8 families that do not purchase any six-packs of diet soda weekly out of 12 randomly selected families.
3. The probability of randomly selecting less than three families that purchase more than three six-packs of beer weekly out of five randomly selected families.
4. The probability of randomly selecting one or more families that prefer the Grand Union out of seven randomly selected families.
5. The probability of randomly selecting between three and five families (inclusive) that favor Paul's Texaco Station out of nine randomly selected families.
6. The probability of randomly selecting between two and four single-person families (inclusive) out of nine families.
7. The probability of randomly selecting 3 families that own less than two cars out of 100 randomly selected families.
8. The probability of randomly selecting 50 or more families that own homes with between three and eight rooms inclusive out of 75 randomly selected families.

PART FOUR

SAMPLING DISTRIBUTIONS

INTRODUCTION

In this chapter we are going to formalize something we have been doing since Chapter 1, drawing conclusions about populations from samples. The process of making a generalization from a sample statistic to a population parameter is called statistical inference. In Chapter 1, we drew a simple random sample of 10 Stat City dwelling units and computed the average number of rooms per dwelling unit to be 6.5. We then inferred that the average number of rooms for all Stat City dwelling units is 6.5. This generalization is a statistical inference.

A major advantage of random sampling is that it allows a researcher to assign a probability that a sample statistic accurately estimates an unknown population parameter. Learning how to assign probability statements to sample statistics will be a major focus of this chapter.

To use statistical inference you must understand the behavior of sample statistics. For example, logic would dictate that if two random samples of size 10 were drawn from Stat City to estimate the average number of rooms per dwelling unit, each sample would have different sample means. This would happen because different samples include different elements of the population. Consequently, you must learn to deal with errors which occur when estimating population parameters due to sample-to-sample variation. Sample-to-sample variation is called sampling error. You will use the probability models you studied in Chapter 7 (binomial, normal, plus others) to better understand sampling error and the problems it causes when estimating population parameters.

The frequency distribution of all possible sample statistics that can be drawn from a population is called a sampling distribution. An understanding of sampling distributions will allow you to make proper statistical inferences about a population from a sample by accounting for sampling error.

Understanding a statistical inference is more complex than a person not trained in statistics would suspect. Consequently, it is imperative that the footnotes in Exhibit 8–1 (or some rendition thereof) appear whenever a statistical inference is being reported in a memorandum. These footnotes serve as a partial disclaimer for the errors which can occur when computing statistics from a particular sample; remember, a sample statistic is subject to sampling and nonsampling error.

This chapter will give you insights into the sampling distributions of the sample mean and proportion. I hope that these insights in conjunction

Exhibit 8–1   Footnotes

[1] The summary statistics reported are estimates derived from a sample survey of _____ (_____) elementary units; drawn via a _____ random sampling plan.

"There are two types of errors possible in an estimate based on a sample survey–sampling and nonsampling. Sampling errors occur because observations are made only on a sample, not on the entire population. Nonsampling errors can be attributed to many sources; for example, inability to obtain information about all cases in the sample, definitional difficulties, differences in the interpretation of questions, inability or willingness to provide correct information on the part of respondents, mistakes in recording or coding the data obtained and other errors of collection, response, processing, coverage, and estimation for missing data. Nonsampling errors also occur in complete censuses. The 'accuracy' of a survey result is determined by the joint effects of sampling and non–sampling errors." (Journal of the American Statistical Association, vol. 70, no. 351, part II [September 1975], p. 6.)

[2] "The particular sample used in this survey is one of a large number of all possible samples of the same size that could have been selected using the same sample design. Estimates derived from the different samples would differ from each other. The difference between a sample estimate and the average of all possible samples is called the sampling deviation. The standard or sampling error of a survey estimate is a measure of the variation among the estimates from all possible samples, and thus is a measure of the precision with which an estimate from a particular sample approximates the average result of all possible samples. The relative standard error is defined as the standard error of the estimate divided by the value being estimated.

"As calculated for this report, the standard error also partially measures the effect of certain nonsampling errors but does not measure any systematic biases in the data. Bias is the difference, averaged over all possible samples, between the estimate and the desired value. Obviously, the accuracy of a survey result depends on both the sampling and nonsampling errors measured by the standard error and the bias and other types of nonsampling error not measured by the standard error." Journal of the American Statistical Association, vol. 70, no. 351, part II [September 1975], p. 6.

with the theory your professor is giving you will make it clear how to make proper statistical inferences.

There will not be any problems in this chapter; there will just be illustrations of how the sampling distributions of the mean and proportion behave under various sets of "real-world circumstances."

GENERAL DISCUSSION OF THE SAMPLING DISTRIBUTION OF THE SAMPLE MEAN (x)

If all possible random samples of size n are drawn from a population of size N and the mean for each sample is computed, the frequency distribution of all the sample means is called the sampling distribution of $\bar{x}$. The shape of the sampling distribution of $\bar{x}$ changes in accordance with certain "real-world conditions." Table 8–1 indicates that there are three basic shapes for the sampling distribution of $\bar{x}$: normal, t, and unknown.

Each of these basic shapes must be thoroughly understood to make proper statistical inferences about μ from $\bar{x}$. Table 8–1 also delineates the relationship between "real-world conditions" and the shape of the sampling distribution of $\bar{x}$. Consequently, once you can isolate which "real-world conditions" are present in your problem, you can determine the basic shape of the sampling distribution of $\bar{x}$ and can make appropriate statistical inferences about μ from your $\bar{x}$. It is important to realize that

Table 8–1

Case	Sample size n (n ≥ 30 is large) (n < 30 is small)	Population standard deviation (σ)	Distribution of original population	Shape of the sampling distribution of x̄		
				Shape 1	Shape 2	Shape 3
1.............	Large	Known	Normal	Normal		
2.............	Large	Known	Unknown	Normal		
3.............	Large	Unknown	Normal	Normal		
4.............	Large	Unknown	Unknown	Normal		
5.............	Small	Known	Normal	Normal		
6.............	Small	Unknown	Normal		t	
7.............	Small	Known	Unknown			Unknown
8.............	Small	Unknown	Unknown			Unknown

sampling distributions are never constructed in practice, they are theoretical abstractions which allow researchers to make appropriate statistical inferences.

Sampling distribution of x̄ when the sample size is large, the population standard deviation is known, and the distribution of the original population is normal (Case 1)

To understand the sampling distribution of x̄ for Case 1, 100 random samples of sizes 2, 5, 10, and 30, were drawn concerning family incomes of Stat City families. For each sample size, all 100 sample means were computed.

Table 8–2 shows the relationship between the sample size and the standard error of the mean for total family income in Stat City, given the population standard deviation is known (σ = \$22,847.12).

Table 8–2

Case	Sample size (n)	Population standard deviation (σ)	Population size (N)	Finite population correction factor $\sqrt{\frac{N-n}{N-1}}$	Standard error of x̄ ($\sigma_{\bar{x}}$)	
					Infinite population $\sigma/\sqrt{n}$	Finite population $(\sigma/\sqrt{n})\sqrt{\frac{N-n}{N-1}}$
1.........	1	\$22,847.12	1,373	1.0	\$22,847.12	\$22,847.12
2.........	2	\$22,847.12	1,373	0.99964	\$16,155.35	\$16,149.46
3.........	5	\$22,847.12	1,373	0.99854	\$10,271.54	\$10,202.64
4.........	10	\$22,847.12	1,373	0.99671	\$ 7,224.89	\$ 7,201.16
5.........	30	\$22,874.12	1,373	0.98938	\$ 4,171.29	\$ 4,126.97

As you can see, the standard error of x̄ decreases as the sample size increases. This simply demonstrates that as *n* is increased, the chances of any given x̄ deviating from μ decreases, because the sample mean is based on a larger proportion of the population. The above rule holds for infinite and finite populations.[1]

Frequency distributions of the 100 sample means for each sample size are shown in Exhibit 8–2. This table graphically illustrates that the larger the sample size, the more closely the relative frequency distribution of the sample means (sampling distributions of x̄) approaches the normal distribution. Further, the mean of the sampling distribution of

---

[1] For all five cases in Table 8–1 the finite population correction factor (fpc) is negligible; $\frac{\sigma}{\sqrt{n}}\sqrt{\frac{N-n}{N-1}} \cong \frac{\sigma}{\sqrt{n}}[1] \cong \frac{\sigma}{\sqrt{n}}$. The fpc is unimportant when [n/N] is less than 10 percent. If [n/N] is greater than 10 percent (as *n* approaches *N*), the fpc becomes a significant modifier of the standard error of x̄.

$\bar{x}$ is μ and the standard deviation (called the standard error of $\bar{x}$) is $\sigma/\sqrt{n}$ if the population is infinite, and $\dfrac{\sigma}{\sqrt{n}} \sqrt{\dfrac{N-n}{N-1}}$ if the population is finite.

The above phenomenon can be demonstrated mathematically and is called the central limit theorem for means.

Exhibit 8–2

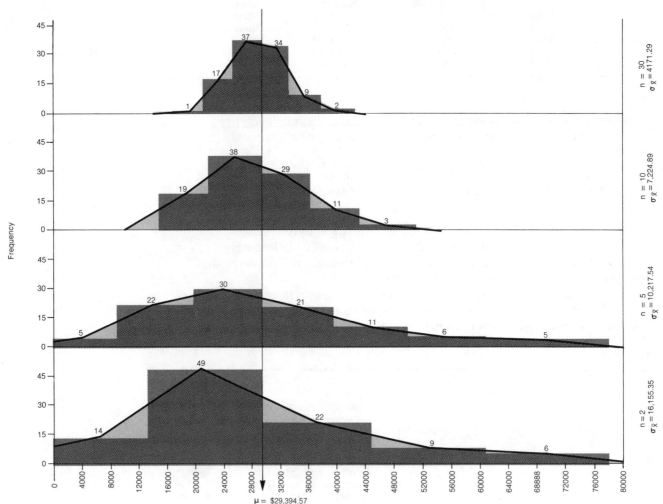

Central limit theorem for means

If random samples of n observations are drawn from a population with a finite mean (μ) and a finite standard deviation (σ), as n becomes large (greater than 30), the distribution of the sample means ($\bar{x}$) will approximate a normal distribution with mean μ and standard error $\sigma/\sqrt{n}$ for infinite populations, and $\dfrac{\sigma}{\sqrt{n}} \sqrt{\dfrac{N-n}{N-1}}$ for finite populations.

Now that it has been established that the frequency distribution of the sample means is approximately normal for large n, you can use what you know about the normal distribution to interpret the ONE SAMPLE MEAN you have in practice.

Remember, you know that approximately 99.73 percent of all data in a normal population lies within three standard deviations (σ) of the population mean (μ).

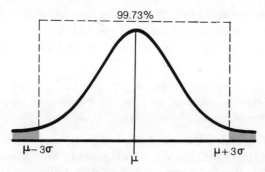

99.73%

$\mu - 3\sigma$ μ $\mu + 3\sigma$

Hence, if the sampling distribution of means is normal, then approximately 99.73 percent of all sample means will fall within three standard errors of the unknown population mean. Consequently, if the standard error is large, 99.73 percent of all sample means will fall within a WIDE range around μ (three standard errors either way of μ) and you cannot have "confidence" that your sample mean will provide a good estimate of μ. However, if the standard error is small, 99.73 percent of all sample means will fall within a small range around μ (three standard errors either way of μ) and you can have "confidence" that your sample mean will provide a good estimate of μ.

Remember, even if the standard error of the mean is small, you never know for certain if your $\bar{x}$ is an accurate estimate of μ. As the standard error decreases, the chances of your $\bar{x}$ deviating from μ decrease, but there is always a finite probability that even with a small standard error, your $\bar{x}$ may be one of the $\bar{x}$s which will not provide a good estimate of μ. However, since you never know if your $\bar{x}$ is a good estimator of μ or not, you will base your evaluation of $\bar{x}$ on the following criteria: assume that $\bar{x}$ is a good estimator of μ if the standard error is small or assume that $\bar{x}$ may be a poor estimator of μ if the standard error is large.

Finally, a way of increasing your confidence in $\bar{x}$ as an estimator of μ is to increase the sample size upon which $\bar{x}$ is based. Remember, the standard error of $\bar{x}$ is:

$$\frac{\sigma}{\sqrt{n}} \text{ for infinite populations, and}$$

$$\frac{\sigma}{\sqrt{n}} \sqrt{\frac{N - n}{N - 1}} \text{ for finite populations.}$$

As n gets larger, the standard error of the mean gets smaller and the $\bar{x}$s are more closely grouped around μ. In other words, the interval in which 99.73 percent of all sample means would fall (three standard errors either way of μ) is reduced as n is increased. Hence, each $\bar{x}$ has a better chance of being a reasonable estimator of μ. Do not foreget that n is usually restricted by budgetary considerations. This precludes researchers from taking a census and makes sampling all the more important.

Sampling distribution of $\bar{x}$ when the sample size is large, regardless of whether the population standard deviation is known or the distribution of the original population is normal (Cases 2, 3, and 4).

The sampling distribution of the $\bar{x}$s is normal when n is large (n is greater than or equal to 30), regardless of whether the population standard deviation is known or the original population is normal. Upon reflection, you can logically accept the foregoing statement so another example is unnecessary.

Sampling distribution of x when the sample size is small, the population standard deviation is known, and the distribution of the original population is normal (Case 5).

The sampling distribution of the $\bar{x}$s is normal when n is small if σ is known and the original population is normal. Upon reflection, you can logically accept the foregoing statement so another example is unnecessary.

Sampling distribution of $\bar{x}$ when the sample size is small, the population standard deviation is unknown, and the distribution of the original population is normal (Case 6).

The sampling distribution of the $\bar{x}$s is distributed according to student's t-distribution (this probability model has yet to be discussed) when: n is small ($n < 30$), σ is unknown, and the original population is normal.

**Central limit theorem
for means based on
small sample sizes**

If random samples of n observations are drawn from a normal population with finite mean (μ) and unknown and finite standard deviation (σ), the distribution of the sample means ($\bar{x}$) will be t-distributed with mean μ and standard error $s/\sqrt{n}$ for infinite populations, and $[s/\sqrt{n}]\sqrt{(N - n)/(N - 1)}$ for finite populations.

The manner in which you will interpret a sample mean is changed only in the percent of sample means that lies within a specified number of standard errors either way of μ. To ascertain the percent of sample means that lies within a specified number of standard errors of μ, you must consult a t-table; see Table B–4 in Appendix B.

Sampling distribution of $\bar{x}$ when the sample size is small, the original population distribution is unknown, and the population standard deviation is known or unknown (Cases 7 and 8).

Statistical inferences about μ cannot be made from $\bar{x}$ if the shape of the sampling distribution of $\bar{x}$ is unknown. Since the shape of the sampling distribution of $\bar{x}$ is unknown, you cannot know with any specificity what percent of the $\bar{x}$s lie within a given range around μ (you could use Chebychev's inequality). Since this percent is unknown, you will be unable to determine if $\bar{x}$ is a good estimator of μ.

GENERAL DISCUSSION OF THE SAMPLING DISTRIBUTION OF THE SAMPLE PROPORTION ($\bar{p}$)

The shape of the sampling distribution of $\bar{p}$ changes in accordance with certain "real-world conditions" as does the sampling distribution of $\bar{x}$. Table 8–3 indicates that there are four basic shapes for the sampling distribution of $\bar{p}$: normal, binomial, Poisson, and hypergeometric (the latter being a probability model we have not yet discussed in this text).

The hypergeometric distribution and the binomial distribution are both concerned with the probability of obtaining x successes in n trials. However, the binomial model assumes that the population under study is infinite while the hypergeometric model assumes that the population under study is finite. The effect of the population being finite is that the probability of success changes from trial to trial; it is not constant.

Table 8–3

Case	Distribution of the original population	Sample size (n)	Proportion between .1 & .9 ($.1 \le \pi \le .9$)	Sampling fraction ($n/N \le .1$)	Comments	Shape of the sampling distribution of $\bar{p}$
1	Infinite Bernoulli process (binomial distribution) or sampling with replacement	Large	Yes	Not relevant for infinite population	$n\pi > 5$ $n(1-\pi) > 5$	Normal
2		Large	No		$n\pi$ and/or $n(1-\pi) < 5$	Poisson
3		Small	Yes		$n\pi$ and/or $n(1-\pi) < 5$	Binomial
4		Small	No		$n\pi$ and/or $n(1-\pi) < 5$	Binomial
5	Finite Bernoulli process (hypergeometric distribution) or sampling without replacement	Large	Yes	No	$n\pi > 5$ $n(1-\pi) > 5$	Normal with fpc
6		Large	Yes	Yes	$n\pi > 5$ $n(1-\pi) > 5$	Normal
7		Large	No	No	$n\pi$ and/or $n(1-\pi) < 5$	Binomial with fpc
8		Large	No	Yes	$n\pi$ and/or $n(1-\pi) < 5$	Binomial
9		Small	Yes	No	$n\pi$ and/or $n(1-\pi) < 5$ and $n/N > .1$	Hypergeometric
10		Small	Yes	Yes	$n\pi$ and/or $n(1-\pi) < 5$ and $n/N \le .1$	Binomial
11		Small	No	No	$n\pi$ and/or $n(1-\pi) < 5$ and $n/N > .1$	Hypergeometric
12		Small	No	Yes	$n\pi$ and/or $n(1-\pi) < 5$ and $n/N \le .1$	Binomial

The formula for the hypergeometric distribution is:

$$P(X|n,N,R) = \frac{\begin{bmatrix} R \\ X \end{bmatrix} \begin{bmatrix} N-R \\ n-X \end{bmatrix}}{\begin{bmatrix} N \\ n \end{bmatrix}}$$

$$= \frac{\left[\dfrac{R!}{X!\,(R-X)!}\right]\left[\dfrac{(N-R)!}{(\{N-R\}-\{n-X\})!(n-X)!}\right]}{\left[\dfrac{N!}{n!\,(N-n)!}\right]}$$

where

n = the sample size
N = the population size
R = the number of successes in the population
$N - R$ = the number of failures in the population
X = the number of successes in the sample
$P(X|n,N,R)$ = the probability of obtaining X successes in n trials, given a finite population of N units containing R successes

Each of the shapes must be understood to make proper statistical inferences about π from $\bar{p}$. Table 8–3 delineates the relationship between "real-world conditions" and the shape of the sampling distribution of $\bar{p}$. Consequently, once you can isolate which "real-world conditions" are present in your problem you can determine the basic shape of the sampling distribution of $\bar{p}$ and can make proper statistical inferences about π from your $\bar{p}$. Remember, sampling distributions are never constructed in practice, they are theoretical abstractions which allow people to make sound statistical inferences.

Sampling distribution of $\bar{p}$ is normal (Cases 1, 5, and 6).

If $n\pi$ and $n(1 - \pi)$ are both greater than 5, the distribution of sample proportions is approximately normal. To demonstrate, 100 samples of size 2, 5, 10, and 30 were drawn concerning types of dwelling units for Stat City families. The sample proportions were computed for all 100 samples of each size.

Table 8–4 shows the relationship between sample size and the standard error of the proportion of houses in Stat City, given that the population

Table 8–4

Sample size (n)	Population standard deviation (σ)	Population size (N)	Finite population correction factor (fpc)	Standard error of $\bar{p}$ ($\sigma_{\bar{p}}$) Infinite pop.	Finite pop.
2	.481	1373	.99964	.340	.340
5	.481	1373	.99854	.215	.215
10	.481	1373	.99671	.152	.151
30	.481	1373	.98938	.088	.087

standard deviation is known ($\sigma = .481 = \sqrt{\pi(1 - \pi)} = \sqrt{.363 \times .637}$). Table 8–4 clearly indicates that the standard error of $\bar{p}$ decreases as the sample size increases. The table demonstrates that as n is increased, the chances of $\bar{p}$ deviating from π decreases because $\bar{p}$ is based on a larger percentage of the population. The above rule works for both infinite and finite populations.[2]

Frequency distributions of the 100 sample proportions for each sample size are shown in Exhibit 8–3. The larger the sample size, the more closely the relative frequency distribution of the sample proportion ($\bar{p}$) approaches the normal distribution. The above phenomena is called the central limit theorem for proportions and only works when $n\pi$ and $n(1 - \pi)$ are both greater than 5 (in this example $n\pi = 10.89$ and $n(1 - \pi) = 19.11$).

**Central limit theorem
for proportions**

If random samples of n observations are drawn from an infinite population, which is a Bernoulli process, as n enlarges (such that $n\pi$ and $n[1 - \pi]$ are greater than 5), the distribution of the sample proportions ($\bar{p}$) will approximate a normal distribution with mean π and standard error $\sqrt{\pi(1 - \pi)/n}$ for infinite populations, and $\sqrt{\pi(1 - \pi)/n}\sqrt{(N - n)/(N - 1)}$ for finite populations.

---

[2] The fpc is inconsequential in this problem because $[n/N]$ is less than 10 percent, or
$$\frac{\sigma}{\sqrt{n}} \sqrt{\frac{N - n}{N - 1}} \cong \frac{\sigma}{\sqrt{n}} \sqrt{1} \cong \frac{\sigma}{\sqrt{n}}.$$

Exhibit 8-3

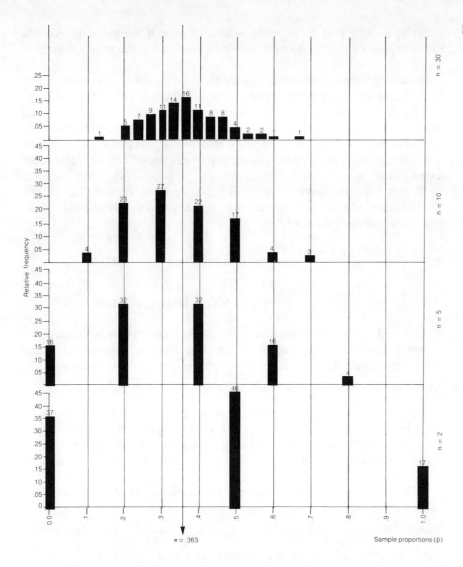

$\pi = .363$

Sample proportions ($\bar{p}$)

Now that it has been established that the frequency distribution of $\bar{p}$s (sampling distribution of the proportions) is approximately normal (if $n\pi$ and $n[1-\pi]$ are both greater than 5), you can use what you know about the normal distribution to interpret the ONE SAMPLE PROPORTION you have in practice.

If the sampling distribution of proportions is normal, then approximately 99.73 percent of all sample proportions will fall within three standard errors of the population proportion. Consequently, if the standard error is large, 99.73 percent of all sample proportions will fall within a WIDE range around π (three standard errors either way of π) and you cannot have "confidence" that your sample proportion will provide a good estimate of π. However, if the standard error is small, 99.73 percent of all sample proportions will fall within a SMALL range around π (three standard errors either way of π) and you can have "confidence" that your sample proportion will provide a good estimate of π.

Remember, even if the standard error of the proportion is small, you can never be certain that your $\bar{p}$ is an accurate estimator of π. As the standard error decreases, the chance of your $\bar{p}$ deviating from π decreases, but there is always a finite probability that even with a small standard error your $\bar{p}$ may be one of the $\bar{p}$s which will not provide a good estimator of π. However, since you never know if your $\bar{p}$ is a good estimator of π or not, you will base your evaluation of $\bar{p}$ on the following

Sampling distributions

111

criteria: assume that $\bar{p}$ is a good estimator of π if the standard error is small or assume that $\bar{p}$ may be a poor estimator of π if the standard error is large.

Finally, a way of increasing your confidence in $\bar{p}$ as an estimator of π is to increase the sample size upon which $\bar{p}$ is based. As n increases, the standard error becomes smaller and the $\bar{p}$s are more closely grouped around π. To elaborate, the interval in which 99.73 percent of all sample proportions would fall (three standard errors either way of π) is reduced as n is increased. Hence, each $\bar{p}$ has a better chance of being a reasonable estimator of π.

If π is unknown and $n\bar{p}$ and $n(1 - \bar{p})$ are both greater than 5, the distributions of sample proportions is also approximately normal. However, the standard error of $\bar{p}$ is:

$$\sqrt{\frac{\bar{p}(1 - \bar{p})}{n}} \text{ assuming an infinite population, and}$$

$$\sqrt{\frac{\bar{p}(1 - p)}{n}} \sqrt{\frac{N - n}{N - 1}} \text{ assuming a finite population.}$$

Everything else in the prior discussion holds true when π is unknown.

Note: The following three sections are directed toward advanced statistics classes or for the more ambitious introductory statistics student.

Sampling distribution of $\bar{p}$ is binomial (Cases 3, 4, 7, 8, 10, 12)

Given the frequency distribution of ps is binomial it is possible to compute the percentage of $\bar{p}$s that fall within a specified number of standard errors of the unknown population proportion. Consequently, probability statements concerning the sample proportion's accuracy can be made.

Finally, if π is unknown $\bar{p}$ can be substituted in its place without any loss of generality.

Sampling distribution of $\bar{p}$ is Poisson (Case 2)

If n is large and π is small (or large), such that $n\pi$ and/or $n(1 - \pi)$ are less than 5, given an infinite population, the distribution of sample proportions is distributed Poisson. Given the frequency distribution of sample proportions is Poisson, it is possible to compute the percentage of $\bar{p}$s that fall within a specified number of standard errors of the unknown population proportion. Hence, a probability statement concerning the sample proportion's accuracy can be made.

Finally, if π is unknown $\bar{p}$ can be substituted in its place without any loss of generality.

Sampling distribution of $\bar{p}$ is hypergeometric (Cases 9 and 11)

If $n\pi$ and/or $n(1 - \pi)$ is less than 5, and the sampling fraction is greater than .1 $(n/N > .1)$, the sampling distribution of the proportion is hypergeometric. Since the frequency distribution of $\bar{p}$s is hypergeometric, probability statements concerning the sample proportion's accuracy can be made. Unfortunately, the hypergeometric distribution is often computationally burdensome.

Finally, if π is unknown $\bar{p}$ can be substituted in its place without any loss of generality.

SUMMARY   Chapter 8 is an exercise in abstract thinking. In this chapter you have been asked to conceptualize the shape of the distribution of all possible sample statistics drawn from a population. This is abstract because sampling distributions are never constructed in practice. Understanding the shapes of sampling distributions (a theoretical abstraction) allows you to state the percentage of possible sample statistics that fall within a specified number of standard errors from the true, but unknown, population parameter (another theoretical abstraction). Being able to state the probability that your sample statistic is within a specified distance of the unknown population parameter lends "confidence" to the use of your sample statistic.

ESTIMATION OF μ AND Π

9

Estimating parameters, such as μ and π, from simple random samples of size n was presented in Chapter 4. In that chapter you noted that different samples of size n yield different estimates, such as $\bar{x}$ and $\bar{p}$, of parameters. This sample-to-sample variability is called sampling error.

Chapter 8 dealt with the sampling distributions of $\bar{x}$ and $\bar{p}$. You learned that if the sampling distribution of $\bar{x}$ (or $\bar{p}$) is known, it is possible to state the percentage of sample means (or proportions) that fall within a specified number of standard errors from the true, but unknown, population mean (or proportion). Consequently, it is possible to state the probability that the ONE SAMPLE MEAN (or proportion) you have computed is within a specified distance of the unknown population mean (or proportion). Being able to make the above probability statement lends "confidence" to the use of your sample mean (or proportion). Remember, you never know for certain if your sample mean (or proportion) is one of the many sample means (or proportions) that is within the specified distance from the unknown population mean (or proportion).

Now you have all the information necessary to estimate μ (or π) while considering sampling error. In this chapter you will take the concepts you learned in Chapters 4 and 8 and combine them so that you can make probability statements about μ (or π) called confidence intervals, from your sample mean (or proportion). You will also be instructed how to compute the sample size required to estimate μ or π in this chapter.

Understanding and interpreting confidence intervals is not a trivial task, especially to a person without formal training in statistics. Consequently, the footnote seen in Exhibit 9–1 (or some version thereof) should be attached to every memorandum in which a confidence interval is reported.

INTRODUCTION

Example 9–1   Continuation of the survey for the Stat City Tax Assessor's Office (see Examples 3–2 and 4–1)

Rewrite the report you prepared for the Stat City Tax Assessor's Office concerning the distribution of rooms per dwelling unit in Zone 2 of Stat City (see Exhibit 9–2). Recall, the tax assessor needed this information to determine new tax assessment rates so that Stat City can obtain a target total tax.

Exhibit 9-1   Footnote

Table 9-1

Random identification number	Number of rooms
198	6
272	9
287	9
131	6
134	7
281	9
280	8
274	7
278	10
230	8
149	8
270	5
255	7
139	9
233	8
168	7
282	9
152	8
204	8
266	7
262	5
132	7
188	8
210	8
171	10
222	8
269	7
283	8
218	7
161	8

[3] "The sample estimate and an estimate of its standard error permit us to construct interval estimates with prescribed confidence that the interval includes the average result of all possible samples (for a given sampling rate).

"To illustrate, if all possible samples were selected, if each of these were surveyed under essentially the same conditions and, if an estimate and its estimated standard error were calculated from each sample, then:

"a. Approximately two thirds of the intervals from one standard error below the estimate to one standard error above the estimate would include the average value of all possible samples. We call an interval from one standard error below the estimate to one standard error above the estimate a two-third confidence interval.

"b. Approximately nine tenths of the intervals from 1.6 standard errors below the estimate to 1.6 standard errors above the estimate would include the average value of all possible samples. We call an interval from 1.6 standard errors below the estimate to 1.6 standard errors above the estimate a 90 percent confidence interval.

"c. Approximately 19/20 of the intervals from two standard errors below the estimate to two standard errors above the estimate would include the average value of all possible samples. We call an interval from two standard errors below the estimate to two standard errors above the estimate a 95 percent confidence interval.

"d. Almost all intervals from three standard errors below the sample estimate to three standard errors above the sample estimate would include the average value of all possible samples.

"The average value of all possible samples may or may not be contained in any particular computed interval. But for a particular sample, one can say with specified confidence that the average of all possible samples is included in the constructed interval." (<u>Journal of the American Statistical Association</u>, vol. 70, no. 351, part II [September 1975], p. 6.)

Include information in the new report on the mean, minimum, maximum, range, and standard deviation of the number of rooms per dwelling unit in Zone 2. Be sure to discuss the sampling error of the mean by using a 95 percent confidence interval for the mean in your report.

Use the same sample that you used in writing your original report to the Stat City Tax Assessor's Office. This simple random sample is repeated in Table 9-1 for your convenience.

The sample statistics are shown below: **Solution**

1. Mean $= \bar{x} = \Sigma x/n = 231/30 = 7.7$
2. Minimum $=$ 5.0
3. Maximum $=$ 10.0
4. Range $=$ maximum-minimum $= 10 - 5 = 5$
5. Standard deviation $= s = 1.236$. See Table 9–2.
6. Standard error of $\bar{x} = [s/\sqrt{n}] \sqrt{\dfrac{N-n}{N-1}} = \left[\dfrac{1.236}{\sqrt{30}}\right] \sqrt{\dfrac{157-30}{157-1}}$

                 $= .2257 (.9023) = .2036$

                 $\approx$ approximately 1/5 of a room

7. Confidence interval for the mean $= \bar{x} \pm Z \left[\dfrac{s}{\sqrt{n}} \sqrt{\dfrac{N-n}{N-1}}\right]$

                 $= 7.7 \pm 1.96 \,(.2036)$

                 $= 7.7 \pm .3991$

                 $\approx 7.7 \pm .4$

                 ≈ 7.3 to 8.1

Table 9–2

Random identification number	Number of rooms (x)	$(x - \bar{x}) = (x - 7.7)$	$(x - \bar{x})^2$
198	6	−1.7	2.89
272	9	1.3	1.69
287	9	1.3	1.69
131	6	−1.7	2.89
134	7	−.7	.49
281	9	1.3	1.69
280	8	.3	.09
274	7	−.7	.49
278	10	2.3	5.29
230	8	.3	.09
149	8	.3	.09
270	5	−2.7	7.29
255	7	−.7	.49
139	9	1.3	1.69
233	8	.3	.09
168	7	−.7	.49
282	9	1.3	1.69
152	8	.3	.09
204	8	.3	.09
266	7	−.7	.49
262	5	−2.7	7.29
132	7	−.7	.49
188	8	.3	.09
210	8	.3	.09
171	10	2.3	5.29
222	8	.3	.09
269	7	−.7	.49
283	8	.3	.09
218	7	−.7	.49
161	8	.3	.09

$$\Sigma(x - \bar{x})^2 = 44.3$$

$$s = \sqrt{\frac{44.3}{30 - 1}} = \sqrt{1.5276}$$

$$= 1.236$$

The memorandum appears in Exhibit 9-2.

Exhibit 9-2

HOWARD S. GITLOW, PH.D.
STATISTICAL CONSULTANT
_____

MEMORANDUM

TO: Stat City Tax Assessor's Office

FROM: Howard Gitlow

DATE: July 21, 1980

RE: Revision of the report on the distribution of rooms
       per dwelling unit in Zone 2 of Stat City, as of January
       1980.

As per your request and using commonly accepted
statistical techniques, I have computed several summary
measures concerning the number of rooms per dwelling unit in
Zone 2.[1] My research indicates that the typical Zone 2
dwelling unit has approximately 7.7 rooms ($\pm$.2 rooms).[2]
Allowing for the errors in the survey estimates, you can have
"95 percent confidence" that the average number of rooms per
dwelling unit in Zone 2 falls in the range from 7.3 rooms to
8.1 rooms.[3] Further, the number of rooms per dwelling unit in
Zone 2 ranges between 5 and 10[4], indicating some degree of
variability in dwelling unit size.[5]

If you have any further questions, please do not hesitate
to call me at 305-999-9999.

Exhibit 9–2 (*continued*)

_____

Footnotes

[1] The summary statistics reported are estimates derived from a sample survey of 30 Zone 2 dwelling units; drawn via a simple random sampling plan.

"There are two types of errors possible in an estimate based on a sample survey—sampling and nonsampling. Sampling errors occur because observations are made only on a sample, not on the entire population. Nonsampling errors can be attributed to many sources; for example, inability to obtain information about all cases in the sample, definitional difficulties, differences in the interpretation of questions, inability or willingness to provide correct information on the part of respondents, mistakes in recording or coding the data obtained and other errors of collection, response, processing, coverage, and estimation for missing data. Nonsampling errors also occur in complete censuses. The 'accuracy' of a survey result is determined by the joint effects of sampling and nonsampling errors." (Journal of the American Statistical Association, vol. 70, no. 351, part II, [September 1975], p. 6).

[2] "The particular sample used in this survey is one of a large number of all possible samples of the same size that could have been selected using the same sample design. Estimates derived from the different samples would differ from each other. The difference between a sample estimate and the average of all possible samples is called the sampling deviation. The standard or sampling error of a survey estimate is a measure of the variation among the estimates from all possible samples, and thus is a measure of the precision with which an estimate from a particular sample approximates the average result of all possible samples. The relative standard error is defined as the standard error of the estimate divided by the value being estimated.

"As calculated for this report, the standard error also partially measures the effect of certain nonsampling errors but does not measure any systematic biases in the data. Bias is the difference, averaged over all possible samples, between the estimate and the desired value. Obviously, the accuracy of a survey result depends on both the sampling and nonsampling errors measured by the standard error and the bias and other types of nonsampling

Exhibit 9–2 (*continued*)

error not measured by the standard error." (<u>Journal of the American Statistical Association</u>, vol. 70, no. 351, part II, [September 1975], p. 6)

[3] "The sample estimate and an estimate of its standard error permit us to construct interval estimates with prescribed confidence that the interval includes the average result of all possible samples (for a given sampling rate).

"To illustrate, if all possible samples were selected, if each of these were surveyed under essentially the same conditions, and if an estimate and its estimated standard error were calculated from each sample, then:

"1. Approximately two thirds of the intervals from one standard error below the estimate to one standard error above the estimate would include the average value of all possible samples. We call an interval from one standard error below the estimate to one standard error above the estimate a two thirds confidence interval.

"2. Approximately nine tenths of the intervals from 1.6 standard errors below the estimate to 1.6 standard errors above the estimate would include the average value of all possible samples. We call an interval from 1.6 standard errors below the estimate to 1.6 standard errors above the estimate a 90 percent confidence interval.

"3. Approximately 19/20 of the intervals from two standard errors below the estimate to two standard errors above the estimate would include the average value of all possible samples. We call an interval from two standard errors below the estimate to two standard errors above the estimate a 95 percent confidence interval.

"4. Almost all intervals from three standard errors below the sample estimate to three standard errors above the sample estimate would include the average value of all possible samples.

The average value of all possible samples may or may not be contained in any particular computed interval. But for a particular sample, one can say with specified confidence that the average of all possible samples is included in the constructed interval." (<u>Journal of the American Standard Association</u>, vol. 70, no. 351, part II, [September 1975] p. 6)

[4] Dwelling units range in size from two bedroom, one bathroom, one kitchen, one living room up to four bedroom, three bathroom, one kitchen, one den, one living room.

[5] The degree of variability in the number of rooms per dwelling unit in Zone 2 of Stat City is also evidenced by the fact that the

Exhibit 9-2 (*concluded*)

standard deviation of the number of rooms per dwelling unit in Zone 2 is 1.236 rooms. This number indicates that approximately 95 percent of Zone 2 dwelling units have between 5.28 and 10.12 rooms. The above statement is true if the sample mean and standard deviation accurately reflect the true, but unknown, population mean and standard deviation and the distribution of rooms in Zone 2 is bell-shaped. I believe that this assumption is tenable; consequently, the standard deviation can be interpreted as stated above.

Do not be confused by the discrepancy between 95 percent of the dwelling units having between 5.28 rooms and 10.12 rooms and that the number of rooms per dwelling unit in Zone 2 ranging between 5 and 10 rooms. The discrepancy can be easily explained by realizing that all the statistics presented are only sample estimates. If the true, but unknown, minimum and maximum number of rooms per dwelling unit in Zone 2 were available, this discrepancy would no longer exist. In other words, the minimum and maximum number of rooms per dwelling unit would be larger than the interval created in which 95 percent of Zone 2 dwelling units would fall.

Point 1 The lengthy footnotes appended to Exhibit 9–2 are necessary if the reader is to be properly informed about the potential errors arising from sample surveys. It is imperative that the footnotes discuss sampling and nonsampling errors, the sampling plan used, the interpretation of standard errors, and the interpretation of confidence intervals.

Point 2 The finite population correction factor (fpc) was used in calculating the standard error because the sampling fraction (n/N) was greater than 0.10 (30/157 = 0.1911). A large sampling fraction indicates that the fpc will be significantly less than 1.0 and must be used to modify the standard error. For Zone 2 of Stat City, the relationship between the sampling fraction and the fpc is shown in Table 9–3.

As you can see, the fpc becomes significant between a sample size of 10 and 30.

Point 3 Reporting only the sample mean of the number of rooms per dwelling unit in Zone 2 of Stat City (7.7 rooms) would be inadequate to properly explain the size of the typical Zone 2 dwelling unit. The above inadequacy stems from the unqualified use of a point estimate ($\bar{x}$) which could impart a false sense of exactness (accuracy) to a reader. Consequently, point estimates (like 7.7 rooms) should always be qualified by their standard errors (like 0.2036 rooms).

Points of interest concerning Example 9–1

Table 9–3

	$N = 157$	
n	n/N	$\sqrt{(N - n)/(N - 1)}$
10	0.0637	0.9707
30	0.1911	0.9023
100	0.6369	0.6045
157	1.000	0.0000

Estimation of μ and π

121

Example 9–2 Survey to estimate the average monthly electric bill in Stat City

Mr. Saul Reisman, chief financial officer of the State City Electric Company, needs to estimate the average monthly electric bill for Stat City families, as of January 1980. Mr. Reisman has retained you to conduct a survey of 25 Stat City families. Prepare a typed memorandum for Mr. Reisman reporting your findings (see Exhibit 9–4). Be sure to report all appropriate statistics: mean, minimum, maximum, range, standard deviation, standard error, and 95 percent confidence interval.

Use the simple random sample of 25 Stat City dwelling units listed in Table 9–4.

Solution The statistics are shown below:

1. Mean $= \bar{x} = \Sigma x/n = 1,476/25 = \$59.04$
2. Minimum $= \$21$
3. Maximum $= \$101$
4. Range $= \$101 - \$21 = \$80$
5. Standard deviation $= s = \$19.58$
6. Standard error $= s/\sqrt{n} = \$3.92$
7. 95 percent confidence interval $= \bar{x} + [t_{(n-1,1-\alpha/2)}][s/\sqrt{n}]$

$$59.04 \pm 2.064\,(3.92)$$
$$59.04 \pm 8.09$$
$$\$50.95 \text{ to } \$67.13$$

The above confidence interval was constructed using t (not Z) standard errors. Consequently, the footnote which interprets confidence intervals (standard footnote 3) must be altered to account for the use of t (not Z) standard errors. Exhibit 9–3 displays the format of a standard footnote (3A) that explains how to interpret confidence intervals based on t standard errors.

Table 9–4

ID	ELEC
14.	101.
42.	65.
45.	91.
151.	72.
174.	80.
196.	78.
204.	74.
236.	79.
257.	59.
289.	21.
302.	65.
433.	85.
458.	47.
495.	47.
515.	44.
648.	43.
699.	67.
793.	43.
804.	45.
886.	49.
988.	50.
1077.	38.
1079.	39.
1227.	51.
1323.	43.

Exhibit 9–3

[3A] The sample estimate and an estimate of its standard error permit us to construct interval estimates with prescribed confidence that the interval includes the average result of all possible samples for a sample of size _____.

To illustrate, if all possible samples of size _____ were selected, if each of these were surveyed under essentially the same conditions, and if an estimate and its estimated standard error were calculated from each sample, then:

a. Approximately two thirds of the intervals from _____ standard errors below the estimate to _____ standard errors above the estimate would include the average value of all possible samples. We call an interval from _____ standard errors below the estimate to _____ standard error(s) above the estimate a thirds confidence interval.

b. Approximately 19/20 of the intervals from _____ standard errors below the estimate to _____ standard errors above the estimate would include the average value of all possible samples. We call an interval from _____ standard errors below the estimate to _____ standard errors above the estimate a 95 percent confidence interval.

The average value of all possible samples may or may not be contained in any particular computed interval. But for a particular sample, one can say with specified confidence that the average of all possible samples is included in the constructed interval.

The memorandum appears in Exhibit 9-4.

HOWARD S. GITLOW, PH.D.
STATISTICAL CONSULTANT

MEMORANDUM

TO: Mr. Saul Reisman, Chief Financial Officer
        Stat City Electric Company

FROM: Howard Gitlow

DATE: July 23, 1980

RE: Survey to estimate the average monthly electric bill
        in Stat City as of January 1980

   I have computed several summary measures concerning the
monthly electric bills in Stat City using commonly accepted
statistical techniques.[1] My research indicates that the
average monthly electric bill for a Stat City family, as of
January 1980, is $59.04 (±$3.92).[2] Allowing for errors in
survey estimates, you can have "95 percent confidence" that
the average monthly electric bill, as of January 1980, is
contained within the range from $50.95 to $67.13.[3] Further,
the monthly electric bills range between $21.00 and
$101.00, indicating some degree of variability.[4]
   If you have any questions, please call me at 305-999-9999.

Exhibit 9-4 (*continued*)

Footnotes

[1] Insert standard footnote 1, using a sample size of 25 drawn via simple random sampling.

[2] Insert standard footnote 2.

[3] The sample estimate and an estimate of its standard error permit us to construct interval estimates with prescribed confidence that the interval includes the average result of all possible samples for a sample of size 25.

To illustrate, if all possible samples of size 25 were selected, if each of these were surveyed under essentially the same conditions, and if an estimate and its estimated standard error were calculated from each sample, then:

<u>a</u>. Approximately two thirds of the intervals from 1 standard error below the estimate to 1 standard error above the estimate would include the average value of all possible samples. We call an interval from 1 standard error below the estimate to 1 standard error above the estimate a two-thirds confidence interval.

<u>b</u>. Approximately 19/20 of the intervals from 2.064 standard errors below the estimate to 2.064 standard errors above the estimate would include the average value of all possible samples. We call an interval from 2.064 standard errors below the estimate to 2.064 standard errors above the estimate a 95 percent confidence interval.

The average value of all possible samples may or may not be contained in any particular computed interval. But for a particular sample, one can say with specified confidence that the average of all possible samples is included in the constructed interval.

[4] The degree of variability in monthly electric bills in Stat City is also evidenced by the fact that the standard deviation is $19.58. This number indicates that approximately 95 percent of Stat City dwelling units have monthly electric bills between $20.66 and $97.42. The above statement is true if the sample mean and standard deviation accurately reflect the true, but unknown, population mean and standard deviation and the distribution of electric bills in Stat City is bell-shaped. I believe that this assumption is tenable; consequently, the standard deviation can be interpreted as stated above.

Table 9-5

ID	PHONE
258.	24.
295.	65.
386.	23.
415.	22.
416.	17.
736.	14.
858.	21.
1016.	19.
1149.	27.
1242.	22.

Example 9-3   Survey to estimate the average monthly telephone bill in Stat City

Mr. Jack Davis, chairman of the Stat City Telephone Company, needs to estimate the average monthly Stat City telephone bill, as of January 1980. Mr. Davis wants to have 95 percent confidence in the estimate. Further, he wants the estimate to be accurate to within $5 of the true average telephone bill.

You have been retained by Mr. Davis to perform the desired survey. The pilot study shown in Table 9-5 should be helpful in your endeavor. Type a memorandum to Mr. Davis reporting your findings (see Exhibit 9-5). Note: This problem requires you to compute the proper sample size!

Solution

The first step is to compute the proper sample size (*n*). Since the population is finite the following formulas are used.

$$n_0 = \frac{Z^2 \hat{s}^2}{e^2}$$

$$n = \frac{n_0}{\left(\dfrac{n_0 + (N - 1)}{N} \right)}$$

where

n = the sample size modified by the finite population correction factor

n_0 = the sample size unmodified by the finite population correction factor

N = the population size (1,373 dwelling units)

z = the number of standard errors required to set the desired level of confidence ($z = 1.96$ for 95 percent confidence)

$\hat{s}^2$ = the pilot study variance. $\hat{s}^2$ is computed by drawing a random sample of, say, 10 Stat City dwelling units and computing the variance from the pilot study (see Table 9–6).

Table 9–6

ID	PHONE
258.	24.
295.	65.
386.	23.
415.	22.
416.	17.
736.	14.
858.	21.
1016.	19.
1149.	27.
1242.	22.

Pilot study $\bar{x}$ = $ 25.40

Pilot study $\hat{s}^2$ = $206.93

e = the tolerable error = $(\bar{x} - \mu)$ = $ 5

Consequently,

$$n_0 = \frac{(1.96)^2(\$206.93)}{(\$5)^2} = 31.8$$

$$n = \frac{31.8}{\left[\frac{(31.8 + 1,372)}{1,373}\right]} = 31.1 = 32$$

The sample size of 32 implies that 32 Stat City families must be randomly sampled to estimate the population mean telephone bill with 95 percent confidence, allowing for a tolerable error of $5.00.

Once the sample size has been determined the random sample can be drawn. Table 9–7 lists the random sample selected in this study.

Finally, the data can be analyzed and conclusions can be drawn. The chart below shows the results of the survey.

Table 9–7

CASE-N	ID	PHONE
1	13.	14.
2	44.	16.
3	60.	15.
4	121.	16.
5	266.	20.
6	285.	21.
7	337.	118.
8	341.	68.
9	358.	98.
10	505.	21.
11	533.	14.
12	536.	55.
13	687.	97.
14	695.	17.
15	698.	19.
16	782.	10.
17	809.	35.
18	825.	14.
19	846.	15.
20	884.	139.
21	887.	80.
22	943.	65.
23	1000.	107.
24	1035.	92.
25	1057.	10.
26	1118.	72.
27	1152.	40.
28	1211.	15.
29	1216.	74.
30	1257.	21.
31	1274.	20.
32	1346.	26.

$n = 32$

$\bar{x} = \$45.13$

$s = \$37.76$

$s_{\bar{x}} = \$6.68$

95 percent confidence interval = $45.13 ± 1.96(6.68)

= $45.13 ± $13.09

= $32.04 to $58.22

The sample indicates that the average monthly telephone bill in Stat City, as of January 1980, is $45.13. The 95 percent confidence interval ranges from $32.04 to $58.22.

The memorandum is shown in Exhibit 9–5.

Exhibit 9–5

```
HOWARD S. GITLOW, PH.D.
STATISTICAL CONSULTANT


                           MEMORANDUM

TO:     Mr. Jack Davis, Chairman
        Stat City Telephone Company

FROM:   Howard Gitlow

DATE:   September 1, 1980

RE:     Estimating the average monthly telephone bill in
        Stat City, as of January 1980.


    Per your request, I have conducted a survey of Stat City
families to estimate the average monthly telephone bill, as
of January 1980.¹ My research indicates that the typical
household's average monthly telephone bill is $45.13
(±$6.68).² Allowing for the errors in survey estimates, you
can have "95 percent confidence" that the average monthly
telephone bill per household in Stat City falls in the range
from $32.04 to $58.22.³ Furthermore, average monthly
telephone bills range between $10 and $139, indicating some
degree of variability in telephone bill size.
    If you have any further questions, please call me at
305–999–9999.


        _____

                           Footnotes


    ¹ Insert standard footnote 1 (insert a sample size of 32 drawn via
simple random sampling).
    ² Insert standard footnote 2.
    ³ Insert standard footnote 3.
```

Example 9–4   Estimating the percentage of childless Zone 4 families

Mr. Hector Chavez, manager of the Grand Union supermarket, believes that the presence of children in a household strongly affects supermarket shopping behavior. Consequently, Mr. Chavez would like to determine the percentage of Zone 4 families that have no children.

Mr. Chavez has agreed that the instance of a single-parent family is rare in Stat City; hence, if a household has two or more people, it is safe to assume that two are adults and the remaining people are children.

You have been retained by Mr. Chavez to determine the percent of childless Zone 4 families. Type a memorandum reporting your findings to Mr. Chavez (see Exhibit 9–6).

Mr. Chavez has provided you with funds to draw a sample of 131 Zone 4 families. Use the simple random sample shown in Table 9–8 in conducting your analysis.

Table 9–8

CASE-N	ID	PEPLE	CASE-N	ID	PEPLE
1	627.	6.	66	952.	4.
2	632.	2.	67	956.	6.
3	643.	6.	68	966.	3.
4	645.	4.	69	972.	1.
5	649.	5.	70	973.	4.
6	653.	6.	71	974.	2.
7	656.	7.	72	980.	4.
8	661.	8.	73	991.	4.
9	662.	1.	74	1008.	4.
10	665.	3.	75	1010.	7.
11	670.	7.	76	1016.	5.
12	672.	5.	77	1018.	4.
13	673.	6.	78	1028.	1.
14	676.	5.	79	1040.	7.
15	682.	4.	80	1044.	5.
16	684.	5.	81	1046.	3.
17	686.	7.	82	1078.	6.
18	691.	1.	83	1082.	6.
19	695.	5.	84	1090.	6.
20	698.	5.	85	1098.	7.
21	700.	5.	86	1102.	4.
22	702.	4.	87	1104.	4.
23	703.	3.	88	1124.	2.
24	705.	4.	89	1127.	4.
25	715.	2.	90	1130.	6.
26	726.	4.	91	1131.	2.
27	737.	6.	92	1132.	5.
28	739.	2.	93	1138.	5.
29	743.	4.	94	1140.	5.
30	758.	8.	95	1143.	4.
31	759.	4.	96	1144.	5.
32	769.	6.	97	1148.	4.
33	770.	4.	98	1149.	4.
34	774.	3.	99	1152.	7.
35	778.	6.	100	1156.	6.
36	780.	4.	101	1161.	6.
37	799.	2.	102	1165.	2.
38	806.	4.	103	1167.	2.
39	819.	3.	104	1171.	5.
40	820.	5.	105	1172.	7.
41	822.	5.	106	1177.	3.
42	825.	4.	107	1188.	7.
43	830.	7.	108	1190.	3.
44	836.	4.	109	1213.	6.
45	847.	7.	110	1218.	1.
46	864.	4.	111	1261.	5.
47	870.	7.	112	1266.	3.
48	877.	5.	113	1272.	5.
49	878.	3.	114	1277.	6.
50	884.	6.	115	1286.	6.
51	891.	3.	116	1289.	5.
52	897.	7.	117	1290.	6.
53	900.	6.	118	1295.	4.
54	903.	7.	119	1302.	2.
55	905.	7.	120	1303.	6.
56	911.	6.	121	1316.	1.
57	917.	4.	122	1323.	6.
58	927.	4.	123	1336.	4.
59	936.	2.	124	1340.	4.
60	937.	8.	125	1348.	4.
61	938.	5.	126	1350.	4.
62	940.	4.	127	1357.	6.
63	943.	7.	128	1362.	8.
64	945.	2.	129	1370.	1.
65	950.	4.	130	1371.	5.
			131	1373.	6.

Solution   The statistics are shown below and in Table 9–9.

Table 9–9

Number of people in household, as of January 1980	Number	Percentage
1 or 2 (single or married—no children)...............	19	14.5
3 or more (married with children).....................	112	85.5
Total	131	100.0

1. Sample proportion $= \bar{p} = .145$
2. Estimated standard error of the proportion

$$= \sqrt{\bar{p}(1 - \bar{p})/n}\sqrt{(N - n)/(N - 1)}$$
$$= .031(.909)$$
$$= .028$$

3. 95 percent confidence interval for the population proportion

$$= \bar{p} \pm Z[\sqrt{\bar{p}(1 - \bar{p})/n}\sqrt{(N - n)/(N - 1)}$$
$$= .145 \pm 1.96[.028]$$
$$= .145 \pm .055$$
$$= .090 \text{ to } .200$$

The memorandum appears in Exhibit 9–6.

Exhibit 9–6

HOWARD S. GITLOW, PH.D.
STATISTICAL CONSULTANT

MEMORANDUM

TO: Mr. Hector Chavez, Manager
        Grand Union

FROM: Howard Gitlow

DATE: July 24, 1980

RE: Estimating the percentage of childless Zone 4
        families, as of January 1980

Using commonly accepted statistical techniques, I have computed several summary measures concerning the percentage of childless[1] Zone 4 families.[2] My research indicates that approximately 14.5 percent (±2.8 percent)[3] of Zone 4 families are childless. Allowing for the errors in survey estimates, you can have "95 percent confidence" that the percentage of Zone 4 families which are childless is contained within the range from 9.0 percent to 20.0 percent.[4]

If you have any further questions, please do not hesitate to call me at 305-999-9999.

Footnotes

[1] A dwelling unit containing one or two persons is considered childless. This determination is reasonable because the instance of single-parent families is rare in Stat City. In other words, it is safe to assume that if a household has two or more people, two are adults and the remainder are children.
[2] Insert standard footnote 1 (insert a sample size of 131 drawn via simple random sampling).
[3] Insert standard footnote 2.
[4] Insert standard footnote 3.

Example 9–5 Estimating the percentage of Stat City families that have their automotive repairs performed by a gas station

Mr. Paul Sugrue, the owner of Paul's Texaco Station, wants to determine the percentage of Stat City families that have their automotive repairs performed in a gas station, as opposed to "dealer" or "do-it-yourself" repairs. Mr. Sugrue wants to have 95 percent confidence in his estimate, with a tolerable error in estimation of 10 percentage points.

You have been retained by Mr. Sugrue to ascertain the desired information. Type a memorandum reporting your results to Mr. Sugrue (see Exhibit 9–7).

Note: This problem requires that you compute the appropriate sample size to meet Mr. Sugrue's specifications.

Solution   The first step is to compute the proper sample size (n). Since the population is finite the following formulas are used.

$$n_0 = \frac{z^2(\hat{\pi})(1 - \hat{\pi})}{e^2}$$

$$n = \frac{n_0}{\left[\dfrac{n_0 + (N - 1)}{N}\right]}$$

where

n = the sample size modified by the finite population correction factor

n_0 = the sample size unmodified by the finite population correction factor

N = the population size (1,373)

$\hat{\pi}$ = the "guesstimated" value of the population parameter ($\hat{\pi}$ = .5 because no better information is available; hence, maximum variation is assumed). $\hat{\pi}$ could also be estimated from a prior or pilot study

Z = the number of standard errors required to set the desired level of confidence (Z = 1.96 for 95 percent confidence)

e = the tolerable error (.10)

Hence,

$$n_0 = \left[\frac{z^2(\hat{\pi})(1 - \hat{\pi})}{e^2}\right] = \frac{(1.96)^2(.5)(.5)}{(.1)^2} = 96.04$$

$$n = \frac{n_0}{\left[\dfrac{n_0 + (N - 1)}{N}\right]} = 89.8 = 90$$

The sample size of 90 implies that 90 Stat City families must be randomly sampled to estimate the population percentage of families that have their automotive repairs performed by a gas station with 95 percent confidence, allowing for a tolerable error of 10 percentage points.

Once the sample size has been computed the random sample can be drawn. Table 9–10 lists the random sample selected in this study.

CASE-N	ID	REPAR	CASE-N	ID	REPAR	Table 9–10
1	4.	0.	46	568.	0.	
2	9.	1.	47	570.	0.	
3	12.	1.	48	572.	0.	
4	27.	0.	49	599.	0.	
5	40.	0.	50	602.	1.	
6	41.	1.	51	615.	0.	
7	49.	0.	52	622.	0.	
8	58.	1.	53	641.	1.	
9	81.	1.	54	664.	1.	
10	90.	0.	55	673.	0.	
11	126.	0.	56	688.	0.	
12	133.	0.	57	771.	1.	
13	180.	0.	58	776.	1.	
14	184.	0.	59	785.	1.	
15	211.	0.	60	817.	0.	
16	215.	0.	61	819.	1.	
17	230.	1.	62	836.	0.	
18	235.	1.	63	897.	1.	
19	263.	0.	64	910.	1.	
20	268.	1.	65	955.	1.	
21	274.	0.	66	990.	1.	
22	278.	0.	67	995.	0.	
23	282.	0.	68	1001.	0.	
24	284.	0.	69	1036.	0.	
25	300.	0.	70	1045.	0.	
26	302.	0.	71	1058.	1.	
27	354.	1.	72	1073.	0.	
28	386.	0.	73	1087.	0.	
29	388.	0.	74	1108.	1.	
30	414.	0.	75	1114.	1.	
31	424.	0.	76	1130.	0.	
32	434.	1.	77	1159.	1.	
33	460.	1.	78	1165.	1.	
34	465.	0.	79	1173.	1.	
35	472.	1.	80	1175.	1.	
36	489.	1.	81	1247.	0.	
37	490.	1.	82	1248.	1.	
38	498.	0.	83	1253.	0.	
39	509.	0.	84	1278.	0.	
40	514.	0.	85	1279.	1.	
41	521.	0.	86	1327.	1.	
42	530.	0.	87	1337.	1.	
43	531.	1.	88	1339.	0.	
44	545.	0.	89	1358.	1.	
45	566.	0.	90	1371.	0.	

Note: Code for repairs: 1 = Done by service station (originally 1).
2 = Done somewhere else (originally 0 or 2).

Finally, the data can be analyzed and conclusions can be drawn. The chart below shows the results of the survey.

$$n = 90$$
$$\bar{p} = .422$$
$$s_{\bar{p}} = .052 = \sqrt{\frac{.422 \times .578}{90}}$$
95 percent confidence interval =
$$.422 \pm 1.96(.052) = .422 \pm .102$$
$$= .320 \text{ to } .524$$

The sample indicates that the proportion of Stat City families that have their automotive repairs performed by a gas station, as of January 1980, is .422. The 95 percent confidence interval ranges from .320 to .524.

The memorandum appears in Exhibit 9–7.

Exhibit 9–7

HOWARD S. GITLOW, PH.D.
STATISTICAL CONSULTANT

MEMORANDUM

TO: Mr. Paul Sugrue, Owner
        Paul's Texaco Station

FROM: Howard Gitlow

DATE: March 9, 1981

RE: Estimating the percentage of Stat City families that
        have their automotive repairs performed by a gas
        station, as of January 1980.

In accordance with our consulting contract I have conducted a survey to estimate the percentage of Stat City families that have their automotive repairs performed by a gas station, as of January 1980.[1] My research, which was performed using commonly accepted statistical techniques, indicates that the percentage of Stat City families that have their automotive repairs performed by a gas station is 42.2 percent (±5.2 percent).[2] Allowing for the errors in survey estimates, you can have "95 percent confidence" that the percentage of Stat City families which have their automotive repairs performed by a gas station is contained within the range from 32.0 percent to 52.4 percent.[3]

If you have any further questions, call me at 305–999–9999.

———————

Footnotes

[1] Insert standard footnote 1 (insert a sample size of 90 drawn via simple random sampling).
[2] Insert standard footnote 2.
[3] Insert standard footnote 3.

ADDITIONAL PROBLEMS

9–6 Need for a promotional campaign to increase supermarket patronage in Stat City The Stat City Supermarket Association would like to estimate the average weekly supermarket bill per family in Stat City, as of January 1980. The association requires 95 percent confidence in the estimate. They have provided you with sufficient funds to draw a random sample of 50 Stat City families. Type a memorandum to the association reporting your findings.

Use the following simple random sample for computing the required statistics.

ID	EAT	ID	EAT	ID	EAT
2.	81.	585.	76.	1043.	69.
14.	155.	604.	85.	1045.	101.
120.	45.	624.	87.	1089.	127.
158.	182.	640.	67.	1104.	80.
159.	99.	670.	161.	1113.	112.
164.	71.	674.	70.	1123.	29.
177.	93.	682.	76.	1146.	70.
223.	60.	688.	86.	1150.	61.
232.	97.	752.	98.	1160.	81.
271.	72.	757.	142.	1220.	53.
314.	153.	813.	101.	1292.	130.
316.	122.	816.	97.	1305.	107.
399.	86.	841.	26.	1313.	90.
474.	144.	875.	87.	1320.	142.
533.	109.	915.	87.	1332.	42.
541.	93.	959.	70.	1361.	72.
579.	112.	1025.	115.		

9–7 Another study to estimate the average monthly electric bill in Stat City Mrs. Jessica Mara, assistant financial officer of the Stat City Electric Company, must estimate the average monthly electric bill in Stat City, as of January 1980. She wants to have 95 percent confidence in the estimate and be accurate to within $10 of the true average electric bill. Unfortunately, she is not aware of the study performed for Mr. Reisman (Example 9–2).

She has retained you to perform the required study. She has provided you with the following data from a pilot survey of eight Stat City families.

ID	ELEC
22.	59.
83.	86.
147.	81.
298.	127.
393.	32.
448.	47.
531.	61.
776.	35.

Type a memorandum to her reporting your findings.

Note: After you compute the proper sample size, your professor will hand out the random sample you should use to compute the required statistics. This is in order that everyone in the class will have the same sample.

9–8 Estimating the percentage of Stat City families who do not use the hospital Ms. Arlene Davis, director of the Stat City Hospital, wants to ascertain the percentage of Stat City families that do not use the Stat City hospital. Ms. Davis wants 90 percent confidence in the estimate. Further, she wants the estimate to be accurate to within 4 percentage points.

You have been retained by Ms. Davis to conduct a survey to compute the desired percentage. Ms. Davis told you that a similar study performed several years ago indicated that 6.7 percent of Stat City families do not use the hospital. Type a memorandum to Ms. Davis reporting your findings.

Estimation of μ and π

133

Note: After you compute the proper sample size, your professor will hand out the random sample you should use to compute the required statistics so that everyone in the class will have the same sample.

COMPUTATIONAL EXERCISES

9–9 Mr. Marc Cooper, manager of the A&P supermarket, is interested in studying several characteristics of Stat City dwelling units. In particular, Mr. Cooper is interested in: the average weekly supermarket bill as of January 1980 (EAT), and the average weekly purchase of six-packs of diet soda (LSODA), regular soda (HSODA), and beer (BEER), as of January 1980. Mr. Cooper has supplied you with enough funds to survey a random sample of 49 Stat City families. Compute the 99 percent confidence interval for each variable.

Use the random sample and summary statistics which accompany this problem for your computations.

Random sample for Exercise 9–9

CASE-N	ID	EAT	LSODA	HSODA	BEER
1	1.	128.	2.	0.	5.
2	30.	79.	1.	0.	2.
3	131.	53.	1.	0.	0.
4	136.	89.	2.	0.	9.
5	202.	156.	1.	0.	0.
6	216.	94.	0.	6.	0.
7	226.	68.	2.	0.	6.
8	261.	68.	0.	5.	7.
9	267.	34.	0.	0.	0.
10	354.	120.	0.	0.	7.
11	373.	66.	0.	4.	0.
12	420.	108.	0.	0.	0.
13	480.	101.	2.	0.	5.
14	498.	117.	0.	0.	2.
15	534.	63.	0.	0.	0.
16	563.	144.	0.	5.	0.
17	591.	72.	1.	0.	3.
18	666.	51.	2.	0.	0.
19	674.	70.	0.	0.	0.
20	703.	65.	0.	0.	0.
21	737.	135.	2.	0.	0.
22	743.	85.	2.	0.	0.
23	786.	134.	0.	4.	0.
24	807.	71.	2.	0.	10.
25	820.	89.	2.	0.	0.
26	847.	138.	0.	5.	10.
27	895.	161.	0.	0.	0.
28	901.	75.	2.	0.	0.
29	909.	59.	0.	0.	6.
30	935.	136.	0.	1.	11.
31	936.	43.	1.	0.	0.
32	937.	126.	2.	0.	0.
33	953.	121.	0.	0.	0.
34	984.	110.	0.	0.	0.
35	987.	89.	0.	0.	0.
36	988.	129.	0.	4.	0.
37	1035.	80.	0.	3.	3.
38	1098.	132.	2.	0.	5.
39	1126.	87.	0.	1.	0.
40	1170.	84.	0.	5.	0.
41	1174.	132.	0.	0.	4.
42	1193.	114.	0.	0.	0.
43	1218.	26.	0.	5.	6.
44	1226.	91.	0.	4.	0.
45	1232.	79.	0.	0.	0.
46	1272.	104.	0.	0.	0.
47	1280.	110.	0.	1.	0.
48	1281.	93.	2.	0.	2.
49	1367.	110.	0.	1.	3.

Summary statistics for Exercise 9–9

```
VARIABLE   EAT         AVERAGE WEEKLY FOOD BILL AS OF 1-80

MEAN          95.694              STD ERROR      4.603           STD DEV    32.224
VARIANCE    1038.384              KURTOSIS       -.666           SKEWNESS    -.013
RANGE        135.000              MINIMUM       26.000           MAXIMUM   161.000
SUM         4689.000

VALID OBSERVATIONS -      49              MISSING OBSERVATIONS -        0

- - - - - - - - - - - - - - - - - - - - - - - - - - - - - - - - - - - - - - - -

VARIABLE   LSODA       AVE WKLY DIET SODA PURCHASE AS OF 1-80

MEAN            .633              STD ERROR       .126           STD DEV      .883
VARIANCE        .779              KURTOSIS     -1.232            SKEWNESS     .810
RANGE          2.000              MINIMUM        .000            MAXIMUM    2.000
SUM           31.000

VALID OBSERVATIONS -      49              MISSING OBSERVATIONS -        0

- - - - - - - - - - - - - - - - - - - - - - - - - - - - - - - - - - - - - - - -

VARIABLE   HSODA       AVE WKLY REG. SODA PURCHASE AS OF 1-80

MEAN           1.102              STD ERROR       .275           STD DEV     1.928
VARIANCE       3.719              KURTOSIS        .346           SKEWNESS    1.431
RANGE          6.000              MINIMUM        .000            MAXIMUM    6.000
SUM           54.000

VALID OBSERVATIONS -      49              MISSING OBSERVATIONS -        0

- - - - - - - - - - - - - - - - - - - - - - - - - - - - - - - - - - - - - - - -

VARIABLE   BEER        AVE WKLY BEER PURCHASE AS OF 1-80

MEAN           2.163              STD ERROR       .465           STD DEV     3.255
VARIANCE      10.598              KURTOSIS        .678           SKEWNESS    1.348
RANGE         11.000              MINIMUM        .000            MAXIMUM   11.000
SUM          106.000

VALID OBSERVATIONS -      49              MISSING OBSERVATIONS -        0
```

9–10 Donna Nelson, a reporter for the *Stat City Beacon,* is writing an article about certain characteristics of Stat City dwelling units. In particular, Ms. Nelson is reporting on: the average yearly heating bill as of January 1980 (HEAT), the average monthly telephone bill as of January 1980 (PHONE), the 1979 total family income (INCOM), the number of automobiles in a household as of January 1980 (CARS), the average bimonthly automobile gasoline bill as of January 1980 (GAS), and the average number of trips to a gas station per month as of January 1980 (GASTR). Ms. Nelson has supplied you with funds to survey a random sample of 10 Stat City dwelling units. Compute the mean, standard deviation, standard error, and 90 percent confidence interval for each of the above variables. Assume that all of the above variables are normally distributed.

Use the accompanying simple random sample.

ID	HEAT	ELEC	PHONE	INCOM	CARS	GAS	GASTR
258.	1136.	73.	24.	116203.	2.	127.	3.
295.	538.	42.	65.	25328.	1.	99.	6.
386.	792.	72.	23.	32667.	2.	160.	6.
415.	413.	37.	22.	26972.	1.	65.	6.
416.	457.	37.	17.	41549.	3.	174.	7.
736.	392.	38.	14.	18404.	2.	120.	12.
858.	256.	21.	21.	17052.	2.	120.	5.
1016.	512.	46.	19.	24800.	3.	185.	5.
1149.	651.	69.	27.	19670.	2.	120.	5.
1242.	471.	47.	22.	24743.	3.	168.	8.

Random sample for Exercise 9–10

9–11 Paul Lund, chairman of the Stat City Chamber of Commerce, is interested in investigating the following attributes of Stat City families as of January 1980: the percentage of Stat City families that did not purchase any six-packs of diet soda (LSODA = 0), the percentage of Stat City families that did not purchase any six-packs of regular soda

Estimation of μ and π

Random sample for Exercise 9–11	ID	LSODA	HSODA	PEPLE	GASTR	REPAR
	6.	0.	0.	2.	3.	1.
	47.	0.	0.	4.	5.	2.
	53.	2.	0.	1.	6.	1.
	68.	1.	0.	1.	4.	1.
	88.	1.	0.	2.	2.	2.
	92.	0.	1.	5.	8.	1.
	164.	0.	1.	3.	3.	1.
	167.	0.	5.	5.	5.	2.
	291.	2.	0.	4.	8.	2.
	459.	0.	0.	6.	6.	0.
	469.	0.	0.	5.	6.	1.
	477.	0.	0.	6.	11.	1.
	490.	0.	0.	4.	6.	1.
	498.	0.	0.	5.	4.	0.
	528.	2.	0.	5.	9.	1.
	538.	2.	0.	7.	5.	2.
	539.	2.	0.	3.	6.	0.
	588.	0.	1.	4.	7.	1.
	597.	0.	4.	4.	2.	1.
	599.	2.	0.	4.	6.	2.
	616.	0.	2.	6.	5.	2.
	657.	0.	0.	4.	2.	0.
	684.	0.	4.	5.	4.	1.
	705.	0.	0.	4.	6.	1.
	724.	0.	0.	2.	6.	1.
	764.	0.	0.	2.	3.	0.
	788.	2.	0.	7.	3.	1.
	795.	2.	0.	5.	4.	1.
	797.	2.	0.	4.	3.	1.
	816.	0.	5.	4.	8.	1.
	835.	0.	1.	3.	7.	0.
	854.	0.	0.	7.	5.	0.
	873.	2.	0.	6.	7.	1.
	897.	0.	0.	7.	4.	1.
	962.	2.	0.	6.	3.	0.
	970.	0.	0.	6.	12.	1.
	996.	0.	0.	4.	4.	1.
	1073.	2.	0.	6.	7.	0.
	1086.	0.	0.	5.	5.	1.
	1092.	0.	0.	4.	6.	2.
	1134.	2.	0.	8.	4.	1.
	1136.	0.	5.	8.	5.	2.
	1192.	0.	0.	7.	2.	1.
	1215.	2.	0.	5.	7.	0.
	1244.	2.	0.	3.	2.	1.
	1279.	0.	1.	2.	6.	1.
	1297.	1.	0.	5.	6.	0.
	1299.	0.	6.	4.	7.	2.
	1320.	0.	0.	7.	4.	1.
	1348.	2.	0.	4.	3.	0.

(HSODA = 0), the percentage of childless Stat City families (PEPLE ≤ 2), the percentage of Stat City families making an average of nine or more trips to the gas station per month (GASTR ≥ 9), and the percentage of Stat City families that have their automotive repairs performed at a dealer (REPAR = 2). Mr. Lund has supplied you with funds to survey 50 randomly selected Stat City dwelling units and to compute percentages, standard errors, and 95 percent confidence intervals for each variable under study.

Use the accompanying random sample for your computations.

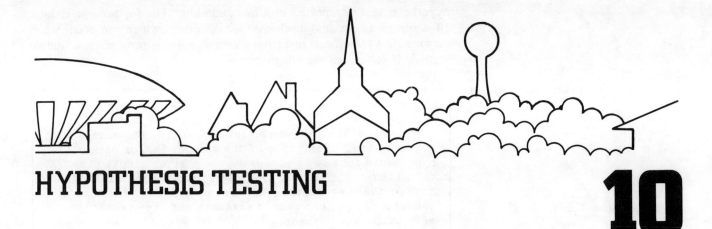

HYPOTHESIS TESTING

10

In Chapter 9 you learned about the branch of statistical inference concerned with making point and interval estimates of parameters. In this chapter you will study another branch of statistical inference: the branch concerned with testing preconceived ideas (hypotheses) about parameters. Generally, if a preconceived idea about a parameter remains tenable after it has been tested the idea gains credibility, and some interest group will pursue a particular course of action. For example, the Stat City Supermarket Association believes that the typical Stat City family purchases less than one six-pack of diet soda per week, on average. If the idea remains credible after a hypothesis test, the association will mount a promotional campaign to increase diet soda sales.

The steps involved in testing a hypothesis are as follows:

1. List the states of nature.
2. List the alternative courses of action.
3. Construct a decision matrix (see Table 10–1).

Table 10–1

States of nature

		State 1	State 2
Alternative courses of action	Act 1		
	Act 2		

4. State the null hypothesis (H_0).
5. State the alternative hypothesis (H_A).
6. State the level of significance (α).
7. State the proper test statistic.
8. Set up the rejection region(s) for the null hypothesis.
9. Compute the sample size (n).
10. Draw the sample and compute the test statistic.
11. Ascertain if the test statistic falls within the rejection region(s).
12. State the statistical decision (reject the H_0 or not).
13. Relate the statistical decision to the appropriate person(s).

Understanding hypothesis testing is more difficult than a person without formal training in statistics would suspect. Hence, it is important that the footnote in Exhibit 10–1 (or some version thereof) appears whenever a hypothesis test is reported in a memorandum. The footnote in Exhibit 10–1 serves as a partial disclaimer for the errors which can occur when performing a hypothesis test from a sample survey; remember, a sample statistic is subject to sampling error.

Exhibit 10–1   Footnote

---

[4] The hypothesized population parameter, the sample estimate, and an estimate of the standard error permit a hypothesis test to be performed for a given sample size.

To illustrate, if all possible samples were selected, if each of these were surveyed under essentially the same conditions, and if a sample estimate and its standard error were calculated, then:

   a.   Approximately nine-tenths of the hypothesis tests conducted at the 90 percent level of confidence (or 10 percent level of significance) would fail to reject the null hypothesis when it was tenable (or 10 percent would reject the null hypothesis when it was tenable).

   b.   Approximately 19/20 of the hypothesis tests conducted of the 95 percent level of confidence (of 5 percent level of significance) would fail to reject the null hypothesis when it was tenable (or 5 percent would reject the null hypothesis when it was tenable).

Once the level of confidence (or significance) has been set, the chances of failing to reject the null hypothesis when it is false (type II error) must be computed. If these probabilities are unacceptably high, then corrective action must be taken.

The formulas presented in this chapter for computing sample sizes for hypothesis testing problems are limited to one sample and one-tailed tests, for means and proportions. The formula for computing the sample size for a hypothesis test concerning a mean is:

$$n = \frac{\hat{s}^2 (Z_\alpha + Z_\beta)^2}{(\mu_0 - \mu_1)^2}$$

where

   n = the sample size[1]
   $\hat{s}^2$ = an estimate of the variance of the variable under study (obtained from a pilot or prior study)
   μ_0 = value of the population mean under the null hypothesis

---

[1] Use of the finite population correction factor in conjunction with the above formula can cause logical inconsistencies.

μ_1 = value of the population mean under the alternative hypothesis

Z_α = Z value for a given α level of significance (omit the sign of Z_α)

Z_β = Z value for a given β risk of a type II error (omit the sign of Z_β)

The formula for computing the sample size for a hypothesis test concerning a proportion is:

$$n = \frac{\{Z_\alpha\sqrt{\pi_0(1-\pi_0)} + Z_\beta\sqrt{\pi_1(1-\pi_1)}\}^2}{(\pi_1 - \pi_0)^2}$$

where,

n = the sample size[1]

π_0 = value of the population proportion under the null hypothesis

π_1 = value of the population proportion under the alternative hypothesis

Z_α = Z value for a given α level of significance (omit sign of Z_α)

Z_β = Z value for a given β risk of a type II error (omit sign of Z_β)

The purpose of this chapter is to develop your hypothesis testing skills. The following examples should be helpful in your gaining a firm grasp of hypothesis testing.

Example 10–1   Applying for HHS funds for Zone 4 families

The U.S. Department of Health and Human Services (HHS) has issued a statement that any urban area with an average 1979 family income below $20,000 is entitled to federal subsidies. Mr. Kaplowitz, the mayor of Stat City, would like to receive federal aid for families living in Zone 4 of Stat City.

The courses of action open to the mayor are twofold: to apply or not to apply for the federal aid. If the mayor applies for the federal aid, when, in fact Stat City is not eligible, he runs the risk of incurring HHS's disfavor. Consequently, Stat City may have an extremely difficult time obtaining other HHS funds in the future. The mayor is far more concerned about applying for the funds when he should not, than he is with refusing to apply for the funds when Stat City may be eligible.

Mr. Kaplowitz has hired you to determine if the average family income in 1979 in Zone 4 was less than $20,000; that is, if he should apply for the funds. Conduct the necessary survey and type a memorandum to the mayor reporting your findings (see Exhibit 10–3).

Solution

BEFORE the survey is performed it is important to specify the hypothesis being tested. The alternative courses of action open to the mayor are: applying for the HHS funds or not applying for the HHS funds. The relevant states of nature are whether or not the average 1979 family income in Zone 4 of Stat City is less than $20,000. The mayor should apply for the funds if the average 1979 family income in Stat City is less than $20,000.

Table 10–2 illustrates the structure of the decision to be made. The box marked A indicates a correct decision; applying for the HHS funds if the average 1979 family income in Zone 4 is less than $20,000 per year. The box marked B also indicates a correct decision; not applying for HHS funds if the average family income in Zone 4 is greater than or equal to $20,000 per year. The box marked C indicates an incorrect decision; not applying for HHS funds if the average 1979 family income in Zone 4 is less than $20,000 per year. Finally, the box marked D also indicates an error; applying for HHS funds if the average 1979 family income in Zone 4 is greater than or equal to $20,000 per year. Recall that this is the error of utmost concern to the mayor.

Table 10–2

	States of nature	
Alternative courses of action	*Average 1979 family income in Zone 4 is less than $20,000 per year* ($\mu < 20,000$)	*Average 1979 family income in Zone 4 is greater than or equal to $20,000 per year* ($\mu \geq 20,000$)
Apply for HHS funds	A	D
Do not apply for HHS funds	C	B

It is customary when working with one-tail tests to designate as the null hypothesis that hypothesis which, if rejected when tenable, leads to the type I error. In this case the null hypothesis is:

$$H_0\text{:}\mu \geq \$20,000$$

Note, $\mu \geq \$20,000$ is designated as the null hypothesis because if it is rejected when it is tenable, the type one error is committed; the mayor applies for HHS funds which Stat City does not deserve and consequently incurs HHS's displeasure. The alternative hypothesis is:

$$H_A\text{:}\mu < \$20,000$$

If the null hypothesis is rejected and the alternative hypothesis is deemed tenable, then action is taken; application is made for the HHS monies.

The problem would be set up as follows:

State of nature	*Alternative courses of action*
$H_0\text{:}\mu \geq \$20,000$	Don't apply for HHS funds
$H_A\text{:}\mu < \$20,000$	Apply for HHS funds

The level of significance for the problem (α) is set to control the more severe error. If $\alpha = 0.05$, then the mayor is willing to reject the null hypothesis when it is tenable 5 percent of the time. In other words, if the mayor were making this decision 100 times he would tolerate being wrong (and incur HHS's displeasure) 5 times. In fact, the mayor only makes this decision once.

A summary of the work done so far and the appropriate statistical test is shown below:

$$H_0\text{: } \mu \geq \$20,000$$
$$H_A\text{: } \mu < \$20,000$$
$$\alpha = 0.05$$
$$z = \frac{\bar{X} - \mu}{\frac{s}{\sqrt{n}}}$$

The next step is to determine the proper sample size (n).

$$n = \frac{\hat{s}^2(Z_\alpha + Z_\beta)^2}{(\mu_0 - \mu_1)^2}$$

where

    n = the sample size[2]
    $\hat{s}^2$ = the pilot study variance. $\hat{s}^2$ is computed by drawing a random sample of, say, five Zone 4 dwelling units and computing the variance from the pilot study.

[2] Use of the finite population correction factor can cause logical inconsistencies in conjunction with this formula.

ID	INCOM
0631.	12,037.
0712.	13,527.
0891.	10,024.
0918.	6,305.
1262.	11,557.

$$\bar{x} = \$10,690$$
$$\hat{s} = \$2,751.66$$
$$\hat{s}^2 = \$7,571,632.76$$

μ_0 = value of the population mean under the null hypothesis
μ_1 = value of the population mean under the alternative hypothesis
Z_α = Z value for a given α level of significance (omit the sign of Z_α)
Z_β = Z value for a given β risk of a type II error (omit the sign of Z_β)

Z_β can be computed after the decision maker states that he/she wishes to have, say, a 90 percent chance (power) of rejecting the null hypothesis (of $20,000) when the population mean is really equal to $19,000 and he/she is willing to permit α to be 5 percent. Hence, $Z_\beta = 1.28$, as shown in Exhibit 10–2.

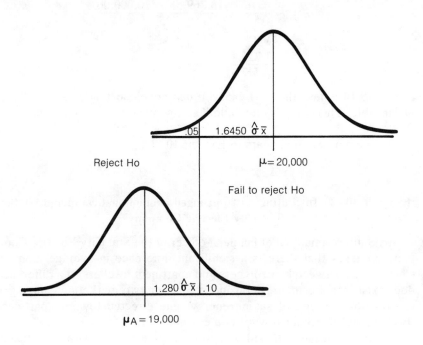

Exhibit 10–2

.05   1.6450 $\hat{\sigma}\bar{x}$

Reject Ho $\mu = 20,000$

Fail to reject Ho

1.280 $\hat{\sigma}\bar{x}$ .10

$\mu_A = 19,000$

Consequently,

$$n = \frac{(\$7,571,632.76)\ (+1.645 + 1.28)^2}{(20,000 - 19,000)^2}$$

$$n = \frac{(7,571,632.76)\ (+2.925)^2}{(1,000)^2} = 64.78 = 65^*$$

* Since $n/N \le .1$, it is not necessary to use the finite population correction factor.

The sample size of 65 implies that 65 Zone 4 families must be randomly sampled if Mr. Kaplowitz is willing to have a 5 percent risk of comitting a type I error and a 90 percent chance of rejecting the null hypothesis (of $20,000) and detecting that the population mean is really $19,000.

Table 10–3

CASE-N	ID	INCOM
1	626.	14081.
2	634.	17987.
3	641.	30195.
4	648.	21242.
5	649.	18425.
6	651.	27781.
7	655.	23675.
8	666.	27030.
9	672.	11480.
10	692.	18489.
11	704.	13755.
12	711.	10961.
13	716.	12364.
14	720.	16777.
15	752.	11187.
16	772.	4322.
17	774.	4886.
18	799.	19071.
19	802.	16371.
20	827.	23872.
21	831.	14573.
22	844.	16144.
23	847.	15130.
24	854.	12097.
25	861.	15055.
26	863.	16916.
27	880.	11445.
28	896.	15904.
29	899.	15823.
30	946.	4087.
31	949.	11867.
32	973.	22127.
33	978.	24772.
34	1000.	20070.
35	1003.	19710.
36	1011.	16007.
37	1013.	20760.
38	1038.	15340.
39	1050.	8664.
40	1060.	7288.
41	1073.	7306.
42	1078.	6894.
43	1102.	19518.
44	1129.	18771.
45	1135.	21870.
46	1136.	23923.
47	1143.	12770.
48	1155.	14018.
49	1161.	11075.
50	1210.	16986.
51	1212.	22910.
52	1221.	24703.
53	1244.	16095.
54	1259.	16406.
55	1272.	13933.
56	1275.	16418.
57	1297.	18922.
58	1303.	18216.
59	1308.	28058.
60	1324.	20395.
61	1325.	13694.
62	1336.	10960.
63	1337.	14338.
64	1338.	11897.
65	1365.	9715.

Once the sample size has been determined the random sample can be drawn. Table 10–3 enumerates the random sample selected in this study.

Finally, the data can be analyzed and conclusions can be drawn. The chart below shows the results of the survey.

$$n = 65$$
$$\bar{x} = \$16,269.55$$
$$s = \$5,851.37$$
$$s_{\bar{x}} = \$725.77$$

The sample indicates that the average income in Zone 4 is $16,269.55. The question arises: Is $16,269.55 below $20,000.00 by a statistically significant amount so that the mayor can decide to apply for the HHS funds and only be wrong 5 percent of the time ($\alpha = 0.05$). The hypothesis test is shown below.

$$H_0: \mu \geq \$20,000$$
$$H_A: \mu < \$20,000$$

$$\alpha = 0.05$$

$$Z = \frac{\bar{x} - \mu}{\frac{s}{\sqrt{n}}} = \frac{16,269.55 - 20,000.00}{725.77}$$

$$= \frac{-3,730.45}{(725.77)} = -5.14$$

Since -5.14 is less than -1.645 you can conclude that $\mu \geq \$20,000$ is not tenable. This conclusion would lead the mayor to apply for the HHS funds.

The memorandum appears in Exhibit 10–3.

Example 10–2 Instituting a two-pronged promotional campaign for the Stat City Electric Company

Mr. Saul Reisman, chief financial officer of the Stat City Electric Company, believes that there is a significant difference in average monthly electric bills between homeowners and apartment dwellers. If a difference does exist, Mr. Reisman is considering having two marketing campaigns to promote the idea of saving energy; one directed toward apartment dwellers and the other toward homeowners.

Mr. Reisman has retained you to determine if there is a statistically significant difference in average monthly electric bills between homeowners and apartment dwellers; that is, if he should spend his stockholders' money to buy the two-pronged campaign. Conduct the necessary survey and type a memorandum to the Stat City Electric Company reporting your findings, at the 5 percent level of significance (see Exhibit 10–4).

Solution

It is important to state the statistical hypothesis to be tested before any data are collected. The alternative courses of action open to the Electric Company are twofold: purchasing the two-pronged marketing campaign or continuing with the present marketing campaign. The relevant states of nature are whether or not the average monthly electric bills of apartment dwellers and homeowners are equal.

Remember that when working with two-tail tests the equality always goes in the null hypothesis. In this case the null hypothesis is:

$$H_0: \mu_{\text{house}} = \mu_{\text{apartment}}$$

Exhibit 10–3

```
┌──────────────────────────────────────────────────────────────┐
│                                                                │
│   HOWARD S. GITLOW, PH.D.                                      │
│   STATISTICAL CONSULTANT                                       │
│   ─────────                                                    │
│                                                                │
│                                                                │
│                       MEMORANDUM                               │
│                                                                │
│                                                                │
│   TO:    Mr. Lee Kaplowitz                                     │
│          Mayor, Stat City                                      │
│                                                                │
│   FROM:  Howard Gitlow                                         │
│                                                                │
│   DATE:  September 17, 1980                                    │
│                                                                │
│   RE:    Applying for HHS funds for Zone 4 families            │
│                                                                │
│                                                                │
│      In accordance with our consulting contract and commonly   │
│   accepted statistical principles, I have computed several     │
│   summary measures concerning the average 1979 family income   │
│   in Zone 4.¹ My research indicates that the average 1979      │
│   family income in Zone 4 was $16,269.55 per year (±$725.77).² │
│   Allowing for the errors in survey estimates you can have "95 │
│   percent confidence" that the average 1979 family income in   │
│   Zone 4 was below the $20,000 guideline.³                     │
│      If you have any further questions, please do not hesitate │
│   to call me at 305–999–9999.                                  │
│                                                                │
│                                                                │
│   ─────────                                                    │
│                                                                │
│                        Footnotes                               │
│                                                                │
│                                                                │
│     ¹ Insert standard footnote 1 (insert a sample size of 65 drawn via │
│   simple random sampling).                                     │
│     ² Insert standard footnote 2.                              │
│     ³ Insert standard footnote 4.                              │
│                                                                │
│                                                                │
└──────────────────────────────────────────────────────────────┘
```

The alternative hypothesis is:

$$H_A: \mu_{house} \neq \mu_{apartment}$$

If the null hypothesis is rejected and the alternative hypothesis is deemed tenable, then the two-pronged campaign will be instituted.

The problem would be set up as follows:

State of nature	Alternative course of action
$H_0: \mu_{house} = \mu_{apartment}$	(Retain the present marketing campaign.)
$H_A: \mu_{house} \neq \mu_{apartment}$	(Institute the new two-pronged marketing campaign.)
Set $\alpha = 0.05$	(If $\alpha = 0.05$, the management of the Stat City Electric Company is willing to reject the null hypothesis when it is tenable 5 percent of the time.)

Table 10–5 Average monthly electric bill, as of January 1980

	House	Apartment
$\bar{x}$	$70.90	$46.39
s	$18.75	$16.79
n	62	88

The next steps are to determine the proper sample size and to draw the sample. For purposes of ease, use the simple random sample shown in Table 10–4 (see next page).

Finally, the data can be analyzed and conclusions can be drawn. Table 10–5 shows the results of the survey.

Table 10–4

CASE-N	ID	ELEC	CASE-N	ID	ELEC	CASE-N	ID	ELEC
1	3.	66.	1	397.	83.	45	890.	39.
2	5.	54.	2	403.	44.	46	905.	35.
3	11.	62.	3	404.	41.	47	922.	95.
4	14.	101.	4	409.	31.	48	940.	38.
5	39.	74.	5	416.	37.	49	966.	38.
6	53.	51.	6	419.	37.	50	967.	47.
7	56.	37.	7	444.	45.	51	970.	47.
8	73.	82.	8	451.	48.	52	982.	26.
9	87.	73.	9	480.	35.	53	1007.	40.
10	88.	42.	10	482.	73.	54	1062.	27.
11	91.	70.	11	502.	63.	55	1069.	60.
12	100.	85.	12	521.	80.	56	1075.	44.
13	102.	66.	13	625.	51.	57	1079.	39.
14	117.	62.	14	632.	24.	58	1086.	20.
15	120.	54.	15	645.	43.	59	1094.	40.
16	132.	66.	16	646.	82.	60	1095.	45.
17	140.	50.	17	647.	71.	61	1104.	62.
18	142.	75.	18	648.	43.	62	1115.	52.
19	149.	72.	19	655.	50.	63	1135.	26.
20	151.	72.	20	660.	68.	64	1140.	51.
21	165.	91.	21	665.	51.	65	1150.	68.
22	172.	72.	22	670.	46.	66	1155.	61.
23	178.	82.	23	679.	54.	67	1165.	24.
24	181.	86.	24	688.	62.	68	1165.	46.
25	205.	77.	25	698.	35.	69	1169.	30.
26	206.	81.	26	722.	40.	70	1172.	38.
27	227.	67.	27	729.	49.	71	1174.	41.
28	230.	67.	28	733.	37.	72	1185.	33.
29	235.	44.	29	745.	39.	73	1192.	52.
30	238.	80.	30	753.	87.	74	1214.	21.
31	279.	87.	31	757.	42.	75	1234.	49.
32	280.	71.	32	759.	73.	76	1241.	58.
33	295.	42.	33	785.	42.	77	1263.	35.
34	301.	96.	34	787.	28.	78	1265.	84.
35	308.	58.	35	799.	21.	79	1266.	53.
36	313.	92.	36	814.	22.	80	1293.	21.
37	326.	65.	37	816.	57.	81	1312.	31.
38	329.	82.	38	825.	44.	82	1335.	27.
39	335.	106.	39	827.	45.	83	1337.	64.
40	340.	33.	40	836.	68.	84	1344.	61.
41	347.	42.	41	856.	44.	85	1348.	65.
42	355.	78.	42	870.	47.	86	1360.	30.
43	360.	67.	43	884.	33.	87	1370.	30.
44	361.	98.	44	889.	36.	88	1373.	38.
45	367.	82.						
46	369.	78.						
47	372.	59.						
48	381.	60.						
49	523.	60.						
50	538.	125.						
51	543.	98.						
52	553.	34.						
53	557.	81.						
54	566.	53.						
55	568.	42.						
56	579.	71.						
57	587.	84.						
58	612.	73.						
59	613.	72.						
60	617.	75.						
61	624.	69.						
62	625.	100.						

Please note the average monthly electric bill for homeowners ($70.90) is quite different from the average monthly electric bill for apartment dwellers ($46.39). The question arises: Is $70.90 different from $46.39 by a statistically significant amount? If there is a statistically significant difference, then the two-pronged promotional campaign should be instituted.

The test statistic is $t = 8.39$. (The following t-test formula may be different from the formula presented in your textbook. Use whichever formula your professor instructs you to use. Either formula will yield the same t-statistic.)

$$H_0: \mu_{\text{house}} = \mu_{\text{apartment}}$$
$$H_A: \mu_{\text{house}} \neq \mu_{\text{apartment}}$$
$$\alpha = 0.05$$

$$t(n_{\text{house}} + n_{\text{apartment}} - 2) = \frac{\bar{x}_{\text{house}} - \bar{x}_{\text{apartment}}}{\sqrt{s_p^2 \left(\frac{1}{n_{\text{house}}} + \frac{1}{n_{\text{apartment}}} \right)}}$$

where

$$s_p^2 = \frac{(n_{house} - 1)s^2_{house} + (n_{apartment} - 1)s^2_{apartment}}{n_{house} + n_{apartment} - 2}$$

$$t(148) = \frac{70.90 - 46.39}{\sqrt{310.61 \left(\frac{1}{62} + \frac{1}{88}\right)}}$$

where

$$s_p^2 = \frac{61(351.56) + 87(281.90)}{148}$$

$$= \frac{21,445.16 + 24,525.30}{148}$$

$$= \frac{45,970.46}{148} = 310.61$$

$$t(148) = \frac{70.90 - 46.39}{2.922} = \frac{24.51}{2.92} = 8.39$$

Since 8.39 is greater than $|1.98|$ $[t(1 - \alpha/2 = .975, 148) \approx t(1 - \alpha/2 = .975, 120) = 1.98]$ $\mu_{house} = \mu_{apartment}$ is not tenable. This conclusion would lead to purchasing the new two-pronged promotional campaign.

The memorandum appears in Exhibit 10–4.

Exhibit 10–4

HOWARD S. GITLOW, PH.D.
STATISTICAL CONSULTANT

MEMORANDUM

TO: Mr. Saul Reisman, Chief Financial Officer
        Stat City Electric Company

FROM: Howard Gitlow

DATE: September 26, 1980

RE: Instituting a two-pronged promotional campaign for
        the Stat City Electric Company

As per your request and using commonly accepted statistical techniques, I have computed several summary measures concerning the average monthly electric bills of Stat City apartment dwellers and homeowners.[1] My research indicates that the average monthly electric bill for a homeowner is $70.90 and $46.39 for an apartment dweller. In other words, the difference in the average monthly electric bills in Stat City between homeowners and apartment dwellers is $24.51 ($\pm$$2.92).[2] Allowing for the errors in survey estimates, you can have "95 percent confidence" that there is a statistically significant difference in average monthly electric bills between apartment dwellers and homeowners.[3]

If you have any further questions, call me at 305-999-9999.

Footnotes

[1] Insert standard footnote 1 (insert a sample size of 150 drawn via simple random sampling).
[2] Insert standard footnote 2.
[3] Insert standard footnote 4.

Example 10–3   Opening an automotive parts outlet in Stat City

"The Pep Boys, Manny, Mo, and Jack" are considering opening an automotive parts outlet in Stat City. They will open the outlet if more than 20 percent of Stat City families perform their own automotive repairs; 20 percent is the break-even percentage to open the automotive parts outlet. The Pep Boys are far more concerned about opening the outlet if there is insufficient demand than they are about not opening the outlet if there is sufficient demand.

The Pep Boys have retained you to determine if more than 20 percent of Stat City dwelling units perform their own automotive repairs as of January 1980; that is, if they should open the outlet in Stat City.

Conduct the necessary survey and type a memorandum to Manny, Mo, and Jack reporting your findings, at the 5 percent level of significance (see Exhibit 10–6).

Solution

BEFORE the survey is performed it is important to specify the hypothesis being tested. The alternative courses of action open to the Pep Boys are: opening the outlet or not opening the outlet. The revelant states of nature are whether or not more than 20 percent of the Stat City families perform their own repairs. The outlet is deemed financially feasible if more than 20 percent of the Stat City families perform their own automotive repairs. Table 10–6 shows the structure of the decision to be made.

Table 10–6

	States of nature	
Alternative courses of action	Less than (or equal to) 20 percent of Stat City families perform their own repairs ($\pi \leq .20$)	More than 20 percent of Stat City families perform their own repairs ($\pi > .20$)
Open outlet	C	B
Do not open outlet	A	D

The box marked A indicates a correct decision; not opening the outlet if less than (or equal to) 20 percent of Stat City families perform their own automotive repairs. The box marked B also indicates a correct decision; opening the outlet if more than 20 percent of the Stat City families perform their own repairs. The cell marked C indicates an incorrect decision; opening the outlet if less than (or equal to) 20 percent of Stat City families perform their own repairs. This error would result in opening an unsuccessful store (an extremely expensive error). The cell marked D also indicates an incorrect decision; not opening the outlet if more than 20 percent of Stat City families perform their own repairs. This error is far less serious to the Pep Boys than the previously stated error. This error (D) represents the opportunity loss of a potentially good business, however, Manny, Mo, and Jack would still have their capital for some other business venture.

Recall, it is customary when working with one-tail tests to designate as the null hypothesis that hypothesis which, if rejected when tenable, leads to the type I error. In this case the null hypothesis is:

$$H_0: \pi \leq .20$$

Note, $\pi \leq .20$ is designated as the null hypothesis because if it is rejected when it is tenable, the type I error is committed; the Pep Boys open the store when demand is not sufficient to warrant opening the store. Remember, opening a store when there is insufficient demand means that the Pep Boys will loose their capital. The alternative hypothesis is:

$$H_A: \pi > .20$$

If the null hypothesis is rejected and the alternative hypothesis is deemed tenable, action is then taken; the outlet would be opened because more than 20 percent of Stat City families perform their own repairs (there is sufficient demand to support an automotive parts outlet).

The problem would be set up as follows:

State of nature	Alternative course of action
$H_0: \pi \leq .20$	(Do not open outlet)
$H_A: \pi > .20$	(Open outlet)

The level of significance for the problem (α) is set to control the more severe error. If $\alpha = 0.05$, then the Pep Boys are willing to reject the null hypothesis, when, in fact, it is tenable 5 percent of the time. In lay language, if the Pep Boys were making this decision 100 times they would tolerate being wrong (and loosing their capital) 5 times. The fact remains, the Pep Boys only make this decision once. A summary of the work done so far and the appropriate statistical test is shown below.

$$H_0: \pi \leq .20$$
$$H_A: \pi > .20$$
$$\alpha = 0.05$$
$$Z = \frac{\bar{p} - \pi}{\hat{\sigma}_{\bar{p}}}$$

where

$$\hat{\sigma}_{\bar{p}} = \sqrt{\frac{\pi(1 - \pi)}{n}}$$

The next step is to compute the proper sample size (n).

$$n = \frac{\{Z_\alpha \sqrt{\pi_0(1 - \pi_0)} + Z_\beta \sqrt{\pi_1(1 - \pi_1)}\}^2}{(\pi_1 - \pi_0)^2}$$

where

n = the sample size,
π_0 = value of the population proportion under the null hypothesis,
π_1 = value of the population proportion under the alternative hypothesis,
Z_α = Z value for a given α level of significance (omit sign of Z_α), and
Z_β = Z value for a given β risk of a type II error (omit sign of Z_β)

Z_β can be computed after the decision maker states that he/she wishes to have, say, an 80 percent chance (power) of rejecting the null hypothesis (of $\pi = 0.20$) when the population proportion is actually, say, 30 percent and he/she is willing to permit α to be 5 percent. Hence, $Z_\beta = 0.84$, as shown in Exhibit 10–5.

Consequently,

$$n = \frac{\{1.645\sqrt{.2 \times .8} + .84\sqrt{.3 \times .7}\}^2}{(.30 - .20)^2}$$
$$= \frac{(.658 + .385)^2}{.1^2} = \frac{1.043^2}{.1^2} = 108.8 \cong 109$$

Exhibit 10–5

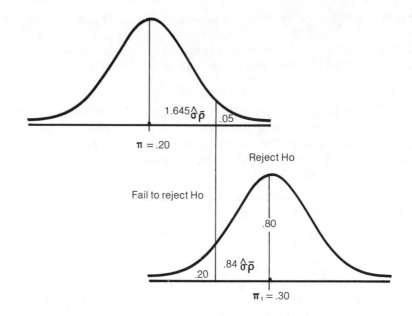

Table 10–7

CASE-N	ID	REPAR	CASE-N	ID	REPAR
1	45.	2.	55	789.	1.
2	79.	2.	56	790.	0.
3	80.	1.	57	792.	0.
4	111.	2.	58	804.	0.
5	124.	0.	59	810.	0.
6	126.	0.	60	822.	0.
7	130.	2.	61	880.	0.
8	145.	2.	62	901.	1.
9	150.	1.	63	904.	1.
10	179.	2.	64	924.	1.
11	186.	2.	65	937.	1.
12	189.	0.	66	951.	1.
13	208.	1.	67	954.	1.
14	212.	2.	68	957.	2.
15	221.	1.	69	967.	1.
16	238.	1.	70	971.	1.
17	276.	0.	71	980.	2.
18	304.	2.	72	983.	0.
19	309.	1.	73	984.	1.
20	331.	2.	74	989.	0.
21	332.	1.	75	994.	2.
22	354.	1.	76	1040.	1.
23	366.	0.	77	1044.	1.
24	391.	2.	78	1051.	1.
25	413.	1.	79	1061.	0.
26	420.	2.	80	1105.	1.
27	424.	2.	81	1119.	2.
28	425.	0.	82	1120.	0.
29	431.	2.	83	1146.	1.
30	435.	0.	84	1155.	1.
31	459.	0.	85	1160.	0.
32	463.	1.	86	1170.	1.
33	465.	2.	87	1180.	1.
34	481.	2.	88	1193.	1.
35	486.	2.	89	1210.	0.
36	496.	2.	90	1211.	2.
37	507.	0.	91	1231.	0.
38	532.	1.	92	1235.	0.
39	559.	1.	93	1240.	2.
40	569.	1.	94	1247.	2.
41	580.	2.	95	1288.	0.
42	608.	2.	96	1291.	1.
43	619.	2.	97	1298.	0.
44	627.	0.	98	1299.	2.
45	628.	0.	99	1305.	2.
46	633.	1.	100	1316.	0.
47	689.	0.	101	1317.	1.
48	704.	1.	102	1318.	0.
49	705.	1.	103	1320.	1.
50	727.	0.	104	1326.	2.
51	753.	1.	105	1335.	1.
52	764.	0.	106	1344.	0.
53	772.	0.	107	1356.	1.
54	779.	1.	108	1362.	0.
			109	1363.	1.

(Since $n/N < .10$, the finite population correction factor is not necessary.[3]) The sample size of 109 implies that 109 Stat City families must be randomly sampled in order for the Pep Boys to have a 5 percent risk of committing a type I error and an 80 percent chance of rejecting the null hypothesis (of 20 percent) while detecting that the population proportion is really 30 percent.

Once the sample size has been determined the random sample can be drawn. Table 10–7 enumerates the random sample selected for this study.

Finally, the data can be analyzed and conclusions can be drawn. Table 10–8 shows the results of the survey.

Table 10–8

Favorite place for automotive repairs as of January 1980	Sample frequency	Estimated percent
Repairs by owner = 0	35	32.1
Repairs performed by gas station = 1	43	39.4
Repairs performed by dealer = 2	31	28.4
Total	109	100.0

The above sample indicates that 32.1 percent of Stat City families perform their own automotive repairs. The question arises: Is 32.1 percent above 20 percent by a statistically significant amount so that the Pep Boys can open the outlet with the understanding that there is only a 5 percent chance of opening the outlet when it should not be opened ($\alpha = 0.05$)? The hypotheses test follows:

$$H_0: \pi \leq 0.20$$
$$H_A: \pi > 0.20$$
$$\alpha = 0.05$$
$$Z = \frac{.321 - .200}{\sqrt{\frac{.2 \times .8}{109}}} = \frac{.121}{.0383}$$
$$= 3.16$$

Since 3.16 is greater than 1.645, you can conclude that $\pi \leq 0.20$ is untenable. This conclusion would lead to opening the automotive parts outlet.

---

[3] Use of the finite population correction factor may cause logical inconsistencies in problems of this type.

The memorandum appears in Exhibit 10-6.

Exhibit 10-6

HOWARD S. GITLOW, PH.D.
STATISTICAL CONSULTANT

MEMORANDUM

TO: Manny Pep, Mo Pep, and Jack Pep

FROM: Howard Gitlow

DATE: September 1, 1980

RE: Opening an automotive parts outlet in Stat City

  As per your request and using commonly accepted
statistical techniques, I have computed several summary
measures concerning the proportion of Stat City families
that perform their own automotive repairs.[1] My research
indicates that approximately 32.1 percent (±3.83 percent)[2]
of Stat City families perform their own automotive repairs.
Allowing for the errors in survey estimates, you can have "95
percent confidence" that the proportion of Stat City
families that perform their own repairs exceeds the
"break-even" percentage of 20 percent.[3]
  If you have any further questions, please do not hesitate
to call me at 305-999-9999.

_____

Footnotes

  [1] Insert standard footnote 1 (insert a sample size of 109 drawn
via simple random sampling).
  [2] Insert standard footnote 2.
  [3] Insert standard footnote 4.

**Example 10–4   Instituting a two-pronged promotional campaign for
Howie's Gulf Station**

Mr. Howie, the owner of Howie's Gulf Station, is considering having two promotional campaigns; one directed toward apartment dwellers and one for homeowners. Mr. Howie will pay for the two campaigns if there is a significant difference between apartment dwellers and homeowners in respect to their favorite gas station.

Mr. Howie has retained you to determine if there is a significant difference between apartment dwellers and homeowners with respect to their choice of gas station; that is, should he consider an expensive promotional campaign. Conduct the necessary survey and type a memorandum to Mr. Howie reporting your findings, at the 1 percent level of significance (see Exhibit 10–7).

Solution

Remember, it is important to state the statistical hypothesis being tested before any data are collected. The alternative courses of action open to Mr. Howie are: instituting a new two-pronged promotional campaign or retaining the present one-pronged promotional campaign. The relevant states of nature are whether or not the percentage of apartment dwellers favoring Howie's Gulf is equal to the percentage of homeowners' favoring Howie's Gulf.

Recall that when working with two-tail tests the equality always goes in the null hypothesis. In this case the null hypothesis is:

$$H_0: \pi_{\text{apartment}} = \pi_{\text{house}}$$

The alternative hypothesis is:

$$H_A: \pi_{\text{apartment}} \neq \pi_{\text{house}}$$

If the null hypothesis is rejected and the alternative hypothesis is deemed tenable, then the two-pronged campaign will be instituted.

The problem would be set up as follows:

State of nature	Alternative course of action
$H_0: \pi_{\text{apartment}} = \pi_{\text{house}}$	(Continue present promotional campaign.)
$H_A: \pi_{\text{apartment}} \neq \pi_{\text{house}}$	(Institute the new two-pronged promotional campaign.)
Set $\alpha = 0.01$	(If $\alpha = 0.01$, Mr. Howie is willing to reject the null hypothesis when in fact it is tenable 1 percent of the time.)

The next steps are to determine the proper sample size and to draw the sample. For purposes of ease, use the simple random sample given in Table 10–9.

Table 10–9

CASE-N	ID	DWELL	FAVGA	CASE-N	ID	DWELL	FAVGA
1	22.	1.	1.	51	785.	0.	1.
2	41.	1.	2.	52	799.	0.	1.
3	48.	1.	2.	53	807.	0.	2.
4	99.	1.	2.	54	833.	0.	1.
5	108.	1.	1.	55	837.	0.	2.
6	125.	1.	1.	56	839.	0.	2.
7	128.	1.	1.	57	841.	0.	1.
8	132.	1.	1.	58	870.	0.	1.
9	133.	1.	2.	59	885.	0.	1.
10	134.	1.	1.	60	888.	0.	2.
11	137.	1.	2.	61	910.	0.	2.
12	142.	1.	2.	62	912.	0.	1.
13	160.	1.	2.	63	913.	0.	2.
14	165.	1.	2.	64	914.	0.	1.
15	181.	1.	1.	65	916.	0.	2.
16	182.	1.	2.	66	934.	0.	2.
17	199.	1.	2.	67	940.	0.	1.
18	205.	1.	2.	68	945.	0.	2.
19	230.	1.	2.	69	977.	0.	2.
20	231.	1.	1.	70	983.	0.	2.
21	249.	1.	2.	71	996.	0.	2.
22	263.	1.	2.	72	1011.	0.	2.
23	266.	1.	2.	73	1031.	0.	2.
24	298.	1.	1.	74	1039.	0.	1.
25	302.	1.	2.	75	1041.	0.	1.
26	335.	1.	1.	76	1045.	0.	2.
27	341.	1.	2.	77	1052.	0.	2.
28	345.	1.	2.	78	1089.	0.	2.
29	349.	1.	1.	79	1095.	0.	2.
30	350.	1.	2.	80	1113.	0.	1.
31	352.	1.	1.	81	1122.	0.	2.
32	372.	1.	2.	82	1129.	0.	2.
33	383.	1.	1.	83	1136.	0.	2.
34	461.	0.	1.	84	1143.	0.	1.
35	482.	0.	1.	85	1148.	0.	1.
36	487.	0.	2.	86	1157.	0.	2.
37	491.	0.	1.	87	1158.	0.	2.
38	515.	0.	2.	88	1159.	0.	1.
39	536.	1.	2.	89	1163.	0.	2.
40	545.	1.	2.	90	1171.	0.	1.
41	565.	1.	1.	91	1176.	0.	2.
42	567.	1.	2.	92	1180.	0.	2.
43	622.	1.	1.	93	1246.	0.	2.
44	644.	0.	2.	94	1247.	0.	2.
45	668.	0.	2.	95	1275.	0.	1.
46	672.	0.	2.	96	1307.	0.	2.
47	691.	0.	2.	97	1321.	0.	2.
48	716.	0.	1.	98	1328.	0.	2.
49	721.	0.	2.	99	1346.	0.	1.
50	732.	0.	2.	100	1356.	0.	2.

Finally, the data can be analyzed and conclusions can be drawn. Table 10–10 shows the results of the survey.

Table 10–10

Type of dwelling	Favorite gas station		Total
	Paul's Texaco	Howie's Gulf	
Apartment	22	40	62
House	15	23	38
Total	37	63	100

The above sample indicates that 64.5 percent (40/62) of apartment dwellers favor Howie's Gulf while 60.5 percent (23/38) of homeowners favor Howie's Gulf. The question arises: Is 64.5 percent different from 60.5 percent by a statistically significant amount? If there is a statistically significant difference, then the two-pronged promotional campaign will be instituted.

The hypothesis test is shown below:

$$H_0: \pi_{apartment} = \pi_{house}$$
$$H_A: \pi_{apartment} \neq \pi_{house}$$
$$\alpha = 0.05$$

$$z \cong \frac{\bar{p}_{apartment} - \bar{p}_{house}}{\sqrt{\bar{p}(1 - \bar{p}) \left(\frac{1}{n_{apartment}} + \frac{1}{n_{house}} \right)}}$$

where

$$\bar{p} = \frac{X_{apartment} + X_{house}}{n_{apartment} + n_{house}}$$

$X_{apartment}$ = number of families' favoring Howie's Gulf out of apartment dwellers

X_{house} = number of families' favoring Howie's Gulf out of homeowners

The assumptions required to perform the above test are: (1) independent random samples, (2) large sample sizes ($n_{apartment}$ and n_{house} are both greater than 30), and (3) the equality of the population standard deviations.

The test statistic is:

$$Z = \frac{0.645 - 0.605}{\sqrt{0.63 \times 0.37 \left(\frac{1}{62} + \frac{1}{38} \right)}} = \frac{0.040}{0.099} = 0.404$$

where

$$\bar{p} = \frac{40 + 23}{62 + 38} = \frac{63}{100} = 0.63$$

Since $Z = 0.404$ is less than $|1.96|$ you can conclude that $\pi_{apartment} = \pi_{house}$ is tenable. This conclusion favors retention of the present promotional campaign.

The above Z-test formula may be different from the formula in your textbook. Use whichever formula your professor instructs you to use. Either formula will yield the same Z-statistic.

The memorandum appears in Exhibit 10–7.

Exhibit 10–7

HOWARD S. GITLOW, PH.D.
STATISTICAL CONSULTANT

MEMORANDUM

TO: Mr. Howie, Owner
        Howie's Gulf Station

FROM: Howard Gitlow

DATE: September 4, 1980

RE: Promotional campaign for Howie's Gulf Station

   In accordance with our consulting contract and using
commonly accepted statistical techniques, I have computed
several summary measures concerning the proportion of Stat
City apartment dwellers and homeowners that prefer Howie's
Gulf Station over Paul's Texaco Station.[1] My research
indicates that approximately 64.5 percent of apartment
dwellers favor Howie's Gulf while 60.5 percent of
homeowners favor Howie's Gulf. In other words, the
difference in the percent of Stat City families that favor
Howie's Gulf between apartment dwellers and homeowners is 4
percent (±9.9 percent).[2] Allowing for the errors in survey
estimates, you can have "99 percent confidence" that there
is no statistically significant difference between the
proportion of apartment dwellers and homeowners that favor
Howie's Gulf.[3]

   If you have any further questions, please do not hesitate
to call me at 305-999-9999.

Footnotes

   [1] Insert standard footnote 1 (insert a sample of 100 drawn via
simple random sampling).
   [2] Insert standard footnote 2.
   [3] Insert standard footnote 4.

ADDITIONAL PROBLEMS

**10–5   Decision to purchase a new promotional campaign to boost diet
soda sales**   The Stat City Supermarket Association believes that the typical Stat City family purchases less than one six-pack of diet soda per
week, on an average. The association would like to test this hypothesis
because if it is tenable they will mount a promotional campaign to
increase diet soda sales. If the hypothesis is not tenable, the association
will continue with their current promotional format. The association is
more concerned with spending funds for the new promotional campaign
when it is not necessary than they are about retaining the present promotional campaign when they should have purchased the new promotional
campaign.

   You have been retained by the association to test their hypothesis and
help them determine if they should purchase the new promotional campaign. The association is willing to have a 10 percent chance of purchasing

the new campaign when it is unnecessary and wants a 90 percent chance of rejecting the notion that the average purchase is one six-pack per week, when in fact, the average weekly purchase is four cans (two thirds of a six-pack) per week.

Use the pilot sample in the margin of five Stat City dwelling units to obtain an estimate of the variance of average weekly purchases of six-packs of diet soda, as of January 1980.

Once you have computed the appropriate sample size, use the accompanying random number table to draw your simple random sample (start in the upper left-hand corner).

ID

0083
1211
1296
0595
0486

2668	7422	4354	4569	9446	8212	3737	2396	6892	3766
6067	7516	2451	1510	0201	1437	6518	1063	6442	6674
4541	9863	8312	9855	0995	6025	4207	4093	9799	9308
6987	4802	8975	2847	4413	5997	9106	2876	8596	7717
0376	8636	9953	4418	2388	8997	1196	5158	1803	5623
8468	5763	3232	1986	7134	4200	9699	8437	2799	2145
9151	4967	3255	8518	2802	8815	6289	9549	2942	3813
1073	4930	1830	2224	2246	1000	9315	6698	4491	3046
5487	1967	5836	2090	3832	0002	9844	3742	2289	3763
4896	4957	6536	7430	6208	3929	1030	2317	7421	3227
9143	7911	0368	0541	2302	5473	9155	0625	1870	1890
9256	2956	4747	6280	7342	0453	8639	1216	5964	9772
4173	1219	7744	9241	6354	4211	8497	1245	3313	4846
2525	7811	5417	7824	0922	8752	3537	9069	5417	0856
9165	1156	6603	2852	8370	0995	7661	8811	7835	5087
0014	8474	6322	5053	5015	6043	0482	4957	8904	1616
5325	7320	8406	5962	6100	3854	0575	0617	8019	2646
2558	1748	5671	4974	7073	3273	6036	1410	5257	3939
0117	1218	0688	2756	7545	5426	3856	8905	9691	8890
8353	1554	4083	2029	8857	4781	9654	7946	7866	2535
1990	9886	3280	6109	9158	3034	8490	6404	6775	8763
9651	7870	2555	3518	2906	4900	2984	6894	5050	4586
9941	5617	1984	2435	5184	0379	7212	5795	0836	4319
7769	5785	9321	2734	2890	3105	6581	2163	4938	7540
3224	8379	9952	0515	2724	4826	6215	6246	9704	1651
1287	7275	6646	1378	6433	0005	7332	0392	1319	1946
6389	4191	4548	5546	6651	8248	7469	0786	0972	7649
1625	4327	2654	4129	3509	3217	7062	6640	0105	4422
7555	3020	4181	7498	4022	9122	6423	7301	8310	9204
4177	1844	3468	1389	3884	6900	1036	8412	0881	6678
0927	0124	8176	0680	1056	1008	1748	0547	8227	0690
8505	1781	7155	3635	9751	5414	5113	8316	2737	6860
8022	8757	6275	1485	3635	2330	7045	2106	6381	2986
8390	8802	5674	2559	7934	4788	7791	5202	8430	0289
3630	5783	7762	0223	5328	7731	4010	3845	9221	5427
9154	6388	6053	9633	2080	7269	0894	0287	7489	2259
1441	3381	7823	8767	9647	4445	2509	2929	5067	0779
8246	0778	0993	6687	7212	9968	8432	1453	0841	4595
2730	3984	0563	9636	7202	0127	9283	4009	3177	4182
9196	8276	0233	0879	3385	2184	1739	5375	5807	4849
5928	9610	9161	0748	3794	9683	1544	1209	3669	5831
1042	9600	7122	2135	7868	5596	3551	9480	2342	0449
6552	4103	7957	0510	5958	0211	3344	5678	1840	3627
5968	4307	9327	3197	0876	8480	5066	1852	8323	5060
4445	1018	4356	4653	9302	0761	1291	6093	5340	1840
8727	8201	5980	7859	6055	1403	1209	9547	4273	0857
9415	9311	4996	2775	8509	7767	6930	6632	7781	2279
2648	7639	9128	0341	6875	8957	6646	9783	6668	0317
3707	3454	8829	6863	1297	5089	1002	2722	0578	7753
8383	8957	5595	9395	3036	4767	8300	3505	0710	6307

Type a memorandum to the Stat City Supermarket Association reporting your findings.

10–6 Establishment of a policy for showing houses to wealthy, prospective Stat City families Ms. Sharon Vigil, chairperson of the Stat City Real Estate Board, believes that Zone 2 dwelling units have more rooms on an average than Zone 1 dwelling units. She would like to test this hypothesis because if a difference exists the board will direct wealthier prospective Stat City families toward homes in Zone 2, as opposed to Zone 1. If little difference exists, the board will direct wealthier prospective Stat City families toward homes in both Zones 1 and 2. Ms. Vigil is far more concerned about directing wealthy families exclusively to Zone 2 when it would be inappropriate than directing wealthy families to both zones when only Zone 2 should be considered.

Ms. Vigil has hired you to determine if the homes in Zone 2 have significantly more rooms than the homes in Zone 1; that is, if the board should direct wealthier prospective Stat City families exclusively to homes in Zone 2. Ms. Vigil wants "95 percent confidence" in the statistics.

Use the following random sample to estimate the required statistics.

ID	ROOMS	ID	ROOMS	ID	ROOMS
8.	9.	113.	8.	181.	10.
42.	8.	118.	8.	187.	10.
45.	9.	141.	7.	191.	5.
58.	11.	150.	7.	202.	11.
70.	6.	161.	8.	207.	6.
71.	8.	168.	8.	217.	5.
80.	9.	170.	7.	223.	9.
111.	8.	176.	8.	233.	8.
				266.	7.

Type a memorandum to Ms. Vigil reporting your findings.

10–7 Federal funding for the Stat City Hospital Mr. Marc Wurgaft, subdirector of the Department of Health and Human Services (HHS), has issued a statement to the Stat City Hospital stating that:

1. If 25 percent or more of Stat City families use the Stat City Hospital at least once per year, federal funding for the hospital will not be changed.
2. If less than 25 percent of Stat City families use the Stat City Hospital at least once per year, federal funding for the hospital will be restricted.

Mr. Wurgaft is more concerned about unnecessary reductions in the Stat City Hospital's federal funding than he is about missing an opportunity to change the Stat City Hospital's federal funding when it is appropriate.

You have been hired by Mr. Wurgaft to determine if 25 percent or more of Stat City families use the hospital; that is, if HHS should leave the Stat City Hospital's funding intact. Mr. Wurgaft is willing to have a 10 percent chance of reducing the hospital's funding when it is not called for and wants a 70 percent chance of rejecting the notion that the percentage of families that use the hospital is 25 percent, when in fact, it is 15 percent.

Conduct the necessary survey and type a memorandum to Mr. Wurgaft reporting your findings. (Note: Use the accompanying random number table to draw your sample; begin at the upper left hand corner of the table.)

9263	7824	1926	9545	5349	2389	3770	7986	7647	6641
7944	7873	7154	4484	2610	6731	0070	3498	6675	9972
5965	7196	2738	5000	0535	9403	2928	1854	5242	0608
3152	4958	7661	3978	1353	4808	5948	6068	8467	5301
0634	7693	9037	5139	5588	7101	0920	7915	2444	3024
2870	5170	9445	4839	7378	0643	8664	6923	5766	8018
6810	8926	9473	9576	7502	4846	6554	9658	1891	1639
9993	9070	9362	6633	3339	9526	9534	5176	9161	3323
9154	7319	3444	6351	8383	9941	5882	4045	6926	4856
4210	0278	7392	5629	7267	1224	2527	3667	2131	7576
1713	2758	2529	2838	5135	6166	3789	0536	4414	4267
2829	1428	5452	2161	9532	3817	6057	0808	9499	7846
0933	5671	5133	0628	7534	0881	8271	5739	2525	3033
3129	0420	9371	5128	0575	7939	8739	5177	3307	9706
3614	1556	2759	4208	9928	5964	1522	9607	0996	0537
2955	1843	1363	0552	0279	8101	4902	7903	5091	0939
2350	2264	6308	0819	8942	6780	5513	5470	3294	6452
5788	8584	6796	0783	1131	0154	4853	1714	0855	6745
5533	7126	8847	0433	6391	3639	1119	9247	7054	2977
1008	1007	5598	6468	6823	2046	8938	9380	0079	9594
3410	8127	6609	8887	3781	7214	6714	5078	2138	1670
5336	4494	6043	2283	1413	9659	2329	5620	9267	1592
8297	6615	8473	1943	5579	6922	2866	1367	9931	7687
5482	8467	2289	0809	1432	8703	4289	2112	3071	4848
2546	5909	2743	8942	8075	8992	1909	6773	8036	0879
6760	6021	4147	8495	4013	0254	0957	4568	5016	1560
4492	7092	6129	5113	4759	8673	3556	7664	1821	6344
3317	3097	9813	9582	4978	1330	3608	8076	3398	6862
8468	8544	0620	1765	5133	0287	3501	6757	6157	2074
7188	5645	3656	0939	9695	3550	1755	3521	6910	0167
0047	0222	7472	1472	4021	2135	0859	4562	8398	6374
2599	3888	6836	5956	4127	6974	4070	3799	0343	1887
9288	5317	9919	9380	5698	5308	1530	5052	5590	4302
2513	2681	0709	1567	6068	0441	2450	3789	6718	6282
8463	7188	1299	8302	8248	9033	9195	7457	0353	9012
3400	9232	1279	6145	4812	7427	2836	6656	7522	3590
5377	4574	0573	8616	4276	7017	9731	7389	8860	1999
5931	9788	7280	5496	6085	1193	3526	7160	5557	6771
2047	6655	5070	2699	0985	5259	1406	3021	1989	1929
8618	8493	2545	2604	0222	5201	2182	5059	5167	6541
2145	6800	7271	4026	6128	1317	6381	4897	5173	5411
9806	6837	8008	2413	7235	9542	1180	2974	8164	8661
0178	6442	1443	9457	7515	9457	6139	9619	0322	3225
6246	0484	4327	6870	0127	0543	2295	1894	9905	4169
9432	3108	8415	9293	9998	8950	9158	0280	6947	6827
0579	4398	2157	0990	7022	1979	5157	3643	3349	7988
1039	1428	5218	0972	2578	3856	5479	0489	5901	8925
3517	5698	2554	5973	6471	5263	3110	6238	4948	1140
2563	8961	7588	9825	0212	7209	5718	5588	0932	7346
1646	4828	9425	4577	4515	6886	1138	1178	2269	4198

10–8 Percentage of families with children in Zones 1 and 2 of Stat City Ms. Sharon Vigil, chairperson of the Stat City Real Estate Board, has a policy of indiscriminately directing wealthy, prospective Stat City families who want to live in an area with children into Zones 1 and 2. Ms. Vigil would like to determine if the above policy is reasonable. Consequently, she needs to determine if the proportion of Stat City families with children in Zones 1 and 2 are equal (three or more people in a Stat City family indicate the presence of children). If the proportion of families with children in Zones 1 and 2 are equal, the board's policy will be maintained. However, if the proportion of families with children in Zones 1 and 2 are different, the board will have to formulate a new policy for directing wealthy, prospective Stat City families into residential housing zones.

The board has retained you to determine if there is a significant difference (at the 5 percent level of significance) between Zone 1 and Zone 2 in respect to the percent of families with children. Conduct the necessary survey and type a memorandum to the Stat City Real Estate Board reporting your findings.

Use the accompanying simple random sample of Stat City dwelling units.

Random sample for Problem 10-8

CASE-N	ID	ZONE	PEPLE	CASE-N	ID	ZONE	PEPLE
1	7.	1.	4.	1	134.	2.	3.
2	16.	1.	4.	2	147.	2.	4.
3	22.	1.	3.	3	154.	2.	4.
4	30.	1.	4.	4	157.	2.	4.
5	48.	1.	3.	5	160.	2.	2.
6	51.	1.	1.	6	163.	2.	4.
7	57.	1.	4.	7	165.	2.	6.
8	59.	1.	4.	8	166.	2.	3.
9	66.	1.	5.	9	169.	2.	5.
10	68.	1.	1.	10	178.	2.	3.
11	74.	1.	3.	11	179.	2.	2.
12	78.	1.	2.	12	183.	2.	3.
13	80.	1.	4.	13	185.	2.	2.
14	84.	1.	3.	14	188.	2.	3.
15	89.	1.	6.	15	189.	2.	2.
16	91.	1.	4.	16	190.	2.	2.
17	94.	1.	3.	17	195.	2.	4.
18	95.	1.	1.	18	196.	2.	4.
19	96.	1.	2.	19	200.	2.	4.
20	99.	1.	4.	20	210.	2.	3.
21	103.	1.	4.	21	213.	2.	3.
22	105.	1.	3.	22	218.	2.	2.
23	108.	1.	4.	23	220.	2.	6.
24	111.	1.	5.	24	222.	2.	2.
25	114.	1.	4.	25	224.	2.	3.
26	117.	1.	3.	26	229.	2.	5.
27	119.	1.	6.	27	230.	2.	3.
28	124.	1.	1.	28	231.	2.	4.
29	126.	1.	2.	29	232.	2.	4.
30	129.	1.	3.	30	233.	2.	3.
				31	234.	2.	2.
				32	238.	2.	4.
				33	239.	2.	6.
				34	250.	2.	5.
				35	252.	2.	4.
				36	255.	2.	2.
				37	259.	2.	4.
				38	260.	2.	5.
				39	264.	2.	6.
				40	282.	2.	5.

ADVANCED SAMPLING PLANS

11

In Chapter 9 you learned how to compute the size of a simple random sample to estimate μ and π, for a given tolerable error (accuracy) and level of confidence (reliability). In this chapter we shall be concerning ourselves with two advanced sampling plans which, under certain conditions, are more "efficient" than simple random sampling. One sampling plan is more efficient than another sampling plan if it provides, at a given level of confidence, a smaller standard error. To rephrase, one sampling plan is more efficient than another if it provides more information per unit of cost.

The advanced sampling plans that will be discussed in this chapter are: stratified random sampling and single-stage cluster sampling. The situations in which both sampling plans provide more efficient estimators than simple random sampling will be discussed.

A *stratified random sample* is obtained by dividing a population into nonoverlapping groups, called strata, and then selecting independent simple random samples from each stratum. For example, if the mayor of Stat City wants to estimate the average income in Stat City and knows that income level varies from zone to zone, a stratified random sample could be used in which each zone is defined as a stratum. This sampling plan should be more efficient than a simple random sampling plan of the same size because it divides the population into homogeneous strata such that the variability of each stratum is smaller than the variability of the entire population. Consequently, a stratified random sample of a given size will provide a smaller standard error, on average, than a simple random sample of the same size (due to the smaller variability in each stratum).

The major reasons why stratified random sampling provides more efficient estimators than simple random sampling are:

1. It takes advantage of prior knowledge concerning the variability present in each stratum of a population.
2. It takes advantage of prior knowledge concerning the cost of sampling in each stratum of a population (for example, if it is less expensive to sample in Zones 3 and 4, than Zones 1 and 2, stratified sampling could factor these cost differentials into the sampling plan).
3. It provides separate estimates of parameters for each stratum without additional sampling.

A *single-stage cluster sample* is obtained by grouping the elements of a population into bunches (called clusters), selecting a probability sample of clusters, and including all elements in the selected clusters into the sample. For example, the mayor of Stat City has hired you to estimate the average age of Stat City residents. If you use simple random or stratified random sampling you will need a frame listing all residents (not dwelling units) in Stat City. Unfortunately, it would be extremely expensive (almost impossible) to construct such a list. However, you could divide Stat City into blocks (see the Stat City map and variable 3), select a simple random sample of blocks, and construct a frame only for the blocks drawn into the sample; this could save a great deal of time and money. Finally, the age of every resident in each selected block would be determined and the desired estimates computed.

The major situations in which single-stage cluster sampling is more efficient than simple random or stratified random sampling are: (1) a good frame is not available or is very costly to obtain, (2) the cost of obtaining observations increases as the distance separating elements increases, and (3) the occurrence of a rare event is being estimated.

The following sections of this chapter will present: (1) the formulas required to estimate μ from a stratified random sample and μ and π from a single-stage cluster sample, (2) applications of stratified and cluster sampling for Stat City interest groups, (3) comparisons of stratified and simple random sampling, and cluster and stratified sampling, and (4) additional examples so that you can develop your sampling skills.

COMPUTING THE SAMPLE SIZE WITH STRATIFIED RANDOM SAMPLING

Recall that the major purpose of stratified sampling is to reduce costs for a given level of precision or increase precision for a given level of cost when estimating a parameter. The type of stratified sampling plan we will discuss assumes that the cost of sampling one unit is the same in each stratum, the number of units in each stratum varies from stratum to stratum, and the variance of the variable understudy is different for each stratum.

The sample size necessary to estimate μ with a prespecified standard error (V) is:[1]

$$n = \left[\frac{\left(\sum_{h=1}^{\mathcal{L}} W_h \hat{S}_h \right)^2}{V^2 + \frac{1}{N} \sum_{h=1}^{\mathcal{L}} W_h \hat{S}_h^2} \right]$$

where

n = the necessary sample size to estimate μ for a given level of precision
$\mathcal{L}$ = the number of strata
N_h = the population size for the hth strata
N = the total population size
$W_h = \left[\dfrac{N_h}{N} \right]$ = the proportion of the population in the hth strata
$\hat{S}_h^2$ = the estimate of the variance in the hth strata from a pilot or prior study
V = the desired standard error

V is set by management and is directly a function of the desired level of confidence and the tolerable error. This can be demonstrated via simple random sampling.

---

[1] The finite population correction factor for stratified sampling is not presented in this text.

Recall, under simple random sampling the sample size to estimate μ is computed as follows:

$$n = \frac{Z^2 \hat{S}^2}{(\bar{X} - \mu)^2}$$

Hence,

$$\left[\frac{\hat{S}^2}{n} \right] = \frac{(\bar{X} - \mu)^2}{Z^2}$$

$$S_{\bar{X}}^2 = \frac{(\bar{X} - \mu)^2}{Z^2}$$

$$S_{\bar{X}} = \frac{(\bar{X} - \mu)}{Z} = \frac{e}{Z} = V$$

You can see that the standard error is equal to the tolerable error divided by the number of standard errors required to achieve $100(1 - \alpha)$ percent confidence.

The necessary sample size in each stratum is:

$$n_h = \left[\frac{N_h \hat{S}_h}{\sum\limits_{h=1}^{\mathscr{L}} (N_h \hat{S}_h)} \right] [n]$$

The estimate of the mean from stratified sampling is:

$$\bar{y}_{st} = \sum_{h=1}^{\mathscr{L}} \left[\frac{N_h \bar{y}_h}{N} \right]$$

where, $\bar{y}_h = $ is the main sample estimate of the mean in the hth strata. The estimate of the standard error of $\bar{y}_{st}$ from stratified sampling is:

$$S(\bar{y}_{st}) = \left\{ \frac{\sum\limits_{h=1}^{\mathscr{L}} \left[N_h (N_h - n_h) \left(\frac{S_h^2}{n_h} \right) \right]}{N^2} \right\}^{1/2}$$

where

$S_h^2 = $ is the main sample estimate of the variance in the hth strata

$$S_h^2 = \frac{\sum\limits_{i=1}^{n_h} (y_{hi} - \bar{y}_h)^2}{n_h - 1}$$

The 95 percent confidence for μ from a stratified sample is:

$$\bar{y}_{st} \pm 2[S(\bar{y}_{st})]$$

Two standard errors are used to approximate the number of standard errors required to compute a 95 percent confidence interval for μ. The exact number of standard errors required for the above computation can be obtained by using the t-distribution, such that the degrees of freedom lie between the smallest $(n_h - 1)$ and n.

Example 11–1 Report on the average 1979 family income in Stat City

Mr. Lee Kaplowitz, the mayor of Stat City, would like to estimate the average 1979 family income in Stat City so that he can tell the federal government whether Stat City is a low, middle, or upper class community. The federal government uses the definition of community class shown in Table 11–1.

Table 11–1

Class	Average 1979 family income
Lower	Under $20,000
Middle	$20,000 to less than $40,000
Upper	$40,000 or more

Advanced sampling plans

The mayor has asked you to draw a random sample of Stat City dwelling units and estimate the average 1979 family income in Stat City (see Exhibit 11–1). Further, the mayor is on a limited budget so you will want to draw the smallest possible sample while providing statistics with "95 percent confidence" and a tolerable error of $980.

Solution *Constructing a 95 percent confidence interval for the mean income in Stat City via stratified sampling.*

STEP 1: Define the strata—the residential zones in Stat City are being used as the strata because the variability of income changes from zone to zone.

STEP 2: Obtain the size of each stratum and an estimate of the standard deviation of income in each stratum from a pilot study. I arbitrarily decided to draw five random dwelling units from each zone (strata) to estimate $\hat{S}_h$; see Table 11–2. The results of the pilot study are shown in Table 11–3.

Table 11–2

CASE-N	ID	ZONE	INCOM
1	5.	1.	40603.
2	11.	1.	54135.
3	14.	1.	28180.
4	76.	1.	46185.
5	104.	1.	41494.

CASE-N	ID	ZONE	INCOM
1	149.	2.	89656.
2	168.	2.	52722.
3	243.	2.	46337.
4	253.	2.	80265.
5	261.	2.	61120.

CASE-N	ID	ZONE	INCOM
1	360.	3.	28224.
2	379.	3.	31622.
3	429.	3.	38844.
4	555.	3.	37546.
5	557.	3.	43040.

CASE-N	ID	ZONE	INCOM
1	724.	4.	17943.
2	966.	4.	9963.
3	989.	4.	22009.
4	1090.	4.	24361.
5	1308.	4.	28058.

Table 11–3

h = zone	N_h	$\hat{S}_h$
$h = 1$	130	$ 9,459
$h = 2$	157	$18,370
$h = 3$	338	$ 5,907
$h = 4$	748	$ 6,926

STEP 3: Decide upon the desired level of precision by setting the size of the standard error, V. In this example we want 95 percent confidence ($Z = 1.96$) and a tolerable error of $980 ($980/1.96 = $500 = V$).

STEP 4: Compute the sample size.

$$n = \left[\frac{\left(\dfrac{130}{1,373} \times 9,459 + \dfrac{157}{1,373} \times 18,370 + \dfrac{338}{1,373} \times 5,907 + \dfrac{748}{1,373} \times 6,926 \right)^2}{500^2 + \dfrac{1}{1,373} \left(\dfrac{130}{1,373} \, 9,459^2 + \dfrac{157}{1,373} \, 18,370^2 + \dfrac{338}{1,373} \times 5907^2 + \dfrac{748}{1,373} \, 6,926^2 \right)} \right]$$

$$n = \frac{(895.6 + 2,100.6 + 1,454.2 + 3,773.2)^2}{250,000 + \left[\dfrac{1}{1,373} (81,782,276.3) \right]}$$

$$n = \frac{(8,223.6)^2}{309,564.7} = 218.5 = 219$$

STEP 5: Allocate the sample size to each stratum; see Table 11–4.

Table 11–4

h	N_h	$\hat{S}_h$	$N_h \hat{S}_h$	$\dfrac{N_h \hat{S}_h}{\Sigma N_h \hat{S}_h}$	n	n_h
1	130	9,459	1,229,670	.109	219	24
2	157	18,370	2,884,090	.255	219	56
3	338	5,907	1,996,566	.177	219	39
4	748	6,926	5,180,648	.459	219	100
			11,290,974			219

STEP 6: Draw the stratified sample; see Table 11–5.

Table 11–5

$$n_1 = 24$$

CASE-N	ID	ZONE	INCOM	CASE-N	ID	ZONE	INCOM
1	1.	1.	56419.	13	74.	1.	43249.
2	5.	1.	40603.	14	75.	1.	28812.
3	11.	1.	54135.	15	76.	1.	46185.
4	12.	1.	57872.	16	77.	1.	59055.
5	14.	1.	28180.	17	79.	1.	41305.
6	15.	1.	23328.	18	82.	1.	17046.
7	25.	1.	63040.	19	85.	1.	39171.
8	32.	1.	31608.	20	100.	1.	28863.
9	34.	1.	44956.	21	104.	1.	41494.
10	37.	1.	36483.	22	112.	1.	32375.
11	43.	1.	38757.	23	118.	1.	51411.
12	69.	1.	33369.	24	123.	1.	56479.

$$n_2 = 56$$

CASE-N	ID	ZONE	INCOM	CASE-N	ID	ZONE	INCOM
1	132.	2.	20234.	8	149.	2.	89656.
2	133.	2.	50001.	9	152.	2.	83484.
3	134.	2.	63330.	10	153.	2.	80217.
4	136.	2.	32648.	11	158.	2.	42314.
5	140.	2.	23576.	12	160.	2.	78086.
6	142.	2.	61106.	13	161.	2.	67110.
7	148.	2.	120030.	14	166.	2.	108774.

Table 11–5 (*continued*)

$$n_2 = 56$$

CASE-N	ID	ZONE	INCOM	CASE-N	ID	ZONE	INCOM
15	168.	2.	52722.	36	246.	2.	111260.
16	176.	2.	60262.	37	247.	2.	141031.
17	177.	2.	60080.	38	253.	2.	80265.
18	183.	2.	77238.	39	254.	2.	40217.
19	186.	2.	73136.	40	256.	2.	121005.
20	188.	2.	101290.	41	258.	2.	116203.
21	190.	2.	61570.	42	261.	2.	61120.
22	193.	2.	87855.	43	262.	2.	52098.
23	195.	2.	60714.	44	263.	2.	70732.
24	201.	2.	96448.	45	265.	2.	71569.
25	203.	2.	76402.	46	267.	2.	113297.
26	207.	2.	34286.	47	272.	2.	66364.
27	217.	2.	72563.	48	273.	2.	68609.
28	223.	2.	91255.	49	274.	2.	57519.
29	225.	2.	30043.	50	275.	2.	81038.
30	228.	2.	80710.	51	276.	2.	27835.
31	230.	2.	66701.	52	281.	2.	109434.
32	233.	2.	70074.	53	282.	2.	58220.
33	241.	2.	47650.	54	285.	2.	50449.
34	243.	2.	46337.	55	286.	2.	50602.
35	245.	2.	140379.	56	287.	2.	43982.

$$n_3 = 39$$

CASE-N	ID	ZONE	INCOM	CASE-N	ID	ZONE	INCOM
1	289.	3.	17624.	21	464.	3.	39873.
2	299.	3.	25780.	22	469.	3.	36907.
3	303.	3.	24357.	23	470.	3.	45300.
4	315.	3.	58703.	24	473.	3.	31398.
5	335.	3.	43767.	25	483.	3.	25687.
6	339.	3.	45259.	26	488.	3.	31762.
7	354.	3.	16117.	27	502.	3.	23004.
8	355.	3.	23584.	28	503.	3.	24262.
9	360.	3.	28224.	29	528.	3.	41557.
10	364.	3.	25067.	30	531.	3.	49726.
11	379.	3.	31622.	31	534.	3.	25329.
12	388.	3.	28381.	32	545.	3.	30647.
13	419.	3.	38626.	33	555.	3.	37546.
14	429.	3.	38844.	34	557.	3.	43040.
15	433.	3.	33033.	35	568.	3.	51522.
16	449.	3.	29863.	36	574.	3.	54133.
17	450.	3.	32865.	37	598.	3.	50684.
18	451.	3.	24255.	38	603.	3.	24343.
19	453.	3.	26582.	39	608.	3.	28593.
20	458.	3.	34061.				

Table 11–5 (*concluded*)

$$n_4 = 100$$

CASE-N	ID	ZONE	INCOM	CASE-N	ID	ZONE	INCOM
1	629.	4.	16014.	51	989.	4.	22009.
2	632.	4.	15867.	52	990.	4.	19681.
3	637.	4.	16917.	53	994.	4.	34964.
4	642.	4.	21554.	54	997.	4.	17201.
5	657.	4.	28604.	55	1008.	4.	20216.
6	658.	4.	22708	56	1011.	4.	16007.
7	661.	4.	28116.	57	1021.	4.	14160.
8	664.	4.	27946.	58	1023.	4.	18889.
9	676.	4.	12755.	59	1030.	4.	15273.
10	677.	4.	17854.	60	1039.	4.	7228.
11	679.	4.	15597.	61	1053.	4.	12661.
12	687.	4.	14967.	62	1069.	4.	8148.
13	689.	4.	14369.	63	1070.	4.	7936.
14	700.	4.	14217.	64	1072.	4.	9005.
15	702.	4.	16006.	65	1090.	4.	24361.
16	711.	4.	10961.	66	1093.	4.	25152.
17	718.	4.	19359.	67	1106.	4.	19336.
18	724.	4.	17943.	68	1113.	4.	19245.
19	731.	4.	17309.	69	1123.	4.	16711.
20	735.	4.	12741.	70	1164.	4.	7369.
21	746.	4.	12769.	71	1182.	4.	15918.
22	760.	4.	10268.	72	1185.	4.	4581.
23	769.	4.	6038.	73	1192.	4.	5786.
24	790.	4.	14757.	74	1194.	4.	8841.
25	792.	4.	12063.	75	1212.	4.	22910.
26	802.	4.	16371.	76	1222.	4.	20390.
27	827.	4.	23872.	77	1224.	4.	25525.
28	831.	4.	14573.	78	1226.	4.	19567.
29	841.	4.	10857.	79	1231.	4.	19576.
30	844.	4.	16144.	80	1235.	4.	15042.
31	850.	4.	18348.	81	1237.	4.	21549.
32	854.	4.	12097.	82	1241.	4.	18150.
33	872.	4.	7465.	83	1254.	4.	13827.
34	875.	4.	13243.	84	1258.	4.	6473.
35	886.	4.	13897.	85	1262.	4.	11557.
36	890.	4.	15787.	86	1269.	4.	4877.
37	895.	4.	11960.	87	1279.	4.	5555.
38	903.	4.	17922.	88	1282.	4.	5326.
39	904.	4.	6095.	89	1285.	4.	6671.
40	917.	4.	11300.	90	1289.	4.	8726.
41	940.	4.	8637.	91	1293.	4.	13723.
42	941.	4.	9726.	92	1298.	4.	19654.
43	949.	4.	11867.	93	1308.	4.	28058.
44	952.	4.	8962.	94	1319.	4.	10349.
45	954.	4.	14832.	95	1328.	4.	18497.
46	956.	4.	11543.	96	1330.	4.	18369.
47	959.	4.	24312.	97	1333.	4.	8807.
48	964.	4.	24560.	98	1343.	4.	23260.
49	965.	4.	35201.	99	1356.	4.	8544.
50	966.	4.	9963.	100	1370.	4.	8648.

Table 11-6

h	N_h	$\bar{y}_h$	$N_h \bar{y}_h$
1......	130	41,424.79	5,385,222.70
2......	157	71,448.75	11,217,453.75
3......	338	33,895.56	11,456,699.28
4......	748	15,415.41	11,530,726.68
			39,590,102.41

$$\bar{y}_{st} = \frac{39,590,102.41}{1,373} = \$28,834.74$$

STEP 7: Compute $\bar{y}_{st}$ from the stratified sample; see Table 11–6.

STEP 8: Compute the estimate of the standard error of $\bar{y}_{st}$ from the stratified sample; see Table 11–7.

Table 11-7

h	N_h	n_h	S_h	$\left[\dfrac{N_h(N_h - n_n)\left(\dfrac{S_h^2}{n_h}\right)}{N^2}\right]$
1	130	24	12,325.90	46,273.70
2	157	56	28,324.05	120,504.32
3	338	39	10,411.52	149,008.31
4	748	100	6,570.35	110,997.33
				426,783.66

$$S(\bar{y}_{st}) = \sqrt{426,783.66} = \$653.29$$

STEP 9: Compute the 95 percent confidence interval for μ from the stratified sample.

$$\bar{y}_{st} \pm 2[S(\bar{y}_{st})]$$
$$\$28,834.74 \pm (2)\$653.29$$
$$\$28,834.74 \pm \$1,306.58$$
$$\$27,528.16 \text{ to } \$30,141.32$$

Finally, we can say with "95 percent confidence" that the mean income in Stat City falls in the range from \$27,528.16 to \$30,141.32.

The memorandum appears in Exhibit 11–1.

An illustrative comparison of simple random and stratified random samples: Estimating the average 1979 Stat City income from a sample of 219 dwelling units.

A simple random sample of 219 Stat City dwelling units was drawn[2] to estimate the mean 1979 income (see Table 11–8 for the random sample). The sample yielded the following 95 percent confidence interval.[3]

$$\bar{y} \pm ZS(\bar{y})$$
$$\$28,044.55 \pm (1.96) \$1,395.69$$
$$\$28,044.55 \pm \$2,735.55$$
$$\$25,309.00 \text{ to } \$30,780.10$$

As you can see, the confidence interval for μ computed from the stratified sample is \$27,528.16 to \$30,141.52, and the confidence interval for μ computed from the simple random sample is \$25,309.00 to \$30,780.10. The confidence interval from the stratified sample of 219 families is \$2,857.74 smaller than the confidence interval from the simple random sample of 219 families:

$$(\$30,780.10 - \$25,309.00 = \$5,471.10 \text{ SRS})$$
$$-(\$30,141.52 - \$27,528.16 = \$2,613.36 \text{ STRATA})$$
$$\$2,857.74 \text{ difference}$$

The above *exemplifies* that a stratified sample is more efficient than a simple random sample of the same size, given that the variances in each strata are dissimilar.

---

[2] The finite population correction factor is not used in this example.

[3] Please note that $\bar{y}_{st}$ and $\bar{y}$ are different. The difference is due to $\bar{y}_{st}$'s consideration of the strata size.

Chapter 11

168

Exhibit 11–1

HOWARD S. GITLOW, PH.D.
STATISTICAL CONSULTANT

MEMORANDUM

TO: Mr. Lee Kaplowitz, Mayor
        Stat City

FROM: Howard Gitlow

DATE: November 21, 1980

RE: Report on the average 1979 family income in Stat
        City.

   As per your request and using commonly accepted
statistical techniques, I have computed several summary
measures concerning the average 1979 family income in Stat
City.[1] My research indicates that the average 1979 family
income in Stat City is $28,834.74 (±$653.29).[2] Allowing for
the errors in survey estimates, you can have "95 percent
confidence" that the average 1979 family income in Stat City
is contained within the range from $27,528.16 to
$30,141.32.[3]
   The above statistics, in conjunction with federal
guidelines, indicates that Stat City is a middle–class
community.
   If you have any further questions, please do not hesitate
to call me at 305–999–9999.

Footnotes

   [1] The summary statistics reported are estimates derived from a
sample survey of 219 Stat City dwelling units, drawn via a
stratified sampling plan (zones are the strata). Insert the
remainder of standard footnote 1.
   [2] Insert standard footnote 2.
   [3] Insert standard footnote 3.

*Constructing a 95 percent confidence interval for the mean income in Stat
City via sample random sampling (for a $500 standard error OR equiva-
lently 95 percent confidence and a $980 tolerable error).*

   The simple random sample size required to estimate μ with 95 percent
confidence and a tolerable error of $980 (equivalent to a standard error of
$500) is computed below.

$$n_0 = \left[\frac{1.96^2(19,755.87)^2}{980^2} \right]$$
$$= 1,561.2$$

where $19,755.87 is the pilot study estimate of the population standard
deviation based on all 20 dwelling units' incomes listed in Table 11–2.

$$n = \frac{n_0}{\left[\frac{n + (N - 1)}{N} \right]} = 730.7 = 731$$

   A simple random sample of 731 Stat City dwelling units was drawn to
estimate the mean income. ($n = 731$ is required to attain a $500 standard

Table 11-8

CASE-N	ID	ZONE	INCOM	CASE-N	ID	ZONE	INCOM	CASE-N	ID	ZONE	INCOM
1	13	1	31098	74	487	3	23527	147	966	4	9963
2	34	1	44956	75	496	3	38874	148	968	4	21518
3	36	1	65173	76	501	3	26669	149	969	4	21071
4	40	1	30600	77	502	3	23004	150	976	4	28466
5	48	1	29375	78	504	3	22306	151	978	4	24772
6	50	1	4575	79	513	3	41286	152	985	4	23211
7	58	1	30630	80	516	3	25410	153	986	4	23533
8	60	1	22026	81	518	3	32456	154	996	4	18157
9	92	1	52546	82	529	3	20543	155	1002	4	20517
10	94	1	36165	83	549	3	44364	156	1005	4	20075
11	114	1	33910	84	557	3	43040	157	1006	4	21561
12	121	1	24839	85	560	3	46247	158	1009	4	19477
13	140	2	23576	86	571	3	30268	159	1018	4	14985
14	141	2	54903	87	572	3	47162	160	1023	4	16889
15	148	2	120030	88	578	3	73546	161	1026	4	13041
16	154	2	70927	89	583	3	41420	162	1027	4	21026
17	158	2	42314	90	589	3	58090	163	1058	4	19917
18	164	2	68773	91	591	3	51039	164	1070	4	7956
19	166	2	108774	92	593	3	49830	165	1077	4	6920
20	174	2	35840	93	613	3	63844	166	1078	4	6894
21	179	2	56944	94	618	3	28485	167	1100	4	20653
22	181	2	64578	95	621	3	45689	168	1105	4	25132
23	191	2	54855	96	623	3	76129	169	1106	4	19336
24	193	2	87855	97	627	4	13735	170	1113	4	19245
25	195	2	60714	98	630	4	18640	171	1119	4	23802
26	203	2	76402	99	640	4	26006	172	1130	4	19307
27	205	2	57598	100	649	4	18425	173	1141	4	25317
28	214	2	67397	101	650	4	25052	174	1144	4	12597
29	235	2	78359	102	657	4	28604	175	1145	4	13043
30	239	2	33411	103	660	4	23927	176	1150	4	12653
31	241	2	47650	104	673	4	17565	177	1161	4	11075
32	253	2	80265	105	680	4	14355	178	1163	4	19534
33	254	2	40217	106	693	4	19419	179	1171	4	7406
34	261	2	61120	107	695	4	11102	180	1175	4	10227
35	270	2	54379	108	715	4	16864	181	1176	4	13035
36	279	2	136427	109	729	4	13937	182	1188	4	7504
37	287	2	43986	110	744	4	11186	183	1189	4	1807
38	294	3	31873	111	745	4	18317	184	1191	4	7520
39	302	3	36110	112	745	4	25044	185	1206	4	13222
40	305	3	17617	113	747	4	7691	186	1210	4	16986
41	312	3	27274	114	749	4	8460	187	1212	4	22910
42	318	3	27684	115	750	4	10754	188	1218	4	40836
43	319	3	34235	116	752	4	11187	189	1219	4	19757
44	320	3	31665	117	759	4	12717	190	1222	4	20390
45	326	3	25287	118	760	4	10268	191	1226	4	19567
46	329	3	25900	119	763	4	8571	192	1230	4	11648
47	331	3	33285	120	769	4	6038	193	1237	4	21549
48	332	3	43656	121	777	4	17565	194	1241	4	18150
49	347	3	42778	122	781	4	8618	195	1247	4	21735
50	348	3	54212	123	787	4	8241	196	1260	4	10392
51	351	3	29330	124	798	4	19876	197	1265	4	23745
52	359	3	18690	125	803	4	26377	198	1273	4	6798
53	390	3	28284	126	825	4	23502	199	1278	4	4664
54	402	3	32598	127	830	4	25456	200	1279	4	5555
55	403	3	31756	128	841	4	10657	201	1286	4	6521
56	407	3	30951	129	845	4	11638	202	1291	4	14120
57	414	3	30457	130	851	4	12642	203	1294	4	14656
58	421	3	38785	131	876	4	8308	204	1302	4	24761
59	426	3	28316	132	886	4	13897	205	1303	4	18216
60	428	3	44682	133	910	4	14508	206	1306	4	29034
61	433	3	33033	134	913	4	7646	207	1308	4	28058
62	441	3	22881	135	915	4	9527	208	1311	4	16962
63	453	3	26582	136	917	4	11300	209	1316	4	12494
64	454	3	37265	137	920	4	9239	210	1327	4	18355
65	460	3	31137	138	921	4	3976	211	1331	4	19738
66	462	3	42231	139	936	4	4642	212	1333	4	8807
67	463	3	56220	140	936	4	4322	213	1336	4	10960
68	464	3	39873	141	939	4	11033	214	1337	4	14338
69	466	3	49898	142	947	4	4148	215	1349	4	22455
70	471	3	33427	143	948	4	10380	216	1350	4	23560
71	473	3	31398	144	949	4	11667	217	1358	4	9207
72	482	3	25538	145	950	4	6182	218	1363	4	5556
73	484	3	29990	146	960	4	19304	219	1366	4	8603

error). See Table 11-9 for the random sample. The sample yielded the following interval:

$$\bar{y} \pm ZS(\bar{y})$$
$$\$29,126.61 \pm 1.96(\$836.025)$$
$$\$29,126.61 \pm \$1,638.61$$
$$\$27,488.00 \text{ to } \$30,765.22$$

The confidence interval for μ computed from the stratified sample based on 219 families is \$27,528.16 to \$30,141.52, and the confidence interval from the simple random sample based on 731 families is \$27,488.00 to \$30,765.22. The stratified sample of 219 families produced a confidence interval \$663.86 smaller $(3,277.22 - 2,613.36 = 663.86)$ than a simple random sample of 731 families. This *exemplifies* that stratified sampling is more efficient than simple random sampling in estimating μ for a given level of precision (if stratified sampling is appropriate).

The next section of this chapter is devoted to discussing single-stage cluster sampling for estimating μ and π and to compare cluster and stratified random sampling.

Table 11–9

CASE-N	ID	ZONE	INCOM	CASE-N	ID	ZONE	INCOM
1	1.	1.	56419.	71	139.	2.	90435.
2	2.	1.	23280.	72	140.	2.	23576.
3	3.	1.	33350.	73	141.	2.	54903.
4	5.	1.	40603.	74	143.	2.	76450.
5	8.	1.	35118.	75	145.	2.	83669.
6	9.	1.	38822.	76	146.	2.	120030.
7	12.	1.	57872.	77	152.	2.	83484.
8	13.	1.	31098.	78	154.	2.	70927.
9	15.	1.	23328.	79	157.	2.	52041.
10	16.	1.	23094.	80	158.	2.	42314.
11	19.	1.	49234.	81	160.	2.	78066.
12	20.	1.	41890.	82	163.	2.	69817.
13	21.	1.	52729.	83	164.	2.	68773.
14	23.	1.	36304.	84	166.	2.	108774.
15	25.	1.	63040.	85	174.	2.	35840.
16	26.	1.	33893.	86	176.	2.	60262.
17	28.	1.	38567.	87	179.	2.	56944.
18	29.	1.	54086.	88	181.	2.	64578.
19	30.	1.	29053.	89	184.	2.	35059.
20	34.	1.	44956.	90	185.	2.	45174.
21	35.	1.	32357.	91	187.	2.	63817.
22	36.	1.	65173.	92	188.	2.	101290.
23	37.	1.	36483.	93	189.	2.	56692.
24	38.	1.	49754.	94	190.	2.	61570.
25	40.	1.	30600.	95	191.	2.	54855.
26	41.	1.	19555.	96	193.	2.	87855.
27	45.	1.	30237.	97	195.	2.	60714.
28	47.	1.	51671.	98	196.	2.	39201.
29	48.	1.	29375.	99	198.	2.	49668.
30	50.	1.	45756.	100	201.	2.	96448.
31	51.	1.	41062.	101	202.	2.	73538.
32	52.	1.	24718.	102	203.	2.	76402.
33	53.	1.	43451.	103	204.	2.	61103.
34	54.	1.	42770.	104	205.	2.	57598.
35	56.	1.	30630.	105	206.	2.	70492.
36	59.	1.	33536.	106	208.	2.	94297.
37	60.	1.	22026.	107	209.	2.	80920.
38	61.	1.	47583.	108	211.	2.	96618.
39	62.	1.	29501.	109	214.	2.	67797.
40	66.	1.	72453.	110	215.	2.	78511.
41	68.	1.	41636.	111	216.	2.	60222.
42	75.	1.	28812.	112	221.	2.	132830.
43	77.	1.	59055.	113	222.	2.	126365.
44	80.	1.	39197.	114	224.	2.	91208.
45	82.	1.	17046.	115	225.	2.	30043.
46	84.	1.	53950.	116	226.	2.	72187.
47	85.	1.	39171.	117	227.	2.	85373.
48	86.	1.	40283.	118	228.	2.	80710.
49	89.	1.	34680.	119	230.	2.	66701.
50	91.	1.	44914.	120	232.	2.	80995.
51	92.	1.	52548.	121	234.	2.	166518.
52	94.	1.	38165.	122	235.	2.	78359.
53	99.	1.	22771.	123	239.	2.	33411.
54	106.	1.	32150.	124	240.	2.	62170.
55	107.	1.	41403.	125	241.	2.	47650.
56	110.	1.	15895.	126	242.	2.	62162.
57	111.	1.	51931.	127	246.	2.	111260.
58	113.	1.	28802.	128	250.	2.	76231.
59	114.	1.	33910.	129	251.	2.	82865.
60	116.	1.	31244.	130	252.	2.	67588.
61	119.	1.	15300.	131	253.	2.	80265.
62	121.	1.	24839.	132	254.	2.	40217.
63	122.	1.	62205.	133	256.	2.	121005.
64	125.	1.	48674.	134	257.	2.	64797.
65	127.	1.	38611.	135	259.	2.	63568.
66	128.	1.	24917.	136	260.	2.	78070.
67	130.	1.	31286.	137	261.	2.	61120.
68	131.	2.	51589.	138	262.	2.	52098.
69	134.	2.	63330.	139	263.	2.	70732.
70	138.	2.	94249.	140	267.	2.	113297.

Table 11–9 (*continued*)

CASE-N	ID	ZONE	INCOM	CASE-N	ID	ZONE	INCOM
141	269.	2.	122969.	211	411.	3.	27132.
142	270.	2.	54379.	212	412.	3.	27804.
143	271.	2.	116346.	213	413.	3.	34889.
144	272.	2.	66364.	214	414.	3.	30457.
145	273.	2.	68609.	215	417.	3.	21512.
146	274.	2.	57519.	216	418.	3.	40912.
147	275.	2.	81038.	217	419.	3.	38626.
148	276.	2.	27835.	218	421.	3.	38785.
149	279.	2.	136427.	219	422.	3.	28757.
150	280.	2.	134443.	220	423.	3.	36459.
151	282.	2.	56220.	221	425.	3.	28988.
152	285.	2.	50449.	222	426.	3.	26316.
153	287.	2.	43982.	223	427.	3.	30535.
154	294.	3.	31873.	224	428.	3.	44682.
155	295.	3.	25328.	225	431.	3.	32195.
156	297.	3.	33745.	226	433.	3.	33033.
157	298.	3.	34378.	227	434.	3.	33632.
158	299.	3.	25780.	228	435.	3.	35841.
159	301.	3.	41160.	229	437.	3.	15515.
160	302.	3.	36110.	230	439.	3.	25971.
161	305.	3.	17617.	231	441.	3.	22081.
162	306.	3.	31669.	232	443.	3.	20404.
163	307.	3.	27342.	233	445.	3.	30474.
164	308.	3.	31791.	234	447.	3.	44565.
165	312.	3.	27274.	235	449.	3.	29863.
166	315.	3.	58703.	236	453.	3.	26582.
167	316.	3.	28784.	237	454.	3.	37265.
168	318.	3.	27684.	238	455.	3.	42136.
169	319.	3.	34235.	239	456.	3.	34392.
170	320.	3.	31665.	240	460.	3.	31137.
171	325.	3.	34853.	241	462.	3.	42231.
172	326.	3.	25287.	242	463.	3.	56220.
173	327.	3.	41507.	243	464.	3.	39873.
174	329.	3.	25900.	244	465.	3.	44327.
175	330.	3.	21138.	245	466.	3.	49896.
176	331.	3.	33285.	246	468.	3.	47583.
177	332.	3.	43656.	247	469.	3.	36907.
178	333.	3.	42494.	248	471.	3.	33427.
179	334.	3.	37439.	249	473.	3.	31398.
180	344.	3.	28264.	250	476.	3.	31567.
181	346.	3.	37524.	251	477.	3.	29803.
182	347.	3.	42778.	252	478.	3.	27430.
183	348.	3.	54212.	253	480.	3.	12976.
184	351.	3.	29330.	254	482.	3.	25536.
185	356.	3.	25999.	255	484.	3.	29990.
186	357.	3.	33578.	256	486.	3.	36237.
187	358.	3.	21129.	257	487.	3.	23527.
188	359.	3.	18890.	258	488.	3.	31762.
189	362.	3.	42293.	259	489.	3.	25193.
190	364.	3.	25067.	260	490.	3.	27809.
191	365.	3.	36839.	261	491.	3.	26430.
192	366.	3.	37566.	262	496.	3.	38874.
193	369.	3.	42447.	263	497.	3.	34964.
194	375.	3.	31398.	264	498.	3.	41587.
195	377.	3.	21821.	265	499.	3.	31801.
196	382.	3.	33414.	266	500.	3.	40799.
197	383.	3.	28119.	267	501.	3.	26669.
198	386.	3.	32667.	268	502.	3.	23004.
199	388.	3.	28381.	269	503.	3.	24262.
200	390.	3.	28284.	270	504.	3.	22306.
201	391.	3.	26049.	271	506.	3.	26626.
202	393.	3.	15598.	272	509.	3.	27390.
203	398.	3.	31103.	273	513.	3.	41266.
204	400.	3.	27444.	274	514.	3.	44547.
205	402.	3.	32598.	275	515.	3.	33009.
206	403.	3.	31758.	276	516.	3.	25410.
207	407.	3.	30951.	277	517.	3.	37596.
208	408.	3.	33300.	278	518.	3.	32458.
209	409.	3.	35005.	279	520.	3.	39430.
210	410.	3.	27779.	280	527.	3.	44147.

Table 11-9 (*continued*)

CASE-N	ID	ZONE	INCOM	CASE-N	ID	ZONE	INCOM
281	529.	3.	20543.	351	660.	4.	23927.
282	531.	3.	49726.	352	661.	4.	28116.
283	532.	3.	44212.	353	665.	4.	21450.
284	534.	3.	25329.	354	666.	4.	27030.
285	537.	3.	34667.	355	667.	4.	8495.
286	542.	3.	40026.	356	668.	4.	12964.
287	543.	3.	36756.	357	669.	4.	6892.
288	549.	3.	44364.	358	671.	4.	15948.
289	550.	3.	45409.	359	672.	4.	11480.
290	554.	3.	44108.	360	673.	4.	17565.
291	557.	3.	43040.	361	677.	4.	17854.
292	560.	3.	46247.	362	680.	4.	14055.
293	564.	3.	50821.	363	682.	4.	14864.
294	565.	3.	43199.	364	685.	4.	16566.
295	570.	3.	44546.	365	690.	4.	17302.
296	571.	3.	30268.	366	692.	4.	18489.
297	572.	3.	47162.	367	693.	4.	19419.
298	574.	3.	54133.	368	695.	4.	14592.
299	575.	3.	48159.	369	696.	4.	16149.
300	576.	3.	31458.	370	699.	4.	11102.
301	575.	3.	73548.	371	700.	4.	14217.
302	579.	3.	45016.	372	702.	4.	16006.
303	581.	3.	41974.	373	703.	4.	23862.
304	582.	3.	33408.	374	704.	4.	13755.
305	583.	3.	41420.	375	711.	4.	10961.
306	584.	3.	60473.	376	713.	4.	9759.
307	587.	3.	45456.	377	715.	4.	18864.
308	588.	3.	44824.	378	719.	4.	17045.
309	589.	3.	58090.	379	720.	4.	16777.
310	590.	3.	44993.	380	725.	4.	15310.
311	591.	3.	51039.	381	727.	4.	15458.
312	592.	3.	54468.	382	729.	4.	13937.
313	593.	3.	49830.	383	730.	4.	16900.
314	596.	3.	40191.	384	732.	4.	10795.
315	597.	3.	48075.	385	734.	4.	20553.
316	598.	3.	50684.	386	736.	4.	18404.
317	600.	3.	19730.	387	739.	4.	11381.
318	602.	3.	12305.	388	740.	4.	11186.
319	604.	3.	37915.	389	742.	4.	18317.
320	606.	3.	44766.	390	745.	4.	25044.
321	607.	3.	54820.	391	746.	4.	12769.
322	608.	3.	28593.	392	747.	4.	7691.
323	611.	3.	18591.	393	749.	4.	8460.
324	612.	3.	41959.	394	750.	4.	10754.
325	613.	3.	63844.	395	751.	4.	10530.
326	616.	3.	27071.	396	752.	4.	11167.
327	618.	3.	28485.	397	759.	4.	12717.
328	621.	3.	45689.	398	760.	4.	10268.
329	622.	3.	28887.	399	761.	4.	10045.
330	623.	3.	76129.	400	763.	4.	8571.
331	625.	3.	36356.	401	764.	4.	8102.
332	627.	4.	13735.	402	766.	4.	8772.
333	629.	4.	16014.	403	767.	4.	7809.
334	630.	4.	18640.	404	769.	4.	6038.
335	631.	4.	12037.	405	770.	4.	10831.
336	635.	4.	20943.	406	771.	4.	7158.
337	637.	4.	16917.	407	772.	4.	4322.
338	640.	4.	26000.	408	773.	4.	8983.
339	642.	4.	21554.	409	775.	4.	3168.
340	643.	4.	27569.	410	776.	4.	5830.
341	646.	4.	22956.	411	777.	4.	17585.
342	647.	4.	23346.	412	778.	4.	6476.
343	649.	4.	18425.	413	780.	4.	6269.
344	650.	4.	25052.	414	781.	4.	8618.
345	651.	4.	27781.	415	782.	4.	8142.
346	654.	4.	29640.	416	784.	4.	9763.
347	655.	4.	23675.	417	785.	4.	9765.
348	656.	4.	29070.	418	787.	4.	6241.
349	657.	4.	28604.	419	789.	4.	2402.
350	659.	4.	28077.	420	792.	4.	12063.

Table 11–9 (*continued*)

CASE-N	ID	ZONE	INCOM	CASE-N	ID	ZONE	INCOM
421	793.	4.	18811.	491	919.	4.	7806.
422	794.	4.	20323.	492	920.	4.	9239.
423	796.	4.	16200.	493	921.	4.	3976.
424	797.	4.	13705.	494	922.	4.	9648.
425	798.	4.	19076.	495	923.	4.	10483.
426	801.	4.	19261.	496	924.	4.	9648.
427	803.	4.	26377.	497	927.	4.	9449.
428	809.	4.	19029.	498	928.	4.	7711.
429	810.	4.	17715.	499	929.	4.	11873.
430	811.	4.	18795.	500	934.	4.	3509.
431	813.	4.	20118.	501	936.	4.	4842.
432	815.	4.	13161.	502	938.	4.	4322.
433	816.	4.	25718.	503	939.	4.	11033.
434	817.	4.	21660.	504	940.	4.	8637.
435	818.	4.	21081.	505	945.	4.	10898.
436	819.	4.	22992.	506	946.	4.	4087.
437	822.	4.	17777.	507	947.	4.	4148.
438	825.	4.	23502.	508	948.	4.	10380.
439	827.	4.	23872.	509	949.	4.	11667.
440	828.	4.	25649.	510	950.	4.	6162.
441	829.	4.	24567.	511	951.	4.	7873.
442	830.	4.	25456.	512	954.	4.	14832.
443	831.	4.	14573.	513	957.	4.	22379.
444	832.	4.	13669.	514	959.	4.	24312.
445	835.	4.	13661.	515	960.	4.	19304.
446	837.	4.	13276.	516	962.	4.	7852.
447	839.	4.	14253.	517	963.	4.	28581.
448	841.	4.	10857.	518	965.	4.	35201.
449	844.	4.	16144.	519	966.	4.	9963.
450	845.	4.	11838.	520	967.	4.	28820.
451	847.	4.	15130.	521	968.	4.	21518.
452	849.	4.	14868.	522	969.	4.	21071.
453	850.	4.	18348.	523	972.	4.	21765.
454	851.	4.	12642.	524	974.	4.	21523.
455	852.	4.	16609.	525	976.	4.	28488.
456	853.	4.	15113.	526	978.	4.	24772.
457	855.	4.	14460.	527	981.	4.	21542.
458	858.	4.	17052.	528	983.	4.	18909.
459	859.	4.	18060.	529	985.	4.	23211.
460	861.	4.	15055.	530	986.	4.	23533.
461	862.	4.	17522.	531	989.	4.	22009.
462	863.	4.	16916.	532	991.	4.	20249.
463	865.	4.	15200.	533	996.	4.	18157.
464	866.	4.	12507.	534	997.	4.	17201.
465	868.	4.	16001.	535	999.	4.	17691.
466	870.	4.	23114.	536	1000.	4.	20070.
467	872.	4.	7465.	537	1002.	4.	20517.
468	874.	4.	11410.	538	1003.	4.	19710.
469	875.	4.	13243.	539	1005.	4.	20075.
470	876.	4.	8308.	540	1006.	4.	21581.
471	880.	4.	11445.	541	1007.	4.	25900.
472	884.	4.	11445.	542	1008.	4.	20216.
473	885.	4.	10673.	543	1009.	4.	19477.
474	886.	4.	13897.	544	1011.	4.	16007.
475	887.	4.	16641.	545	1013.	4.	20760.
476	891.	4.	10024.	546	1016.	4.	24800.
477	892.	4.	15678.	547	1018.	4.	14985.
478	896.	4.	15904.	548	1019.	4.	15404.
479	898.	4.	14765.	549	1023.	4.	18889.
480	899.	4.	15823.	550	1024.	4.	12822.
481	901.	4.	15262.	551	1026.	4.	13041.
482	903.	4.	17922.	552	1027.	4.	21026.
483	908.	4.	10614.	553	1028.	4.	15240.
484	910.	4.	14508.	554	1029.	4.	12212.
485	912.	4.	16601.	555	1032.	4.	18651.
486	913.	4.	7846.	556	1034.	4.	15387.
487	915.	4.	9527.	557	1035.	4.	17928.
488	916.	4.	3714.	558	1038.	4.	15340.
489	917.	4.	11300.	559	1040.	4.	13837.
490	918.	4.	6305.	560	1041.	4.	10987.

Table 11–9 (*continued*)

CASE-N	ID	ZONE	INCOM	CASE-N	ID	ZONE	INCOM
561	1043.	4.	8987.	631	1188.	4.	7504.
562	1044.	4.	11892.	632	1189.	4.	1807.
563	1046.	4.	11424.	633	1191.	4.	7520.
564	1048.	4.	9099.	634	1194.	4.	8841.
565	1050.	4.	8664.	635	1202.	4.	4739.
566	1052.	4.	8618.	636	1203.	4.	5302.
567	1056.	4.	6620.	637	1206.	4.	13222.
568	1058.	4.	19917.	638	1208.	4.	14475.
569	1061.	4.	1399.	639	1209.	4.	10477.
570	1062.	4.	7921.	640	1210.	4.	16986.
571	1063.	4.	7717.	641	1211.	4.	21289.
572	1064.	4.	7941.	642	1212.	4.	22910.
573	1065.	4.	5469.	643	1213.	4.	21847.
574	1066.	4.	8048.	644	1214.	4.	13461.
575	1067.	4.	10811.	645	1215.	4.	21394.
576	1070.	4.	7936.	646	1216.	4.	29141.
577	1072.	4.	9005.	647	1217.	4.	20175.
578	1073.	4.	7306.	648	1218.	4.	40836.
579	1074.	4.	11770.	649	1219.	4.	19757.
580	1077.	4.	6920.	650	1220.	4.	27651.
581	1078.	4.	8894.	651	1222.	4.	20390.
582	1079.	4.	8246.	652	1226.	4.	19567.
583	1085.	4.	28679.	653	1230.	4.	11648.
584	1086.	4.	20539.	654	1231.	4.	19576.
585	1088.	4.	32365.	655	1233.	4.	13477.
586	1091.	4.	35832.	656	1235.	4.	15042.
587	1092.	4.	26576.	657	1236.	4.	19745.
588	1100.	4.	20653.	658	1237.	4.	21549.
589	1102.	4.	19518.	659	1239.	4.	13326.
590	1105.	4.	25132.	660	1241.	4.	18150.
591	1106.	4.	19336.	661	1243.	4.	17542.
592	1109.	4.	20826.	662	1244.	4.	16895.
593	1110.	4.	23902.	663	1245.	4.	22413.
594	1111.	4.	23180.	664	1246.	4.	24855.
595	1113.	4.	19245.	665	1247.	4.	21735.
596	1116.	4.	20851.	666	1250.	4.	16145.
597	1119.	4.	23802.	667	1253.	4.	11296.
598	1121.	4.	22564.	668	1254.	4.	13627.
599	1124.	4.	14333.	669	1255.	4.	9187.
600	1127.	4.	18341.	670	1256.	4.	7492.
601	1130.	4.	19307.	671	1258.	4.	6473.
602	1132.	4.	19179.	672	1260.	4.	10392.
603	1134.	4.	17395.	673	1263.	4.	18385.
604	1138.	4.	18481.	674	1264.	4.	12860.
605	1141.	4.	25317.	675	1265.	4.	23745.
606	1142.	4.	20648.	676	1269.	4.	4877.
607	1144.	4.	12597.	677	1270.	4.	9179.
608	1145.	4.	13043.	678	1273.	4.	6798.
609	1147.	4.	14790.	679	1278.	4.	4664.
610	1150.	4.	12653.	680	1279.	4.	5555.
611	1153.	4.	12575.	681	1282.	4.	5326.
612	1154.	4.	19199.	682	1283.	4.	7080.
613	1160.	4.	17381.	683	1284.	4.	7920.
614	1161.	4.	11075.	684	1286.	4.	8521.
615	1163.	4.	19334.	685	1288.	4.	3383.
616	1165.	4.	11763.	686	1289.	4.	8726.
617	1166.	4.	9336.	687	1291.	4.	14120.
618	1168.	4.	8998.	688	1294.	4.	14656.
619	1169.	4.	11561.	689	1295.	4.	31619.
620	1170.	4.	9609.	690	1297.	4.	18922.
621	1171.	4.	7406.	691	1298.	4.	19654.
622	1172.	4.	7983.	692	1300.	4.	23132.
623	1173.	4.	9844.	693	1301.	4.	20073.
624	1175.	4.	10227.	694	1302.	4.	24761.
625	1177.	4.	19315.	695	1303.	4.	18216.
626	1178.	4.	13035.	696	1306.	4.	29034.
627	1180.	4.	11156.	697	1308.	4.	28358.
628	1182.	4.	15918.	698	1309.	4.	24029.
629	1184.	4.	6549.	699	1311.	4.	16962.
630	1186.	4.	6677.	700	1313.	4.	6971.

Table 11–9 (*concluded*)

CASE-N	ID	ZONE	INCOM
701	1316.	4.	12494.
702	1318.	4.	14303.
703	1320.	4.	19193.
704	1322.	4.	16204.
705	1327.	4.	18355.
706	1328.	4.	18497.
707	1331.	4.	19738.
708	1332.	4.	8999.
709	1333.	4.	8807.
710	1334.	4.	9257.
711	1335.	4.	8907.
712	1336.	4.	10960.
713	1337.	4.	14336.
714	1339.	4.	12668.
715	1343.	4.	23260.
716	1347.	4.	13560.
717	1348.	4.	19695.
718	1349.	4.	22455.
719	1350.	4.	23560.
720	1353.	4.	15445.
721	1354.	4.	8312.
722	1356.	4.	8544.
723	1358.	4.	9207.
724	1361.	4.	8998.
725	1362.	4.	5743.
726	1363.	4.	5556.
727	1364.	4.	7145.
728	1365.	4.	9715.
729	1366.	4.	8603.
730	1372.	4.	7732.
731	1373.	4.	4279.

COMPUTING THE SAMPLE SIZE TO ESTIMATE μ WITH SINGLE-STAGE CLUSTER SAMPLING

Remember that the major reasons for applying single-stage cluster sampling occur when (1) a good frame is not available or is very costly to obtain, (2) the cost of obtaining observations increases as the distance separating the elements increases, and (3) the occurrence of a rare event is being estimated.

The number of clusters necessary to estimate μ with a prespecified standard error is:

$$n = \frac{N\,S_c^2}{ND + S_c^2}$$

where

n = the number of clusters selected in the main sample

N = the number of clusters in the population

$$S_c^2 = \sum_{}^{n_p} (y_i - ym_i)^2/(n_p - 1)$$

$$\bar{y} = \sum_{}^{n_p} y_i \Big/ \sum_{}^{n_p} m_i$$

m_i = the number of elements in cluster i

y_i = the total of all observations in cluster i

n_p = the number of clusters selected in the pilot study

$$D = (B^2 \bar{m}^2)/4$$

$B = 2\sqrt{V(\bar{y})} = 2 \times$ (standard error) = the tolerable error

$\bar{m} = \sum_{}^{n_p} m_i/n_p$ = the average cluster size in the pilot sample

The estimate of the mean from the cluster sample is:

$$\bar{y}_{cl} = \sum_{}^{n} y_i \Big/ \sum_{}^{n} m_i$$

The estimate of the standard error of $\bar{y}_{cl}$ from cluster sampling is:

$$s(\bar{y}_{cl}) = \left[\frac{N-n}{Nn\,\bar{m}^2}\right]\left[\frac{\sum^{n}(y_i - \bar{y}m_i)^2}{n-1}\right]$$

The $100(1 - \alpha)$ percent confidence interval for μ from the cluster sample is:

$$\bar{y}_{cl} \pm Z S(\bar{y}_{cl})$$

The following examples illustrate the strengths and weaknesses of single-stage cluster sampling. Further, single-stage cluster sampling is contrasted with stratified sampling via a Stat City example.

Example 11-2   Report on the average weekly purchase of beer in Stat City

The Stat City Supermarket Association wants to estimate the average weekly purchase of six-packs of beer in Stat City, as of January 1980.

You have been hired by the association to provide the required information (see Exhibit 11-2). The association has a limited budget so you should draw the smallest possible sample while still providing statistics with "95 percent confidence" and a tolerable error of one half of a six-pack.

Note: Imagine that a frame of Stat City families does not exist and would be extremely expensive to obtain.

Solution

Before any information is collected, the purpose of the study should be quantitatively stated. The purpose of this study is to compute point and interval estimates of the average weekly purchase of six-packs of beer in Stat City, as of January 1980.

Cluster sampling will be more efficient than simple random sampling or stratified sampling in this problem because of the (imaginary) absence of an extremely expensive frame. The advantage of cluster sampling is that a frame must be constructed only for those clusters selected into the sample rather than for the entire population, as in simple random and stratified random sampling.

STEP 1: Define the clusters—The blocks in Stat City (variable 3 in the data base) are being defined as clusters. There are 120 blocks in Stat City.

STEP 2: Conduct a pilot study of, say, 3 randomly selected blocks (clusters) to estimate the information required to determine the number of blocks to be included in the main sample (see Tables 11-10 and 11-11).

Table 11–10 Pilot sample of three random blocks

Block 55		Block 70		Block 117	
	Beer		Beer		Beer
1	5	1	0	1	4
2	1	2	0	2	0
3	0	3	0	3	1
4	5	4	0	4	7
5	9	5	1	5	0
6	11	6	10	6	3
7	0	7	0	7	2
8	2	8	0	8	0
$y_1 =$	33	9	2	9	9
		10	3	10	0
		11	0	11	0
		12	9	12	0
		13	0	13	3
		14	7	14	0
		15	0	15	9
		16	0	16	10
		17	0	17	0
		18	3	18	0
		19	3	19	0
		20	3	20	3
		21	0	21	0
		22	2	$y_3 =$	51
		23	1		
		24	0		
		25	0		
		26	0		
		27	0		
		28	0		
		29	6		
		30	4		
		31	3		
		32	0		
		33	0		
		34	5		
		35	0		
		36	2		
		37	0		
		38	4		
		39	2		
		40	0		
		41	0		
		42	2		
		$y_2 =$	72		

Table 11–11 Pilot sample of three blocks

Cluster (block) i	Number of dwelling units (m_i)	Total beer purchases in cluster (block) i (y_i)	$\bar{y}m_i$	$y_i - \bar{y}m_i$	$(y_i - \bar{y}m_i)^2$
55	8	33	17.6	15.4	237.16
70	42	72	92.4	−20.4	416.16
117	21	51	46.2	4.8	23.04

$$\sum_{}^{n_p} m_i = 71$$

$$\sum_{}^{n_p} y_i = 156$$

$$\bar{m} = \frac{71}{3} = 23.7$$

$$\bar{y} = \sum_{}^{n_p} y \Big/ \sum_{}^{n_p} m = \frac{156}{71} = 2.2$$

$$\sum_{}^{n_p} (y_i - \bar{y}m_i)^2 = 676.36$$

$$S_c^2 = \frac{\sum_{}^{n_p} (y_i - \bar{y}m_i)^2}{n_p - 1} = \frac{676.36}{2} = 338.18.$$

STEP 3: Set the desired level of confidence and tolerable error. In this case the desired level of confidence is 95 percent [yielding a Z value of approximately 2 (actually 1.96)] and the tolerable error is one half of a six-pack. As you know, the tolerable error divided by the Z value yields the desired level of accuracy; in this case a standard error of .25 ($e/Z = [0.50/2.00] = .25 = \sqrt{V(\hat{p})}$).

$$N = 120 \text{ blocks}$$
$$B = 2\sqrt{V(\hat{p})} = 2(.25) = .50$$

STEP 4: Compute the number of clusters to be sampled.

$$D = \frac{(.50)^2(23.7)^2}{4} = 35.11$$

$$n = \frac{120(338.18)}{120(35.11) + 338.18} = \frac{40,581.60}{4,551.38} = 8.916 = 9$$

Thus 9 out of 120 blocks must be sampled to estimate the average weekly purchase of six-packs of beer in Stat City with a standard error of approximately one quarter of a six-pack (.25 six-packs).

STEP 5: Construct a frame for the selected clusters, draw the main sample, and construct an analysis table (see Table 11–12).

Table 11–12

Random block number between 001 and 120	m_i	y_i	$\bar{y}m_i$	$(y_i - ym_i)$	$(y_i - \bar{y}m_i)^2$
043	6	16	13.62	2.38	5.6644
081	6	16	13.62	2.38	5.6644
110	21	49	47.67	1.33	1.7689
069	8	31	18.16	12.84	164.8656
050	6	14	13.62	.38	.1444
100	41	97	93.07	3.93	15.4449
116	21	24	47.67	−23.67	560.2689
048	3	5	6.81	−1.81	3.2761
119	21	50	47.67	2.33	5.4289

STEP 6: Compute the required statistics.

$$\sum_{}^{n} m_i = 133$$

$$\bar{m} = \frac{133}{9} = 14.78 \text{ (In this example } \bar{M} \text{ is being estimated by } \bar{m}.)$$

$$\sum_{}^{n} y_i = 302$$

$$\bar{y}_{cl} = \frac{302}{133} = 2.27$$

$$\sum_{}^{n} (y_i - \bar{y}m_i)^2 = 762.5265$$

$$S_c^2 = 762.5265/(9 - 1) = 95.32$$
$$s^2(\bar{y}_{cl}) = [(120 - 9)/((120)(9)(14.78)^2)][95.32]$$
$$= .00047(95.32) = .045$$
$$s(\bar{y}_{cl}) = \sqrt{.045} = .212$$
$$\bar{y}_{cl} \pm 1.96(.212) = 2.27 \pm 0.42$$
$$= 1.85 \leq \mu \leq 2.69$$

Hence, the 95 percent confidence interval ranges from 1.85 six-packs of beer to 2.69 six-packs of beer.

The memorandum appears in Exhibit 11–2.

Exhibit 11–2

HOWARD S. GITLOW, PH.D.
STATISTICAL CONSULTANT
_____

MEMORANDUM

TO: Stat City Supermarket Association

FROM: Howard Gitlow

DATE: March 18, 1981

RE: Average weekly purchase of beer in Stat City, as of
        January 1980

   In accordance with our consulting contract and using
commonly accepted statistical techniques, I have computed
several summary measures concerning the average weekly
purchase of six–packs of beer in Stat City, as of January
1980.[1] My research indicates that the average weekly
purchase of six–packs of beer is 2.27 six packs (±.212
six–packs).[2] Allowing for the errors in survey estimates,
you can have "95 percent confidence" that the average
weekly purchase of six–packs of beer in Stat City (as of
January 1980) is contained within the range from 1.85 to
2.69 six–packs.[3]
   If you have any questions, please call me at
305–999–9999.

_____

Footnotes

   [1] The summary statistics reported are estimates derived from a
sample survey of the families residing on nine randomly selected
blocks in Stat City. The sample was drawn via a single–stage cluster
sampling plan. Complete standard footnote 1.
   [2] Insert standard footnote 2.
   [3] Insert standard footnote 3.

An illustrative comparison of stratified random and single-stage cluster sampling: Estimating the average 1979 Stat City income (see Example 11–1).

This section is devoted to exemplifying what can happen if cluster sampling is used when stratified sampling is appropriate.

Rework Example 11–1 but this time use a single-stage cluster sampling plan. (Note that there is no reason for using a cluster sample in this case.)

STEP 1: Define the clusters—the blocks in Stat City are being used as clusters.

STEP 2: Conduct a pilot study of, say, five blocks to estimate the information required to determine the number of blocks to include in the main sample (see Tables 11–13 and 11–14).

Table 11–13 Pilot sample
of five clusters

Block = 32		Block = 39		Block = 83	
1	63,040	1	42,314	1	15,340
2	33,693	2	68,809	2	31,869
3	44,956	3	47,989	3	46,033
4	32,357	4	88,572	$y_5 =$	93,242
5	50,588	5	65,255		
6	30,237	6	64,578		
$y_4 =$	254,871	$y_1 =$	377,517		

Block = 90		Block = 115	
1	33,796	1	5,780
2	41,160	2	5,660
3	27,684	3	10,145
4	34,235	4	10,803
5	35,425	5	10,116
6	29,141	6	11,296
$y_2 =$	201,441	7	13,827
		8	9,187
		9	7,492
		10	14,385
		11	6,473
		12	16,406
		13	10,392
		14	10,925
		15	11,557
		16	18,385
		17	12,860
		18	23,745
		19	20,685
		20	8,466
		21	24,872
		$y_3 =$	263,457

Table 11–14 Pilot sample
of five clusters

Cluster i	Number of dwelling units m_i	Total income in cluster i y_i	$\bar{y}m_i$	$y_i - \bar{y}m_i$
39	6	377,517	170,075.4	207,441.6
90	6	201,441	170,075.4	31,365.6
115	21	263,457	595,263.9	−331,806.9
32	6	254,871	170,075.4	84,795.6
83	3	93,242	85,037.7	8,204.3

$$\sum^{n_p} m_i = 42$$

$$\sum^{n_p} y_i = 1,190,528.$$

$$\bar{m} = \frac{42}{5} = 8.4 \,(\text{estimate of } \bar{M} \text{ from pilot survey})$$

$$\bar{y} = \frac{\sum^{n_p} y_i}{\sum^{n_p} m_i} = \frac{1,190,528.}{42} = 28,345.90$$

$$\sum^{n_p} (y_i - \bar{y}m_i)^2 = 1.613692415 \text{ E } 11*$$

* Scientific notation is being used because the numbers are so large.

STEP 3: Set the desired level of accuracy, in this case a standard error of $500 $\left(\sqrt{V(\bar{y}_{cl})} = \frac{e}{Z} = \frac{980}{1.96} = \$500 \right)$.

$$N = 120 \text{ blocks}$$
$$B = 2\sqrt{V(\bar{y}_{cl})} = 2(\$500) = \$1,000$$

STEP 4: Compute the number of clusters to be drawn in the main sample.

$$D = \frac{(1,000^2)\ (8.4)^2}{4} = 17,640,000.$$

$$S_c^2 = 4.034231037\text{E}10^*$$

$$n = \frac{120\ (4.03\text{E}10)^*}{120\ (17,640,000) + (4.03\text{E}10)^*}$$

$$n = 114.017397 = 114$$

* Scientific notation is being used because the numbers are so large.

Thus 114 out of 120 clusters must be sampled to estimate the average 1979 family income in Stat City with a standard error of $500. Cluster sampling is not as cost/effective as stratified sampling to estimate 1979 family income in Stat City because the stratified random sample required 219 dwelling units while the cluster sample required almost the entire population (114 out of 120 blocks).

COMPUTING THE SAMPLE SIZE TO ESTIMATE Π WITH A SINGLE-STAGE CLUSTER SAMPLE

The sample size necessary to estimate π with a prespecified standard error is:

$$n = \frac{NS_c^2}{ND + S_c^2}$$

where

n = the number of clusters in the main sample

$$D = \frac{B^2\overline{m}}{4}$$

$$S_c^2 = \frac{\sum\limits_{i=1}^{n_p} (a_i - pm_i)^2}{n_p - 1}$$

N = the number of clusters in the population

$p = \left[\sum\limits_{i=1}^{n_p} a_i\right] \Big/ \left[\sum\limits_{i=1}^{n_p} m_i\right]$ = estimate of the proportion from the pilot study

m_i = the number of elements in cluster i

a_i = the number of elements that contain the characteristic of interest in cluster i

n_p = the number of clusters selected in the pilot study

$B = 2\sqrt{V(\hat{p})}$ = desired level of error in estimation = tolerable error

$\overline{m} = \sum\limits_{i=1}^{n_p} m_i/n_p$ = the average cluster size in the pilot sample

The estimate of the proportion from the cluster sample is:

$$\hat{p} = \sum\limits_{i=1}^{n} a_i \Big/ \sum\limits_{i=1}^{n} m_i$$

The estimate of the standard error of $\hat{p}$ from cluster sampling is:

$$s(\hat{p}_{c1}) = \sqrt{[(N - n)/(Nn\overline{m}^2)] \left[\sum\limits_{i=1}^{n} (a_i - \hat{p}m_i)^2/(n - 1)\right]}$$

The $100(1 - \alpha)$ percent confidence interval for π from the cluster sample is $\hat{p} \pm Zs(\hat{p}_{c1})$.

Example 11–3 Report on the proportion of Stat City families with average monthly telephone bills of $90 or more

Mr. Jack Davis, chairman of the Stat City Telephone Company, needs to estimate the percentage of Stat City families that have average monthly telephone bills of $90 or more.

You have been retained by Mr. Davis to obtain the required information (see Exhibit 11–3). Mr. Davis believes that the percentage of Stat City families with average monthly telephone bills of $90 or more is very small; that is, a rarity.

Solution

Before any information is collected the purpose of the study should be quantitatively stated. This study's purpose is to construct point and interval estimates of the percent of Stat City families that have average monthly telephone bills of $90 or more.

Cluster sampling is appropriate in this study because the task at hand requires estimating the frequency of occurrence of a rare event. Recall that cluster sampling is most useful when: (1) a good frame is not available or is very costly to obtain, (2) the cost of obtaining observations increases as the distance separating the elements increases, or (3) as in this case, THE OCCURRENCE OF A RARE EVENT IS BEING ESTIMATED.

The first step is to conduct a pilot study of, for example, five blocks (clusters) to estimate the information required to determine the number of blocks to be included in the main sample (see Table 11–15).

Table 11–15

Cluster (block) i	Size of cluster (block) m_i	Number of dwelling units with average monthly telephone bills of $90 or more a_i	pm_i	$(a_i - pm_i)$	$(a_i - pm_i)^2$
32	6	0	.402	−.402	.161604
39	6	0	.402	−.402	.161604
84	6	0	.402	−.402	.161604
90	6	1	.402	+.598	.357604
115	21	2	1.407	+.593	.351649

$$\Sigma(a_i - pm_i)^2 = 1.194065$$
$$\Sigma m_i = 45$$
$$\Sigma a_i = 3$$
$$p = \frac{3}{45} = .067$$
$$S_c^2 = \frac{1.194065}{4} = .29851625$$

The second step is to set the desired level of confidence and tolerable error. In this case a tolerable error of 2 percent and a 95 percent level of confidence is desired; this yields a standard error of 1 percent.

$$N = 120 \text{ blocks}$$
$$B = 2\sqrt{V(\hat{p})} = .02$$

So

$$\sqrt{V(\hat{p})} = .01 = 1\%$$

The third step is to compute the number of clusters to be sampled.

$$\overline{M} = \frac{1{,}373}{120} = 11.44 \ (\overline{M} \text{ is a known parameter})$$

$$D = \frac{(.02)^2(11.44)^2}{4} = .01308736$$

$$n = \left[\frac{120(.29851625)}{120(.01308736) + .29851625} \right] = 19.17$$

$$n = 20$$

Thus 20 out of 120 clusters must be sampled to estimate the proportion of Stat City families with average monthly telephone bills of $90 or more within a boundary of 2 percent on the error of estimation and 95 percent confidence.[4]

The fourth step is to draw the main sample; see Table 11–16.

Table 11–16

Random block numbers between 001–120	m_i	a_i	$\hat{p}m_i$	$(a_i - \hat{p}m_i)$	$(a_i - \hat{p}m_i)^2$
104	21	2	1.449	.551	.303601
094	41	3	2.829	.171	.029241
103	21	3	1.449	1.551	2.405601
071	21	1	1.449	−.449	.201601
023	6	0	.414	−.414	.171396
010	6	1	.414	.586	.343396
070	21	2	1.449	.551	.303601
024	6	0	.414	−.414	.171396
007	6	0	.414	−.414	.171396
053	4	0	.276	−.276	.076176
005	6	0	.414	−.414	.171396
097	41	3	2.829	.171	.029241
089	6	0	.414	−.414	.171396
069	8	0	.552	−.552	.304704
042	6	0	.414	−.414	.171396
047	1	0	.069	−.069	.004761
075	3	0	.207	−.207	.042849
003	6	0	.414	−.414	.171396
015	7	1	.483	.517	.267289
062	8	1	.552	.448	.200704

The fifth step is to compute the desired point and interval estimates.

$$\sum^n m_i = 245$$

$$\sum^n a_i = 17$$

$$\hat{p} = \frac{17}{245} = .0694 \cong .069$$

$$S_c^2 = \frac{5.712537}{19} \cong .30$$

$$s^2(\hat{p}_{cl}) = \left[\frac{N - n}{Nn\overline{\overline{M}}^2}\right] S_c^2 =$$

$$\left[\frac{120 - 20}{120 \times 20 \times 11.44^2}\right] (.30)$$

$$= \left(\frac{100}{314096.64}\right) (.30) = .000096$$

$$\hat{p} \pm 1.96\sqrt{s(\hat{p}_{cl})} = .069 \pm 1.96(.0098)$$
$$= .069 \pm .0192$$
$$= .0498 \text{ to } .0882$$

95 percent confidence interval = 4.98 percent to 8.82 percent

[4] The finite population correction factor is not presented for cluster sampling in this text.

The memorandum appears in Exhibit 11–3.

Exhibit 11–3

```
┌─────────────────────────────────────────────────────────────┐
│                                                               │
│   HOWARD S. GITLOW, PH.D.                                     │
│   STATISTICAL CONSULTANT                                      │
│   ─────────                                                   │
│                                                               │
│                      MEMORANDUM                               │
│                                                               │
│   TO:    Mr. Jack Davis, Chairman                             │
│          Stat City Telephone Company                          │
│                                                               │
│   FROM:  Howard Gitlow                                        │
│                                                               │
│   DATE:  December 1, 1980                                     │
│                                                               │
│   RE:    Report on the proportion of Stat City families with  │
│          average monthly telephone bills of $90 or more.      │
│                                                               │
│     As per your request and using commonly accepted           │
│   statistical techniques, I have computed several summary     │
│   measures concerning the proportion of Stat City families    │
│   with average monthly telephone bills of $90 or more.[1] My  │
│   research indicates that the proportion of Stat City         │
│   families with average monthly telephone bills of $90 or more│
│   is 6.9 percent (±0.98 percent).[2] Allowing for the errors in│
│   survey estimates, you can have "95 percent confidence" that │
│   the proportion of Stat City families with average monthly   │
│   telephone bills of $90 or more is contained within the range│
│   from 4.98 percent to 8.82 percent.[3]                       │
│     If you have any further questions, please do not hesitate │
│   to call me at 305–999–9999.                                 │
│                                                               │
│   ─────────                                                   │
│                      Footnotes                                │
│                                                               │
│   [1] The summary statistics reported are estimates derived   │
│   from a sample survey of the families residing on 20         │
│   randomly selected blocks in Stat City consisting of 245     │
│   Stat City families. The sample was drawn via a single–stage │
│   cluster sampling plan. Complete standard footnote 1.        │
│   [2] Insert standard footnote 2.                             │
│   [3] Insert standard footnote 3.                             │
│                                                               │
└─────────────────────────────────────────────────────────────┘
```

CONCLUSION

This chapter introduced two advanced sampling plans that are more efficient than simple random sampling under certain conditions; stratified random sampling and single-stage cluster sampling. You saw that the conditions under which stratified random sampling is more efficient than simple random sampling are: (1) when definable strata exist with different variances, (2) when the costs of sampling vary from stratum to stratum, and (3) when it is desirable to obtain estimates of parameters in each stratum. Further, you saw that single-stage cluster sampling is more efficient than simple random sampling when: (1) a good frame is not available or is very costly to obtain, (2) the cost of obtaining observations increases as the distance between the elements increases, and (3) the occurrence of a rare event is being estimated.

Remember, a major goal of statistics is to provide the maximum amount of information per unit of cost. Consequently, if you are presented with a situation in which an advanced sampling plan is appropriate, it behooves you as an informed user of statistics to use the most efficient sampling plan and to maximize the information per unit of cost.

ADDITIONAL PROBLEMS

11–4   Ms. Arlene Davis, director of the Stat City Hospital, needs to estimate the percentage of Stat City families that average nine or more trips to the hospital per year. (Note: Averaging nine or more trips per year to the Stat City Hospital is considered a rare event.) She wants "95 percent confidence" in her estimate and will tolerate an error of 3 percentage points.

You have been retained by Ms. Davis to obtain the required information. Conduct a survey and report your findings to Ms. Davis.

Use the random number table which accompanies this problem to draw a random sample of three blocks for your pilot study. Begin on line 1, column 1 and use the first three digits. Once you have computed the required sample size, continue down column 1, then column 2—using the first three digits and so on until you have drawn your sample.

Random number table for Problem 11–4

Line/Col.	(1)	(2)	(3)	(4)	(5)	(6)	(7)	(8)	(9)	(10)	(11)	(12)	(13)	(14)
1	10480	15011	01536	02011	81647	91646	69179	14194	62590	36207	20969	99570	91291	90700
2	22368	46573	25595	85393	30995	89198	27982	53402	93965	34095	52666	19174	39615	99505
3	24130	48360	22527	97265	76393	64809	15179	24830	49340	32081	30680	19655	63348	58629
4	42167	93093	06243	61680	07856	16376	39440	53537	71341	57004	00849	74917	97758	16379
5	37570	39975	81837	16656	06121	91782	60468	81305	49684	60672	14110	06927	01263	54613
6	77921	06907	11008	42751	27756	53498	18602	70659	90655	15053	21916	81825	44394	42880
7	99562	72905	56420	69994	98872	31016	71194	18738	44013	48840	63213	21069	10634	12952
8	96301	91977	05463	07972	18876	20922	94595	56869	69014	60045	18425	84903	42508	32307
9	89579	14342	63661	10281	17453	18103	57740	84378	25331	12566	58678	44947	05585	56941
10	85475	36857	43342	53988	53060	59533	38867	62300	08158	17983	16439	11458	18593	64952
11	28918	69578	88231	33276	70997	79936	56865	05859	90106	31595	01547	85590	91610	78188
12	63553	40961	48235	03427	49626	69445	18663	72695	52180	20847	12234	90511	33703	90322
13	09429	93969	52636	92737	88974	33488	36320	17617	30015	08272	84115	27156	30613	74952
14	10365	61129	87529	85689	48237	52267	67689	93394	01511	26358	85104	20285	29975	89868
15	07119	97336	71048	08178	77233	13916	47564	81056	97735	85977	29372	74461	28551	90707
16	51085	12765	51821	51259	77452	16308	60756	92144	49442	53900	70960	63990	75601	40719
17	02368	21382	52404	60268	89368	19885	55322	44819	01188	65255	64835	44919	05944	55157
18	01011	54092	33362	94904	31273	04146	18594	29852	71585	85030	51132	01915	92747	64951
19	52162	53916	46369	58586	23216	14513	83149	98736	23495	64350	94738	17752	35156	35749
20	07056	97628	33787	09998	42698	06691	76988	13602	51851	46104	88916	19509	25625	58104
21	48663	91245	85828	14346	09172	30168	90229	04734	59193	22178	30421	61666	99904	32812
22	54164	58492	22421	74103	47070	25306	76468	26384	58151	06646	21524	15227	96909	44592
23	32639	32363	05597	24200	13363	38005	94342	28728	35806	06912	17012	64161	18296	22851
24	29334	27001	87637	87308	58731	00256	45834	15398	46557	41135	10367	07684	36188	18510
25	02488	33062	28834	07351	19731	92420	60952	61280	50001	67658	32586	86679	50720	94953
26	81525	72295	04839	96423	24878	82651	66566	14778	76797	14780	13300	87074	79666	95725
27	29676	20591	68086	26432	46901	20849	89768	81536	86645	12659	92259	57102	80428	25280
28	00742	57392	39064	66432	84673	40027	32832	61362	98947	96067	64760	64584	96096	98253
29	05366	04213	25669	26422	44407	44048	37937	63904	45766	66134	75470	66520	34693	90449
30	91921	26418	64117	94305	26766	25940	39972	22209	71500	64568	91402	42416	07844	69618
31	00582	04711	87917	77341	42206	35126	74087	99547	81817	42607	43808	76655	62028	76630
32	00725	69884	62797	56170	86324	88072	76222	36086	84637	93161	76038	65855	77919	88006
33	69011	65797	95876	55293	18988	27354	26575	08625	40801	59920	29841	80150	12777	48501
34	25976	57948	29888	88604	67917	48708	18912	82271	65424	69774	33611	54262	85963	03547
35	09763	83473	73577	12908	30883	18317	28290	35797	05998	41688	34952	37888	38917	88050
36	91567	42595	27958	30134	04024	86385	29880	99730	55536	84855	29080	09250	79656	73211
37	17955	56349	90999	49127	20044	59931	06115	20542	18059	02008	73708	83517	36103	42791
38	46503	18584	18845	49618	02304	51038	20655	58727	28168	15475	56942	53389	20562	87338
39	92157	89634	94824	78171	84610	82834	09922	25417	44137	48413	25555	21246	35509	20468
40	14577	62765	35605	81263	39667	47358	56873	56307	61607	49518	89656	20103	77490	18062
41	98427	07523	33362	64270	01638	92477	66969	98420	04880	45585	46565	04102	46880	45709
42	34914	63976	88720	82765	34476	17032	87589	40836	32427	70002	70663	88863	77775	69348
43	70060	28277	39475	46473	23219	53416	94970	25832	69975	94884	19661	72828	00102	66794
44	53976	54914	06990	67245	68350	82948	11398	42878	80287	88267	47363	46634	06541	97809
45	76072	29515	40980	07391	58745	25774	22987	80059	39911	96189	41151	14222	60697	59583
46	90725	52210	83974	29992	65831	38857	50490	83765	55657	14361	31720	57375	56228	41546
47	64364	67412	33339	31926	14883	24413	59744	92351	97473	89286	35931	04110	23726	51900
48	08962	00358	31662	25388	61642	34072	81249	35648	56891	69352	48373	45578	78547	81788
49	95012	68379	93526	70765	10593	04542	76463	54328	02349	17247	28865	14777	62730	92277
50	15664	10493	20492	38391	91132	21999	59516	81652	27195	48223	46751	22923	32261	85653

11–5 Mr. Saul Reisman, chief financial officer of the Stat City Electric Company, needs to know the average number of rooms per dwelling unit in Stat City. He believes that there is a strong relationship between residential zone and the number of rooms per dwelling unit. Mr. Reisman wants "95 percent confidence" in his estimate and will tolerate an error of one room.

You have been retained by Mr. Reisman to ascertain the required information. Conduct a survey and report your findings to Mr. Reisman.

Use the random number table which accompanies this problem to draw a random sample of three families from each zone for your pilot study. Begin on line 51, column 1 and use the first four digits. Continue down column 1, then column 2—using the first four digits and so on, until you have drawn your pilot sample. Once you have computed the required sample size, continue down the random number table (picking up where you left off) and draw the main sample.

Random number table for Problem 11–5

Line/Col.	(1)	(2)	(3)	(4)	(5)	(6)	(7)	(8)	(9)	(10)	(11)	(12)	(13)	(14)
51	16408	81899	04153	53381	79401	21438	83035	92350	36693	31238	59649	91754	72772	02338
52	18629	81953	05520	91962	04739	13092	97662	24822	94730	06496	35090	04822	86772	98289
53	73115	35101	47498	87637	99016	71060	88824	71013	18735	20286	23153	72924	35165	43040
54	57491	16703	23167	49323	45021	33132	12544	41035	80780	45393	44812	12515	98931	91202
55	30405	83946	23792	14422	15059	45799	22716	19792	09983	74353	68668	30429	70735	25499
56	16631	35006	85900	98275	32388	52390	16815	69298	82732	38480	73817	32523	41961	44437
57	96773	20206	42559	78985	05300	22164	24369	54224	35083	19687	11052	91491	60383	19746
58	38935	64202	14349	82674	66523	44133	00697	35552	35970	19124	63318	29686	03387	59846
59	31624	76384	17403	53363	44167	64486	64758	75366	76554	31601	12614	33072	60332	92325
60	78919	19474	23632	27889	47914	02584	37680	20801	72152	39339	34806	08930	85001	87820
61	03931	33309	57047	74211	63445	17361	62825	39908	05607	91284	68833	25570	38818	46920
62	74426	33278	43972	10119	89917	15665	52872	73823	73144	88662	88970	74492	51805	99378
63	09066	00903	20795	95452	92648	45454	09552	88815	16553	51125	79375	97596	16296	66092
64	42238	12426	87025	14267	20979	04508	64535	31355	86064	29472	47689	05974	52468	16834
65	16153	08002	26504	41744	81959	65642	74240	56302	00033	67107	77510	70625	28725	34191
66	21457	40742	29820	96783	29400	21840	15035	34537	33310	06116	95240	15957	16572	06004
67	21581	57802	02050	89728	17937	37621	47075	42080	97403	48626	68995	43805	33386	21597
68	55612	78095	83197	33732	05810	24813	86902	60397	16489	03264	88525	42786	05269	92532
69	44657	66999	99324	51281	84463	60563	79312	93454	68876	25471	93911	25650	12682	73572
70	91340	84979	46949	81973	37949	61023	43997	15263	80644	43942	89203	71795	99533	50501
71	91227	21199	31935	27022	84067	05462	35216	14486	29891	68607	41867	14951	91696	85065
72	50001	38140	66321	19924	72163	09538	12151	06878	91903	18749	34405	56087	82790	70925
73	65390	05224	72958	28609	81406	39147	25549	48542	42627	45233	57202	94617	23772	07896
74	27504	96131	83944	41575	10573	08619	64482	73923	36152	05184	94142	25299	84387	34925
75	37169	94851	39117	89632	00959	16487	65536	49071	39782	17095	02330	74301	00275	48280
76	11508	70225	51111	38351	19444	66499	71945	05422	13442	78675	84081	66938	93654	59894
77	37449	30362	06694	54690	04052	53115	62757	95348	78662	11163	81651	50245	34971	52924
78	46515	70331	85922	38329	57015	15765	97161	17869	45349	61796	66345	81073	49106	79860
79	30986	81223	42416	58353	21532	30502	32305	86482	05174	07901	54339	35909	81250	54238
80	63798	64995	46583	09765	44160	78128	83991	42865	92520	83531	80377	35909	81250	54238
81	82486	84846	99254	67632	43218	50076	21361	64816	51202	88124	41870	52689	51275	83556
82	21885	32906	92431	09060	64297	51674	64126	62570	26123	05155	59194	52799	28225	85762
83	60336	98782	07408	53458	13564	59089	26445	29789	85205	41001	12535	12133	14645	23541
84	43937	46891	24010	25560	86355	33941	25786	54990	71899	15475	95434	98227	21824	19585
85	97656	63175	89303	16275	07100	92063	21942	18611	47348	20203	18534	03862	78095	50136
86	03299	01221	05418	38982	55758	92237	26759	86367	21216	98442	08303	56613	91511	75928
87	79626	06486	03574	17668	07785	76020	79924	25651	83325	88428	85076	72811	22717	50585
88	85636	68335	47539	03129	65651	11977	02510	26113	99447	68645	34327	15152	55230	93448
89	18039	14367	61337	06177	12143	46609	32989	74014	64708	00533	35398	58408	13261	47908
90	08362	15656	60627	36478	65648	16764	53412	09013	07832	41574	17639	82163	60859	75567
91	79556	29068	04142	16268	15387	12856	66227	38358	22478	73373	88732	09443	82558	05250
92	92608	82674	27072	32534	17075	27698	98204	63863	11951	34648	88022	56148	34925	57031
93	23982	25835	40055	67006	12293	02753	14827	22235	35071	99704	37543	11601	35503	85171
94	09915	96306	05908	97901	28395	14186	00821	80703	70426	75647	76310	88717	37890	40129
95	50937	33300	26695	62247	69927	76123	50842	43834	86654	70959	79725	93872	28117	19233
96	42488	78077	69882	61657	34136	79180	97526	43092	04098	73571	80799	76536	71255	64239
97	46764	86273	63003	93017	31204	36692	40202	35275	57306	55543	53203	18098	47625	88684
98	03237	45430	55417	63282	90816	17349	88298	90183	36600	78406	06216	95787	42579	90730
99	86591	81482	52667	61583	14972	90053	89534	76036	49199	43716	97548	04379	46370	28672
100	38534	01715	94964	87288	65680	43772	39560	12918	86537	62738	19636	51132	25739	56947

Reprinted with permission from "A Table of 14,000 Random Units" in *Standard Mathematical Tables*, 15th ed., Samuel Selby, ed., pp. 566–68. Copyright The Chemical Rubber Co., CRC Press, Inc., Boca Raton, Florida.

Line/Col.	(1)	(2)	(3)	(4)	(5)	(6)	(7)	(8)	(9)	(10)	(11)	(12)	(13)	(14)
101	13284	16834	74151	92027	24670	36665	00770	22878	02179	51602	07270	76517	97275	45960
102	21224	00370	30420	03883	96648	89428	41583	17564	27395	63904	41548	49197	82277	24120
103	99052	47887	81085	64933	66279	80432	65793	83287	34142	13241	30590	97760	35848	91983
104	00199	50993	98603	38452	87890	94624	69721	57484	67501	77638	44331	11257	71131	11059
105	60578	06483	28733	37867	07936	98710	98539	27186	31237	80612	44488	97819	70401	95419
106	91240	18312	17441	01929	18163	69201	31211	54288	39296	37318	65724	90401	79017	62077
107	97458	14229	12063	59611	32249	90466	33216	19358	02591	54263	88449	01912	07436	50813
108	35249	38646	34475	72417	60514	69257	12489	51924	86871	92446	36607	11458	30440	52639
109	38980	46600	11759	11900	46743	27860	77940	39298	97838	95145	32378	68038	89351	37005
110	10750	52745	38749	87365	58959	53731	89295	59062	39404	13198	59960	70408	29812	83126
111	36247	27850	73958	20673	37800	63835	71051	84724	52492	22342	78071	17456	96104	18327
112	70994	66986	99744	72438	01174	42159	11392	20724	54322	36923	70009	23233	65438	59685
113	99638	94702	11463	18148	81386	80431	90628	52506	02016	85151	88598	47821	00265	82525
114	72055	15774	43857	99805	10419	76939	25993	03544	21560	83471	43989	90770	22965	44247
115	24038	65541	85788	55835	38835	59399	13790	35112	01324	39520	76210	22467	83275	32286
116	74976	14631	35908	28221	39470	91548	12854	30166	09073	75887	36782	00268	97121	57676
117	35553	71628	70189	26436	63407	91178	90348	55359	80392	41012	36270	77786	89578	21059
118	35676	12797	51434	82976	42010	26344	92920	92155	58807	54644	58581	95331	78629	73344
119	74815	67523	72985	23183	02446	63594	98924	20633	58842	85961	07648	70164	34994	67662
120	45246	88048	65173	50989	91060	89894	36063	32819	68559	99221	49475	50558	34698	71800
121	76509	47069	86378	41797	11910	49672	88575	97966	32466	10083	54728	81972	58975	30761
122	19689	90332	04315	21358	97248	11188	39062	63312	52496	07349	79178	33692	57352	72862
123	42751	35318	97513	61537	54955	08159	00337	80778	27507	95478	21252	12746	37554	97775
124	11946	22681	45045	13964	57517	59419	58045	44067	58716	58840	44557	96345	33271	53464
125	96518	48688	20996	11090	48396	57177	83867	86464	14342	21545	46717	72364	86954	55580
126	35726	58643	76869	84622	39098	36083	72505	92265	23107	60278	05822	46760	44294	07672
127	39737	42750	48968	70536	84864	64952	38404	94317	65402	13589	01055	79044	19308	83623
128	97025	66492	56177	04049	80312	48028	26408	43591	75528	65341	49044	95495	81256	53214
129	62814	08075	09788	56350	76787	51591	54509	49295	85830	59860	30883	89660	96142	18354
130	25578	22950	15227	83291	41737	79599	96191	71845	86899	70694	24290	01551	80092	82118
131	68763	69576	88991	49662	46704	63362	56625	00481	73323	91427	15264	06969	57048	54149
132	17900	00813	64361	60725	88974	61005	99709	30666	26451	11528	44323	34778	60342	60388
133	71944	60227	63551	71109	05624	43836	58254	26160	32116	63403	35404	57146	10909	07346
134	54684	93691	85132	64399	29182	44324	14491	55226	78793	34107	30374	48429	51376	09559
135	25946	27623	11258	65204	52832	50880	22273	05554	99521	73791	85744	29276	70326	60251
136	01353	39318	44961	44972	91766	90262	56073	06606	51826	18893	83448	31915	97764	75091
137	99083	88191	27662	99113	57174	35571	99884	13951	71057	53961	61448	74909	07322	80960
138	52021	45406	37945	75234	24327	86978	22644	87779	23753	99926	63898	54886	18051	96314
139	78755	47744	43776	83098	03225	14281	83637	55984	13300	52212	58781	14905	46502	04472
140	25282	69106	59180	16257	22810	43609	12224	25643	89884	31149	85423	32581	34374	70873
141	11959	94202	02743	86847	79725	51811	12998	76844	05320	54236	53891	70226	38632	84776
142	11644	13792	98190	01424	30078	28197	55583	05197	47714	68440	22016	79204	06862	94451
143	06307	97912	68110	59812	95448	43244	31262	88880	13040	16458	43813	89416	42482	33939
144	76285	75714	89585	99296	52640	46518	55486	90754	88932	19937	57119	23251	55619	23679
145	55322	07589	39600	60866	63007	20007	66819	84164	61131	81429	60676	42807	78286	29015
146	78017	90928	90220	92503	83375	26986	74399	30885	88567	29169	72816	53357	15428	86932
147	44768	43342	20696	26331	43140	69744	82928	24988	94237	46138	77426	39039	55596	12655
148	25100	19336	14605	86603	51680	97678	24261	02464	86563	74812	60069	71674	15478	47642
149	83612	46623	62876	85197	07824	91392	58317	37726	84628	42221	10268	20692	15699	29167
150	41347	81666	82961	60413	71020	83658	02415	33322	66036	98712	46795	16308	28413	05417

Line/Col.	(1)	(2)	(3)	(4)	(5)	(6)	(7)	(8)	(9)	(10)	(11)	(12)	(13)	(14)
151	38128	51178	75096	13609	16110	73533	42564	59870	29399	67834	91055	89917	51096	89011
152	60950	00455	73254	96067	50717	13878	03216	78274	65863	37011	91283	33914	91303	49326
153	90524	17320	29832	96118	75792	25326	22940	24904	80523	38928	91374	55597	97567	38914
154	49897	18278	67160	39408	97056	43517	84426	59650	20247	19293	02019	14790	02852	05819
155	18494	99209	81060	19488	65596	59787	47939	91225	98768	43688	00438	05548	09443	82897
156	65373	72984	30171	37741	70203	94094	87261	30056	58124	70133	18936	02138	59372	09075
157	40653	12843	04213	70925	95360	55774	76439	61768	52817	81151	52188	31940	54273	49032
158	51638	22238	56344	44587	83231	50317	74541	07719	25472	41602	77318	15145	57515	07633
159	69742	99303	62578	83575	30337	07488	51941	84316	42067	49692	28616	29101	03013	73449
160	58012	74072	67488	74580	47992	69482	58624	17106	47538	13452	22620	24260	40155	74716
161	18348	19855	42887	08279	43206	47077	42637	45606	00011	20662	14642	49984	94509	56380
162	59614	09193	58064	29086	44385	45740	70752	05663	49081	26960	57454	99264	24142	74648
163	75688	28630	39210	52897	62748	72658	98059	67202	72789	01869	13496	14663	87645	89713
164	13941	77802	69101	70061	35460	34576	15412	81304	58757	35498	94830	75521	00603	97701
165	96656	86420	96475	86458	54463	96419	55417	41375	76886	19008	66877	35934	59801	00497
166	03363	82042	15942	14549	38324	87094	19069	67590	11087	68570	22591	65232	85915	91499
167	70366	08390	69155	25496	13240	57407	91407	49160	07379	34444	94567	66035	38918	65708
168	47870	36605	12927	16043	53257	93796	52721	73120	48025	76074	95605	67422	41646	14557
169	79504	77606	22761	30518	28373	73898	30550	76684	77366	32276	04690	61667	64798	66276
170	46967	74841	50923	15339	37755	98995	40162	89561	69199	42257	11647	47603	48779	97907
171	14558	50769	35444	59030	87516	48193	02945	00922	48189	04724	21263	20892	92955	90251
172	12440	25057	01132	38611	28135	68089	10954	10097	54243	06460	50856	65435	79377	53890
173	32293	29938	68653	10497	98919	46587	77701	99119	93165	67788	17638	23097	21468	36992
174	10640	21875	72462	77981	56550	55999	87310	69643	45124	00349	25748	00844	96831	30651
175	47615	23169	39571	56972	20628	21788	51736	33133	72696	32605	41569	76148	91544	21121
176	16948	11128	71624	72754	49084	96303	27830	45817	67867	18062	87453	17226	72904	71474
177	21258	61092	66634	70335	92448	17354	83432	49608	66520	06442	59664	20420	39201	69549
178	15072	48853	15178	30730	47481	48490	41436	25015	49932	20474	53821	51015	79841	32405
179	99154	57412	09858	65671	70655	71479	63520	31357	56968	06729	34465	70685	04184	25250
180	08759	61089	23706	32994	35426	36666	63988	98844	37533	08269	27021	45886	22835	78451
181	67323	57839	61114	62192	47547	58023	64630	34886	98777	75442	95592	06141	45096	73117
182	09255	13986	84834	20764	72206	89393	34548	93438	88730	61805	78955	18952	46436	58740
183	36304	74712	00374	10107	85061	69228	81969	92216	03568	39630	81869	52824	50937	27954
184	15884	67429	86612	47367	10242	44880	12060	44309	46629	55105	66793	93173	00480	13311
185	18745	32031	35303	08134	33925	03044	59929	95418	04917	57596	24878	61733	92834	64454
186	72934	40086	88292	65728	38300	42323	64068	98373	48971	09049	59943	36538	05976	82118
187	17626	02944	20910	57662	80181	38579	24580	90529	52303	50436	29401	57824	86039	81062
188	27117	61399	50967	41399	81636	16663	15634	79717	94696	59240	25543	97989	63306	90946
189	93995	18678	90012	63645	85701	85269	62263	68331	00389	72571	15210	20769	44686	96176
190	67392	89421	09623	80725	62620	84162	87368	29560	00519	84545	08004	24526	41252	14521
191	04910	12261	37566	80016	21245	69377	50420	85658	55263	68667	78770	04533	14513	18099
192	81453	20283	79929	59839	23875	13245	46808	74124	74703	35769	95588	21014	37078	39170
193	19480	75790	48539	23703	15537	48885	02861	86587	74539	65227	90799	58789	96257	02708
194	21456	13162	74608	81011	55512	07481	93551	72189	76261	91206	89941	15132	37738	59284
195	89406	20912	46189	76376	25538	87212	20748	12831	57166	35026	16817	79121	18929	40628
196	09866	07414	55977	16419	01101	69343	13305	94302	80703	57910	36933	57771	42546	03003
197	86541	24681	23421	13521	28000	94917	07423	57523	97234	63951	42876	46829	09781	58160
198	10414	96941	06205	72222	57167	83902	07460	69507	10600	08858	07685	44472	64220	27040
199	49942	06683	41479	58982	56288	42853	92196	20632	62045	78812	35895	51851	83534	10689
200	23995	68882	42291	23374	24299	27024	67460	94783	40937	16961	26053	78749	46704	21983

REGRESSION AND CORRELATION

PART FIVE

12

Chapter 5 introduced tools for understanding statistical relationships: regression and correlation analysis, and contingency table analysis. This chapter continues the discussion of regression and correlation analysis directed toward the statistical importance of relationships. Formulas will not be presented in this chapter because there are no standard notations for computing regression-related statistics. Consequently, computations will be presented which you can adapt to the formulas presented in your textbook.

Note that the slope of a population regression line is frequently symbolized by β_1 or β and the Y-intercept of is frequently symbolized by β_0 or α. Further, the slope of a sample regression line is frequently symbolized by $\hat{\beta}_1$, b_1, or b and the y-intercept is frequently symbolized by $\hat{\beta}_0$, b_0, or a.

The assumptions required to make statistical inferences about simple linear regressions are:

1. A linear model is appropriate for the data.
2. The conditional probability distributions of Y given X are normal.
3. The conditional standard deviations of Y given X are all equal (this is called homoscedasticity).
4. The random error terms are independent.
5. The values of the independent variable are set at fixed levels.

Multiple regression analysis is an extension of simple linear regression analysis in which two or more independent variables are included in the model. Once again, formulas will not be presented due to a lack of notational uniformity. Note that the slope of the ith independent variable is frequently symbolized by β_i for a population model and $\hat{\beta}_i$ or b_i for a sample model.

Two more assumptions are required to make statistical inferences about a multiple linear regression. They are:

6. The number of observations must be larger than the number of regression coefficients to be estimated.
7. There must not be an exact linear relationship between any of the independent variables (this is called multicollinearity).

INTRODUCTION

Regression and correlation

191

There are several common errors associated with the use of regression analysis that should be pointed out:

1. Estimating the dependent variable outside the observed range of the independent variable.
2. Assuming a cause and effect relationship exists if a statistically significant regression model is computed.
3. Assuming that conditions present in the past will remain constant in the future.
4. Estimating spurious relationships between variables (finding variables that are mathematically related by chance but have no practical relationship).

Estimating the dependent variable outside the observed range of the independent variable can cause errors because the linear relationship that exists between Y and X in the observed range of X may not be valid outside the observed range or X. An example of this type of error can be seen in Additional Problem 5–4. In that example you were asked to compute the regression line of CARS on PEPLE and to forecast the number of CARS in a household with 10 PEPLE. The relevant statistics from the problem are given in Table 12–1.

Table 12–1

PEPLE (x)		CARS (y)
$\Sigma x = 38$		$\Sigma y = 24$
$\bar{x} = 3.8$		$\bar{y} = 2.4$
$s_x^2 = 1.96$		$s_y^2 = 1.16$
Minimum = 1		Minimum = 2
Maximum = 4	$n = 10$	Maximum = 6
	Slope = .67	
	$y -$ intercept $= -.15$	
	CARS $= -.15 + .67$(PEPLE).	
	CARS $= -.15 + .67(10) = 6.55$.	

The above forecast is subject to error because the relationship between PEPLE and CARS is valid only in the observed range of CARS (in this example the observed range of cars is between one and four). For example, the forecasted number of cars for a family of zero people is -0.15 cars (rather than zero cars), which is clearly absurd.

It is important to realize that statistics cannot imply cause and effect relationships. Three conditions must be present to imply cause and effect relationships: (1) a correlation must exist between two variables, (2) a sequence of events must exist between two variables—the causal factor must occur first, and (3) no other causal factors must exist (no other factors can explain the relationship being investigated). To clarify, the existence of a correlation between two variables is only one of the three requirements necessary to imply that cause and effect relationship exists.

Users of regression analysis must be careful not to assume that conditions which existed in the past will remain constant in the future. The above caveat is especially important if the independent variable is time. For example, if the purchase of automobiles is being forecasted over time and a war is declared, the old relationship between automobile purchases and time will be invalid due to the war effort.

Finally, it is important that users of statistics do not give too much importance to spurious mathematical correlations. For example, prior to 1940 there was a .99 correlation between the number of mules in the United States and the number of Ph.D.s in the United States. This correlation, although humorous, was spurious; it was an accident and should be treated as such. It is obviously ridiculous to attempt to increase the number of Ph.D.s in the United States by increasing the mule population.

The following sections of this chapter are devoted to Stat City related examples which are worked out in detail, and additional problems. Keep the assumptions and the caveats of regression and correlation analysis in mind when you work on the cases in this chapter.

Example 12–1   Forecasting the average number of CARS in a household of four people

Ms. Sharon Lowe, division head of the Stat City Department of Traffic, received your memorandum on forecasting the number of cars in a household from the number of people in a household (see Additional Problem 5–4). She liked the general format of the memo but wants you to rewrite the document to include information on the statistical significance of your findings and add any footnotes that are needed to fully explain your work. Further, she has decided that she is only interested in forecasting the average number of cars for a family with four people (see Exhibit 12–1).

The random sample that was drawn and several summary statistics are listed in Table 12–2 for your convenience.

Table 12–2

ID	CARS (y)	PEPLE (x)
58.	4.	6.
127.	1.	2.
584.	3.	4.
615.	2.	3.
767.	2.	4.
806.	3.	4.
1121.	4.	6.
1160.	1.	3.
1365.	2.	2.
1366.	2.	4.

$$\Sigma x = 38 \qquad \Sigma y = 24 \qquad \Sigma(xy) = 103$$
$$\Sigma(x)^2 = 162 \qquad \Sigma(y)^2 = 68 \qquad \Sigma[(x - \bar{x})(y - \bar{y})] = 11.8$$
$$\Sigma(x - \bar{x})^2 = 17.6 \qquad \Sigma(y - \bar{y})^2 = 10.4$$

Solution

To meet the new requirements of the Stat City Department of Traffic the following items had to be added to the original memorandum:

1. A test of the statistical significance of the regression line.
2. Construction of point and interval estimates of the average number of cars for a family with four people.
3. Insertion of appropriate footnotes as to the dangers of statistical surveys.

Statistical significance of the regression line   The statistical significance of the regression line is tested below:

$$H_0: \beta_1 = 0$$
$$H_A: \beta_1 \neq 0$$
$$\alpha = .05$$
$$n = 10$$
$$\text{slope} = .67$$
$$\text{standard error of slope} = .13$$
$$y\text{-intercept} = -.15$$
$$\text{CARS} = -.15 + .67(\text{PEPLE})$$
$$r = .872$$
$$r^2 = .761$$

$$t = \frac{.67 - 0}{.13} = 5.043$$
$$t(1 - \alpha/2 = .975, 8) = 2.306$$
or
$$F = 25.432$$
$$F(1 - \alpha = .95, 1, 8) = 5.32$$

Since, $|5.043| > 2.306$ (or $25.432 > 5.32$), you can conclude that the null hypothesis (H_0: $\beta_1 = 0$) is not tenable. Consequently, β_1 is not equal to zero and PEPLE is a statistically significant predictor of CARS.

Point and interval estimates of the average number of cars for a family of four people A family of four people will have an average of 2.53 cars; $CARS = -0.15 + .67(4) = 2.53$. The 95 percent confidence interval for the average number of cars for a family of four people is constructed below.

$$\hat{y}_h \pm t(1 - \alpha/2, n - 2)s(\hat{y}_h)$$

where

$\hat{y}_h$ = the mean value of y at the hth level of x
$s(\hat{y}_h)$ = the standard error of the mean value of y at the hth level of x($x = 4$ in this example)

$$= \sqrt{\frac{\Sigma(y - \hat{y})^2}{n - 2}} \sqrt{\frac{1}{n} + \frac{(x_h - \bar{x})^2}{\Sigma (x^2) - \frac{(\Sigma x)^2}{n}}}$$

$$= \sqrt{\frac{2.489}{8}} \sqrt{\frac{1}{10} + \frac{(4 - 3.8)^2}{162 - \frac{382}{10}}} = .55783(.3198)$$

$$= .1784$$

If the above formula is different from the formula in your textbook, use the formula your professor presents in class. Either formula will yield the same $s(\hat{y}_h)$.

$$2.53 \pm 2.306(.1784)$$
$$2.53 \pm .4111$$
$$2.12 \text{ to } 2.94$$
$$2.12 \leq E(\hat{y}_h) \leq 2.94$$

Appropriate footnotes If the memorandum is to meet professional standards it must include footnotes explaining:

1. The sampling plan used.
2. Sampling and nonsampling errors.
3. Standard errors.
4. The interpretation of confidence intervals.

The memorandum appears in Exhibit 12–1.

Exhibit 12–1

HOWARD S. GITLOW, PH.D.
STATISTICAL CONSULTANT
———

MEMORANDUM

TO: Ms. Sharon Lowe, Division Head
        Stat City Department of Traffic

FROM: Howard Gitlow

DATE: August 13, 1980

RE: Predicting the average number of cars in a household
        of four people, as of January 1980

   As per our consulting contract and using commonly
accepted statistical techniques, I have investigated the
relationship between "the number of cars in a household as of
January 1980 (CARS)" and "the number of people in a household
as of January 1980 (PEPLE)."[1] I have found that 76 percent of
the variation in CARS is explained by PEPLE.[2] This strong
relationship[3] between CARS and PEPLE is expressed in the
following equation:

$$CARS = -0.15 + .67 \ (PEPLE)$$

The above equation can be used to estimate the average number
of CARS for different-sized families. An average family of
four people (PEPLE = 4) will have 2.53 cars;

$$CARS = -.015 + .67 \ (4) = 2.53 \ [CARS]$$

Allowing for the errors in survey estimates, you can have "95
percent confidence" that the average number of cars for a

Exhibit 12–1 (*continued*)

family with four people is contained in the range from 2.12 cars to 2.94 cars.[4,5,6]

I hope the above information meets your needs. If you have any questions, please call me at 305-999-9999.

Footnotes

[1] Insert standard footnote 1 (insert a sample size of 10 drawn via simple random sampling).

[2] The coefficient of determination (r^2) is 76 percent. This indicates that the variability in CARS is reduced by 76 percent when PEPLE is considered.

[3] Insert standard footnote 2.

[4] Insert standard footnote 3A. (Insert the following numbers into the footnote: 10, 10, 1, 1, 1, 1, 2.306, 2.306, 2.306, 2.306.)

[5] The standard error of the slope of PEPLE (.67 cars) is 0.13 CARS. The "95 percent confidence" interval for the estimated slope of PEPLE ranges from 0.37 cars to 0.97 cars. Since the confidence interval does not contain zero, you can say that PEPLE is a statistically significant predictor of cars at the 95 percent level of confidence.

[6] The standard error of the predicted number of CARS for a family with four PEPLE is 0.1784 cars. The "95 percent confidence interval" for the predicted number of cars for a family with four people ranges from 2.12 to 2.94 cars.

Example 12–2   Development of a model to forecast gas consumption from electric consumption

Ms. Shelly Dimmerman, division head of the Stat City Department of Energy, believes that families which consume large quantities of electricity are also large consumers of automobile gasoline—these are called "fuelish families." You have been retained by Ms. Dimmerman to ascertain if her "fuelish family" hypothesis is tenable and to develop an equation to forecast gas consumption from electric consumption (see Exhibit 12–2). Ms. Dimmerman has asked you to use the following operational definitions for gas consumption and electric consumption:

GAS = Family's average bimonthly automobile gas bill, as of January 1980.

ELEC = Family's average monthly electric bill, as of January 1980.

Ms. Dimmerman wants to forecast GAS from ELEC because GAS is difficult to measure while ELEC is metered monthly and can be obtained from the Stat City Electric Company.

Use Table 12–3 for a random sample of Stat City families.

Table 12–3

CASE-N	ID	ELEC	GAS	CASE-N	ID	ELEC	GAS
1	19.	84.	185.	71	905.	35.	120.
2	26.	86.	206.	72	1013.	36.	167.
3	28.	62.	159.	73	1016.	46.	185.
4	39.	74.	180.	74	1027.	67.	152.
5	56.	37.	65.	75	1037.	50.	120.
6	62.	76.	184.	76	1039.	21.	65.
7	65.	57.	123.	77	1044.	47.	120.
8	77.	90.	204.	78	1047.	21.	120.
9	81.	87.	193.	79	1076.	48.	218.
10	101.	71.	164.	80	1077.	38.	120.
11	104.	53.	136.	81	1087.	65.	120.
12	106.	85.	200.	82	1106.	36.	65.
13	119.	57.	65.	83	1114.	39.	65.
14	138.	65.	139.	84	1119.	54.	168.
15	142.	75.	202.	85	1136.	43.	120.
16	148.	84.	156.	86	1140.	51.	208.
17	180.	48.	65.	87	1149.	69.	120.
18	181.	86.	199.	88	1186.	45.	120.
19	201.	75.	142.	89	1189.	47.	120.
20	223.	83.	150.	90	1218.	35.	65.
21	228.	63.	105.	91	1242.	47.	168.
22	258.	73.	127.	92	1246.	47.	195.
23	260.	85.	195.	93	1256.	29.	120.
24	274.	66.	144.	94	1272.	36.	65.
25	295.	42.	99.	95	1289.	48.	120.
26	298.	127.	120.	96	1300.	37.	110.
27	330.	60.	175.	97	1307.	48.	218.
28	383.	85.	207.	98	1348.	65.	65.
29	386.	72.	160.	99	1352.	38.	120.
30	390.	111.	220.	100	1362.	45.	65.
31	402.	52.	202.				
32	410.	51.	182.				
33	411.	46.	158.				
34	415.	37.	65.				
35	416.	37.	174.				
36	435.	72.	155.				
37	439.	62.	145.				
38	452.	63.	156.				
39	476.	35.	119.				
40	492.	76.	130.				
41	498.	44.	231.				
42	500.	51.	216.				
43	521.	80.	152.				
44	533.	88.	204.				
45	537.	75.	175.				
46	549.	97.	196.				
47	565.	60.	130.				
48	573.	66.	145.				
49	600.	65.	120.				
50	616.	84.	217.				
51	624.	69.	208.				
52	640.	65.	148.				
53	651.	46.	155.				
54	653.	45.	217.				
55	693.	22.	120.				
56	720.	39.	65.				
57	736.	38.	120.				
58	748.	61.	120.				
59	751.	48.	120.				
60	764.	19.	120.				
61	779.	38.	120.				
62	786.	42.	65.				
63	790.	51.	120.				
64	821.	50.	65.				
65	845.	45.	120.				
66	848.	50.	120.				
67	858.	21.	120.				
68	872.	19.	65.				
69	878.	45.	120.				
70	892.	68.	65.				

The following summary statistics were computed from the sample data: **Solution**

$y = GAS$	$x = ELEC$
$\Sigma y = 14{,}118$	$\Sigma x = 5{,}674$
$\bar{y} = 141.18$	$\bar{x} = 56.74$
$s_y = 47.26$	$s_x = 20.62$

$$r = .49$$
$$r^2 = .24$$

y − intercept $= \$77.45$	Standard error of the slope $= .20$
Slope $= \$ 1.12$	Standard error of the estimate $= 41.40$
Mean square error $= 1{,}714.12$	

The sample regression line is:

$$GAS = \$77.45 + \$1.12(ELEC)$$

The regression line is significant at the 5 percent level of significance (or 95 percent level of confidence); see test below.

$$H_0: \beta_1 = 0$$
$$H_A: \beta_1 > 0$$
$$\alpha = .05$$

$$t = \frac{1.12 - 0}{.20} = 5.6$$
$$t(.95, 98) = 1.66$$
or
$$F = 31.36$$
$$F(.90, 1, 98) = 2.76$$

Since $|5.6| > 1.66$ (or $31.36 > 2.76$), the null hypothesis ($\beta_1 = 0$) is not tenable. Consequently, ELEC is a statistically significant predictor of GAS at the 5 percent level of significance.

The "fuelish family" theory has validity because β_1 is significantly greater than zero. This means that GAS increases as ELEC increases. It is important to realize that a one-tail test was called for here to test the "fuelish family" hypothesis. Consequently, all 5 percent of α is in the upper tail so the t-statistic is 1.66, not 1.99 (the two-tail t-statistic). Be sure you note that the qualitative statement of the "fuelish family" hypothesis has a quantitative, statistically testable, counterpart:

Families with large electric bills have large gas bills

or

GAS is directly related to ELEC

or

There is a positive correlation between GAS and ELEC ($\rho > 0.0$)

or

$\beta_1 > 0$ which implies GAS $= \beta_0 + \beta_1$ ELEC, that is, ELEC is useful in forecasting GAS.

The memorandum appears in Exhibit 12-2.

Exhibit 12-2

HOWARD S. GITLOW, PH.D.
STATISTICAL CONSULTANT

MEMORANDUM

TO: Ms. Shelly Dimmerman, Division Head
       Stat City Department of Energy

FROM: Howard Gitlow

DATE: August 12, 1980

RE: Development of a model to forecast gas consumption
       from electric consumption, as of January 1980.

As per your request and using commonly accepted
statistical techniques, I have investigated the
relationship between "average bimonthly automobile gas
bills, as of January 1980 (GAS)" and average monthly
electric bills, as of January 1980 (ELEC)"[1]. I have found
that 24 percent of the variation in GAS is explained by ELEC[2]
The relationship between GAS and ELEC is expressed in the
following equation:[3]

$$GAS = \$77.45 + \$1.12 \, (ELEC)$$

The above forecasting equation can be used to estimate
average bimonthly gas bills from average monthly electric
bills. Further, the forecasting equation is statistically
significant at the "95 percent level of confidence."[4,5] The
practical implication of the statistical significance of
the forecasting equation is that it lends support to the
"fuelish family" hypothesis.[6]

Exhibit 12–2 (*continued*)

If you have any questions, please call me at
305–999–9999.

_____

Footnotes

[1] Insert standard footnote 1 (insert a sample of 100 drawn via simple random sampling).

[2] The coefficient of determination (r^2) is 24 percent. This indicates that the variability in GAS is reduced by 24 percent when ELEC is considered.

[3] Insert standard footnote 2.

[4] Insert standard footnote 3A. (Insert the following numbers into the footnote: 100, 100, 1, 1, 1, 1, 1.66, 1.66, 1.66, 1.66.)

[5] The standard error of the slope of ELEC ($1.12) is $0.20. The 90 percent confidence interval for the estimated slope of ELEC ranges from $0.72 and $1.52. Since the confidence interval does not contain zero, ELEC is a statistically significant predictor of GAS at the 90 percent level of confidence.

[6] If the regression line had not been statistically significant the forecasting equation would have been GAS = $77.45. To restate, ELEC would have no value in forecasting GAS (it is not in the equation). However, since the forecasting equation includes an ELEC term, and the sign of that term is positive, the "fuelish family" hypothesis is supported; big gas users are also big electricity users.

Example 12–3   Development of a model to forecast electric consumption in Stat City

Mr. Saul Reisman, chief financial officer of the Stat City Electric Company, would like to develop an equation to forecast a dwelling unit's average monthly electric bill. You have been retained by Mr. Reisman to construct the equation. In your discussions with energy experts you have isolated two categories of variables which may influence electric consumption: "fuelishness" of a household and size of a household. The energy experts have told you that the "fuelishness" of a household is measured vis-à-vis a household's yearly heating bill and bimonthly automobile gas bill. The size of a dwelling unit is measured by the number of rooms in the household.

Mr. Reisman has asked you to use the following operational definitions for the relevant variables:

ELEC = Average monthly electric bill as of January 1980.

HEAT = Average yearly heating bill as of January 1980.

GAS = Average bimonthly automobile gas bill as of January 1980.

ROOMS = Number of rooms in a dwelling unit as of January 1980.

Mr. Reisman has provided you with sufficient funds to draw a simple random sample of 144 Stat City dwelling units. Conduct the desired survey and type a memorandum to Mr. Reisman (see Exhibit 12–3).

Use the simple random sample of Stat City families given in Table 12–4.

CASE-N	ID	ROOMS	HEAT	ELEC	GAS	Table 12–4
1	5.	6.	674.	54.	133.	
2	9.	9.	1101.	85.	177.	
3	21.	8.	983.	69.	146.	
4	41.	4.	259.	27.	120.	
5	43.	7.	876.	64.	183.	
6	54.	7.	802.	64.	124.	
7	55.	9.	1065.	83.	187.	
8	63.	7.	857.	65.	134.	
9	67.	9.	1060.	81.	202.	
10	76.	11.	1253.	100.	219.	
11	93.	9.	1019.	82.	231.	
12	100.	9.	1061.	85.	170.	
13	123.	4.	483.	32.	65.	
14	126.	5.	628.	40.	139.	
15	128.	9.	1097.	84.	150.	
16	135.	7.	987.	58.	180.	
17	143.	9.	1265.	91.	162.	
18	161.	8.	1053.	75.	173.	
19	173.	8.	1152.	74.	179.	
20	174.	9.	1309.	80.	211.	
21	175.	9.	1272.	79.	156.	
22	193.	9.	1357.	85.	175.	
23	203.	5.	656.	40.	74.	
24	213.	7.	953.	66.	148.	
25	218.	7.	989.	59.	112.	
26	223.	9.	1249.	83.	150.	
27	224.	7.	923.	62.	130.	
28	238.	8.	1201.	80.	159.	
29	249.	7.	1018.	63.	136.	
30	261.	6.	780.	53.	125.	
31	263.	9.	1256.	80.	162.	
32	267.	7.	990.	54.	105.	
33	283.	8.	1162.	82.	155.	
34	285.	7.	916.	57.	177.	
35	295.	5.	538.	42.	99.	
36	300.	8.	786.	73.	177.	
37	324.	7.	687.	53.	168.	
38	328.	8.	871.	73.	205.	
39	359.	5.	524.	40.	65.	
40	365.	10.	949.	96.	198.	
41	376.	8.	779.	75.	164.	
42	378.	8.	848.	69.	164.	
43	379.	7.	698.	58.	159.	
44	395.	10.	1035.	97.	120.	
45	401.	9.	859.	78.	169.	
46	405.	8.	678.	63.	132.	
47	408.	6.	588.	48.	165.	
48	433.	10.	835.	85.	124.	
49	438.	6.	584.	48.	120.	
50	446.	6.	562.	47.	120.	
51	451.	6.	528.	48.	172.	
52	475.	6.	524.	51.	120.	
53	484.	5.	427.	41.	132.	
54	485.	8.	799.	64.	179.	
55	489.	8.	693.	76.	142.	
56	504.	8.	729.	62.	138.	
57	519.	5.	464.	41.	116.	
58	526.	11.	1105.	109.	120.	
59	528.	8.	779.	71.	199.	
60	555.	8.	809.	74.	146.	
61	574.	7.	625.	62.	166.	
62	585.	8.	825.	75.	136.	
63	598.	9.	952.	81.	205.	
64	604.	7.	730.	65.	159.	
65	623.	8.	850.	75.	203.	
66	629.	5.	395.	37.	120.	
67	635.	8.	637.	67.	215.	
68	639.	8.	621.	64.	120.	
69	652.	6.	481.	41.	197.	
70	667.	6.	470.	44.	65.	

Table 12–4 (*continued*)

CASE-N	ID	ROOMS	HEAT	ELEC	GAS
71	669.	7.	598.	54.	65.
72	697.	5.	422.	32.	120.
73	714.	6.	521.	48.	120.
74	738.	6.	489.	45.	120.
75	740.	7.	523.	62.	120.
76	752.	5.	393.	38.	65.
77	760.	5.	384.	40.	65.
78	778.	5.	391.	38.	65.
79	809.	7.	538.	63.	120.
80	829.	4.	268.	30.	65.
81	835.	7.	601.	59.	120.
82	838.	5.	391.	33.	120.
83	844.	7.	556.	54.	120.
84	846.	5.	374.	41.	120.
85	866.	5.	436.	41.	65.
86	871.	5.	386.	37.	120.
87	878.	6.	533.	45.	120.
88	879.	6.	529.	47.	65.
89	889.	5.	396.	36.	65.
90	894.	6.	483.	53.	120.
91	902.	4.	334.	30.	120.
92	908.	6.	451.	48.	120.
93	938.	7.	569.	56.	65.
94	941.	7.	585.	55.	120.
95	945.	3.	280.	18.	120.
96	947.	5.	388.	37.	120.
97	957.	7.	576.	45.	184.
98	958.	6.	553.	49.	165.
99	969.	8.	616.	68.	126.
100	979.	5.	369.	37.	65.
101	981.	9.	721.	77.	142.
102	987.	6.	503.	48.	185.
103	999.	5.	397.	38.	120.
104	1003.	5.	393.	35.	120.
105	1016.	6.	512.	46.	185.
106	1019.	7.	549.	54.	120.
107	1024.	7.	568.	53.	120.
108	1031.	5.	404.	35.	65.
109	1049.	5.	411.	41.	65.
110	1051.	8.	571.	63.	65.
111	1052.	5.	423.	37.	120.
112	1070.	4.	361.	30.	120.
113	1099.	9.	719.	76.	190.
114	1100.	6.	498.	43.	206.
115	1136.	6.	458.	43.	120.
116	1148.	8.	609.	64.	120.
117	1158.	4.	260.	29.	65.
118	1168.	5.	452.	38.	120.
119	1181.	7.	608.	55.	120.
120	1182.	6.	479.	43.	120.
121	1184.	5.	433.	35.	120.
122	1187.	4.	312.	31.	120.
123	1194.	3.	220.	19.	120.
124	1204.	3.	210.	22.	65.
125	1217.	5.	384.	39.	77.
126	1222.	5.	402.	36.	85.
127	1224.	6.	488.	46.	169.
128	1231.	7.	586.	59.	65.
129	1233.	10.	832.	79.	65.
130	1250.	7.	554.	60.	65.
131	1258.	5.	395.	37.	120.
132	1272.	5.	422.	36.	65.
133	1273.	5.	411.	32.	120.
134	1288.	6.	487.	42.	65.
135	1292.	6.	503.	48.	65.
136	1294.	6.	503.	40.	120.
137	1296.	5.	382.	38.	177.
138	1297.	8.	647.	70.	65.
139	1315.	5.	385.	38.	65.
140	1317.	5.	423.	40.	65.
141	1320.	6.	470.	46.	120.
142	1335.	5.	398.	36.	65.
143	1345.	5.	342.	38.	65.
144	1373.	5.	374.	38.	65.

The bivariate correlation matrix for all four variables used in this study is shown in Table 12–5.

The correlation matrix reveals an extremely high correlation between ROOMS and ELEC ($r = .97$); approximately 94 percent ($r^2 = .94$) of the variation in ELEC is explained by ROOMS. Further, ELEC is highly correlated with HEAT and GAS, and ROOMS is highly correlated with HEAT and GAS. All of the above indicates redundancy among the independent variables; each independent variable is a good predictor of ELEC and the independent variables are also good predictors of each other.[1] Consequently, it is both efficient and expedient to use only the best independent variable to predict ELEC and to delete the others from the model. The other variables should be eliminated because they bring little or no new information which would enhance the model's forecasting ability.

Just for your information, the full regression model is:

$$ELEC = -\$11.73 + \$0.016(HEAT) - \$0.005(GAS) + \$8.65(ROOMS)$$
$$(.002) \qquad (.009) \qquad (.349)$$

The interpretation of the above regression coefficients is as follows: the y-intercept ($-\$11.73$) indicates the average monthly electric bill if the average yearly heating bill is zero, the average bimonthly automobile gasoline bill is zero, and the number of rooms in a dwelling unit is zero (this is obviously an absurd situation and occurs because a forecast is being made outside the ranges of the independent variables). The slope of the average yearly heating bill ($\$0.016$) indicates that for a dwelling unit with a given average biweekly automobile gasoline bill and number of rooms, the average monthly electric bill will increase by $\$0.016$ for every dollar spent on heat. The slope of the average bimonthly automobile gasoline bill ($-\$0.005$) indicates that for a dwelling unit with a given average yearly heating bill and number of rooms, the average monthly electric bill will decrease by $\$0.005$ for every dollar spent on gasoline. The slope of the number of rooms ($\$8.65$) indicates that for a dwelling unit with a given average yearly heating bill and an average bimonthly automobile gasoline bill, the average monthly electric bill will increase by $\$8.65$ for every additional room.

The numbers in brackets under the regression coefficients are the standard errors of the regression coefficients. It is extremely important that t-tests are not performed to test the statistical significance of each individual regression coefficient. The warning is appropriate whenever there is any correlation between the independent variables. If correlations do exist between the independent variables (this is called multicollinearity), then it is impossible to extract the portion of the variability in the dependent variable that is uniquely attributable to a particular independent variable. Consequently, t-tests are meaningless in a multiple regression context in which multicollinearity is present. The bivariate correlation matrix in this problem clearly indicates the large degree of multicollinearity present in this example; hence, t-tests should not be conducted to test the statistical significance of the independent variables. Instead, an analysis of variance table can be computed which allows all of the regression coefficients to be tested for significance simultaneously (H_0: $\beta_1 = \beta_2 = \beta_3 = 0$). Unfortunately, if the null hypothesis is rejected you will not know which of the regression coefficients are in fact statistically different from zero. This is the method you must use to circumvent the problem of multicollinearity

Solution

Table 12–5

	ELEC	HEAT	GAS	ROOMS
ELEC	1.00	.89	.54	.97
HEAT		1.00	.58	.85
GAS			1.00	.54
ROOMS				1.00

---

[1] High correlations between the independent variables is called multicollinearity and can cause severe problems in regression analysis. One way to deal with multicollinearity is to drop all but one of the variables which are highly correlated out of the model so that only the best predictor variable(s) remain.

and still be able to test the statistical significance of your regression model. The ANOVA table for this problem is shown in Table 12–6.

Table 12–6

Source	df	SS	MS	F
Regression	3	48,304.44	16,101.48	1,159.24
Error	140	1,944.55	13.89	
Total	143			

The coefficient of multiple determination measures the percent of variation in the dependent variable that is explained by the independent variables. In this example, the coefficient of multiple determination is 0.96 (96 percent of the variation in ELEC is explained by ROOMS, HEAT, and GAS). The problem of multicollinearity can also be seen when examining the coefficient of multiple determination. Table 12–7 depicts the change in the coefficient of multiple determination as independent variables are added into the model.

Table 12–7

Independent variable	r^2	Change in r^2 due to the addition of another independent variable
ROOMS	.9473	.9473
HEAT	.9612	.0139
GAS	.9613	.0001

Once ROOMS has been included in the model none of the other independent variables makes a substantive contribution to explaining the variation in ELEC.

The best strategy would be to revise the model and to use only ROOMS as a predictor of ELEC. To reiterate, the energy experts were correct concerning the importance of the "fuelish" family and size of dwelling unit variables; however, they failed to warn you about the interrelationships between the independent variables. Consequently, the experts wasted the Electric Company's money by requiring you to collect unnecessary data. Sounder planning on their part would have ensured a more cost/effective study.

The reduced regression model is shown below.

$$ELEC = -\$15.86 + \$10.72(ROOMS)$$
$$r^2 = .94$$

The statistical significance of the regression line is tested below:

$$H_0: \beta_1 = 0$$
$$H_A: \beta_1 \neq 0$$
$$\alpha = .05$$

$$\left. \begin{array}{l} t = \dfrac{10.72 - 0}{.212} = 50.57 \\ t(.975, 142) = 1.97 \end{array} \right\} \text{ or } \left\{ \begin{array}{l} F = 2,556.92 \\ F(.95, 1, 142) = 3.88 \end{array} \right.$$

Since $|50.57| > 1.97$ (or $2,556.92 > 3.88$), you can conclude that H_0 is not tenable. Consequently, $\beta_1 \neq 0$ and ROOMS is a statistically significant predictor of ELEC.

One final note, the regression model computed in this problem affords a perfect example of the dangers of forecasting the dependent variable outside the range of the independent variable. Case in point, a forecast of the electric bill for a one-room dwelling unit yields a negative electric bill (ELEC = −$15.86 + $10.72(1) = −$5.14). Obviously, electric bills cannot be negative (except in the case of a rebate or some other atypical practice).

The memorandum appears in Exhibit 12–3.

Exhibit 12–3

```
HOWARD S. GITLOW, PH.D.
   STATISTICAL CONSULTANT
   _____


                      MEMORANDUM


TO:    Mr. Saul Reisman, Chief Financial Officer
       Stat City Electric Company

FROM:  Howard Gitlow

DATE:  August 19, 1980

RE:    Development of a model to forecast electric
       consumption in Stat City, as of January 1980


   As per your request and using commonly accepted
statistical techniques, I have developed an equation to
forecast a dwelling unit's average monthly electric bill as
of January 1980 (ELEC).[1,2] I have found that approximately 94
percent of the variation in ELEC is explained by the number
of rooms in a dwelling unit as of January 1980 (ROOMS).[3,4] The
relationship between ELEC and ROOMS is expressed in the
following equation:

           ELEC = −$15.86 + $10.72(ROOMS).

The above forecasting equation can be used to estimate the
average monthly electric bill for a dwelling unit (as of
January 1980) from the number of rooms in a dwelling unit (as
of January 1980). Further, the forecasting equation is
statistically significant as the "95 percent level of
confidence."[5,6]
   If you have any questions, please call me at
305-999-9999.
```

Exhibit 12–3 (*continued*)

Footnotes

[1] The original model to forecast ELEC included the "fuelishness" of a family and the size of a dwelling unit as measured by the following independent variables:

> HEAT = Average yearly heating bill as of January 1980.
> GAS = Average bimonthly automobile gas bill as of January 1980.
> ROOMS = Number of rooms in a dwelling unit as of January 1980.

All of the above variables were suggested by the Stat City Electric Company energy experts. HEAT and GAS were surrogate measures of the "fuelishness" of a family. ROOMS was a measure of the size of a dwelling unit.

Preliminary correlation analysis revealed that approximately 94 percent of the variation in ELEC was explained by ROOMS. Further, ELEC was highly correlated with HEAT and GAS, and ROOMS was highly correlated with HEAT and GAS. All of the above indicated a redundacy among the predictor variables; each independent variable was a good predictor of ELEC and the independent variables were also good predictors of each other. Consequently, a simple model was constructed to forecast ELEC which used only ROOMS.

[2] Insert standard footnote 1 (include a sample of 144 drawn via simple random sampling).

[3] The coefficient of determination (r^2) is 94 percent. This indicates that the variability in ELEC is reduced by 94 percent when ROOMS is considered.

[4] Insert standard footnote 2.

[5] Insert standard footnote 3A. (Insert the following numbers into the footnote: 144, 144, 1, 1, 1, 1, 1.97, 1.97, 1.97, 1.97.)

[6] The standard error of the slope of ROOMS (10.72) is 0.212. The 95 percent confidence interval for the estimated slope of ROOMS ranges from $10.30 to $11.14. Since the confidence interval does not contain zero, you can say that ROOMS is a statistically significant predictor of ELEC at the 95 percent level of confidence.

SUMMARY   This chapter dealt with many of the issues necessary to properly use regression and correlation analysis. Primarily, this chapter introduced the concept of statistical inference in regression analysis problems. The assumptions required to make statistical inferences in a regression analysis were enumerated. Finally, several caveats concerning the proper use of regression analysis were discussed.

ADDITIONAL PROBLEMS

12–4   Forecasting heating bills from the number of rooms in a dwelling unit   Ms. Shelly Dimmerman, division head of the Stat City Department of Energy, needs to forecast the yearly heating bill of a dwelling unit from the number of rooms in the dwelling unit. You have been asked by Ms. Dimmerman to conduct a survey and construct the needed forecasting equation. Ms. Dimmerman has asked you to use the following operational definitions for heating bill and rooms:

HEAT = Average yearly heating bill as of January 1980 (for all types of heat).

ROOMS = Number of rooms in a dwelling unit as of January 1980.

Use the following simple random sample of Stat City families:

ID	ROOMS	HEAT
111.	8.	973.
550.	8.	851.
554.	8.	812.
640.	8.	661.
776.	5.	385.
777.	8.	691.
813.	6.	505.
1192.	6.	466.
1253.	5.	419.
1257.	8.	667.

12–5 Developing an equation to forecast monthly mortgage payments from the assessed value of a house   Ms. Sharon Vigil, chairperson of the Stat City Real Estate Board, is trying to develop an equation for forecasting monthly mortgages in Stat City. Ms. Vigil has decided that using the assessed value of a house may provide an accurate indicator of monthly mortgages. Assessed value is a convenient predictor to use because it is publicly available and current as of January 1980.

You have been retained by Ms. Vigil to develop an equation for forecasting monthly mortgages from assessed values. Ms. Vigil has provided you with funds to draw a random sample of 10 Stat City dwelling units. Please use the following simple random sample of Stat City dwelling units.

ID	HCOST	ASST
91.	510.	58524.
129.	606.	70174.
137.	880.	94070.
162.	785.	95599.
218.	785.	87717.
286.	621.	76856.
394.	144.	27898.
537.	222.	44747.
561.	514.	51378.
575.	493.	49758.

12–6 Development of an equation to forecast the number of CARS in a household   Ms. Sharon Lowe, division head of the Stat City Department of Traffic, would like to be able to forecast the number of cars in a household. You have been hired by her to develop the desired forecasting equation.

Ms. Lowe believes that the number of people in a household (PEPLE), the average bimonthly automobile gasoline bill (GAS), and the average monthly telephone bill (PHONE) are all important predictors of the number of cars in a household (CARS).

Ms. Lowe has provided you with the necessary funds to draw a simple random sample of 100 Stat City families. Conduct the desired survey and type a memorandum to Ms. Lowe.

Use the accompanying simple random sample of Stat City families.

Random sample table for Problem 12–6	CASE-N	ID	PHONE	PEPLE	CARS	GAS
	1	20.	35.	2.	2.	137.
	2	54.	55.	3.	2.	124.
	3	58.	63.	6.	4.	217.
	4	69.	59.	5.	4.	210.
	5	70.	44.	3.	2.	149.
	6	72.	39.	3.	2.	131.
	7	91.	18.	4.	3.	177.
	8	93.	22.	5.	5.	231.
	9	94.	23.	3.	2.	158.
	10	129.	21.	3.	2.	122.
	11	137.	18.	3.	2.	135.
	12	143.	85.	4.	2.	162.
	13	162.	23.	4.	2.	157.
	14	175.	21.	4.	2.	156.
	15	180.	12.	1.	1.	65.
	16	208.	20.	3.	2.	122.
	17	214.	22.	2.	1.	104.
	18	218.	67.	2.	1.	112.
	19	234.	93.	2.	1.	111.
	20	268.	15.	3.	2.	129.
	21	286.	71.	2.	1.	104.
	22	291.	68.	4.	3.	176.
	23	320.	36.	5.	3.	200.
	24	337.	118.	7.	2.	120.
	25	342.	63.	6.	2.	120.
	26	353.	21.	5.	4.	209.
	27	375.	78.	5.	3.	184.
	28	388.	20.	6.	5.	235.
	29	394.	45.	2.	1.	105.
	30	411.	59.	4.	2.	158.
	31	421.	54.	5.	3.	193.
	32	435.	22.	4.	2.	155.
	33	437.	46.	3.	2.	120.
	34	449.	16.	6.	2.	120.
	35	452.	13.	4.	2.	156.
	36	478.	30.	6.	4.	219.
	37	483.	16.	3.	2.	139.
	38	498.	19.	5.	5.	231.
	39	503.	17.	5.	4.	208.
	40	516.	13.	4.	2.	155.
	41	547.	17.	6.	4.	218.
	42	554.	22.	4.	2.	152.
	43	578.	64.	5.	3.	187.
	44	598.	15.	5.	4.	205.
	45	607.	22.	5.	4.	209.
	46	616.	48.	6.	4.	217.
	47	677.	17.	3.	2.	120.
	48	700.	20.	5.	2.	120.
	49	735.	27.	8.	1.	65.
	50	741.	86.	3.	1.	65.
	51	745.	77.	2.	1.	100.
	52	768.	37.	1.	2.	120.
	53	813.	17.	5.	4.	201.
	54	826.	16.	6.	2.	120.
	55	837.	39.	3.	2.	120.
	56	850.	23.	5.	2.	120.
	57	868.	66.	5.	2.	120.
	58	879.	71.	5.	1.	65.
	59	886.	18.	5.	1.	65.
	60	896.	17.	1.	2.	120.
	61	918.	17.	3.	2.	120.
	62	919.	29.	6.	2.	120.
	63	937.	12.	8.	2.	120.
	64	940.	76.	4.	2.	120.
	65	947.	25.	4.	2.	120.
	66	961.	13.	4.	3.	178.
	67	984.	19.	6.	2.	120.
	68	1020.	31.	5.	2.	120.
	69	1022.	18.	4.	2.	120.
	70	1040.	23.	7.	1.	65.

CASE-N	ID	PHONE	PEPLE	CARS	GAS
71	1046.	98.	3.	1.	65.
72	1052.	59.	6.	2.	120.
73	1067.	44.	1.	2.	120.
74	1073.	71.	6.	2.	120.
75	1078.	91.	6.	1.	65.
76	1081.	17.	3.	2.	120.
77	1123.	102.	1.	2.	120.
78	1136.	105.	8.	2.	120.
79	1152.	40.	7.	1.	65.
80	1157.	29.	3.	2.	65.
81	1169.	51.	1.	2.	120.
82	1176.	85.	4.	1.	65.
83	1177.	41.	3.	2.	120.
84	1187.	18.	2.	2.	120.
85	1188.	107.	7.	2.	120.
86	1195.	13.	3.	2.	120.
87	1225.	53.	3.	2.	128.
88	1230.	19.	2.	2.	120.
89	1239.	22.	3.	2.	120.
90	1249.	20.	4.	2.	120.
91	1264.	42.	3.	2.	120.
92	1269.	18.	7.	1.	65.
93	1270.	62.	6.	1.	65.
94	1271.	13.	3.	2.	120.
95	1282.	11.	6.	2.	120.
96	1291.	23.	3.	2.	120.
97	1294.	14.	9.	2.	120.
98	1296.	34.	4.	3.	177.
99	1304.	26.	5.	3.	199.
100	1359.	11.	6.	2.	120.

12–7 An empirical test of the "extra familial nonpersonal family size" hypothesis Mr. Brennan (ID = 1041), a sociologist, believes that a family's level of extrafamilial, nonpersonal communication (nonface-to-face communication) increases with family size. The sociologist would like to be able to forecast a family's level of nonpersonal communication from their family size.

You have been employed by the sociologist to conduct a study and test out his hypothesis. Sufficient funds have been allocated to draw a simple random sample of 10 Stat City families. Further, the sociologist has instructed you to use the following operational definitions for his study:

Please use the following simple random sample of Stat City families.

ID	PHONE	PEPLE
3.	62.	4.
227.	69.	2.
300.	15.	4.
367.	57.	5.
714.	21.	6.
759.	77.	4.
821.	14.	7.
1338.	23.	3.
1347.	18.	3.
1372.	45.	5.

Actual variable	Operational variable
Level of extra-familial non-personal communication	PHONE = Average monthly telephone bill as of January 1980.
Family size	PEPLE = Number of people in a family, as of January 1980.

12–8 Developing a model to forecast a family's total 1979 income Mr. Lee Kaplowitz, the mayor of Stat City, would like to be able to forecast a family's income (INCOM) to set donation guidelines for the City Benevolent Association. The mayor is convinced that he will have a difficult time obtaining information on total family income (INCOM) because of his purpose. He would like to see if the number of people in a family (PEPLE) and a family's rent/mortgage (HCOST) is correlated with family income (INCOM). He believes that it will be easy to obtain information on the number of people in a family (PEPLE) and the amount of a family's rent/mortgage (HCOST). Consequently, if a relationship exists

Regression and correlation

between HCOST, PEPLE, and INCOM he can use the easily obtainable variables (HCOST and PEPLE) to forecast his variable of interest (INCOM).

The mayor has engaged you to conduct a survey and develop a model to forecast INCOM from HCOST and PEPLE. Sufficient funds have been allocated to draw a simple random sample of 100 Stat City dwelling units.

Please use the accompanying simple random sample of Stat City families.

Random sample table from Problem 12–8

CASE-N	ID	HCOST	INCOM	PEPLE	CASE-N	ID	HCOST	INCOM	PEPLE
1	41.	263.	19505.	1.	51	687.	547.	14967.	5.
2	52.	775.	24718.	6.	52	688.	572.	16856.	4.
3	63.	653.	35468.	3.	53	700.	323.	14217.	5.
4	67.	667.	55787.	5.	54	714.	353.	12078.	6.
5	81.	840.	33260.	5.	55	735.	313.	12741.	8.
6	118.	549.	51411.	4.	56	741.	564.	16674.	3.
7	119.	389.	15300.	6.	57	744.	330.	4405.	4.
8	136.	884.	32648.	5.	58	762.	455.	8857.	4.
9	148.	956.	120030.	4.	59	776.	343.	5830.	4.
10	161.	672.	67110.	4.	60	782.	327.	8142.	5.
11	164.	827.	68773.	3.	61	790.	416.	14757.	5.
12	173.	767.	69621.	4.	62	792.	528.	12063.	3.
13	186.	766.	73136.	4.	63	794.	430.	20323.	5.
14	201.	714.	96448.	3.	64	799.	252.	19071.	2.
15	205.	675.	57598.	3.	65	807.	540.	24995.	4.
16	207.	631.	34286.	2.	66	816.	603.	25718.	4.
17	214.	714.	67797.	2.	67	828.	361.	25649.	6.
18	227.	674.	85373.	2.	68	840.	501.	14305.	7.
19	230.	818.	66701.	3.	69	864.	386.	15236.	4.
20	237.	950.	61602.	4.	70	868.	450.	16001.	5.
21	241.	768.	47650.	3.	71	873.	485.	11509.	6.
22	262.	418.	52098.	1.	72	894.	448.	16307.	4.
23	273.	689.	68609.	1.	73	906.	460.	17915.	5.
24	290.	250.	30236.	2.	74	911.	411.	5375.	6.
25	306.	297.	31869.	2.	75	917.	303.	11300.	4.
26	315.	515.	58703.	5.	76	933.	382.	16051.	5.
27	340.	193.	18485.	4.	77	972.	232.	21765.	1.
28	348.	468.	54212.	5.	78	978.	446.	24772.	6.
29	360.	470.	28224.	3.	79	1005.	406.	20075.	3.
30	375.	419.	31998.	5.	80	1018.	404.	14985.	4.
31	386.	276.	32667.	4.	81	1043.	531.	8987.	4.
32	396.	298.	52318.	1.	82	1070.	307.	7936.	1.
33	400.	474.	27444.	6.	83	1071.	549.	5813.	5.
34	405.	591.	38353.	3.	84	1083.	560.	31908.	5.
35	410.	391.	27779.	4.	85	1102.	640.	19518.	4.
36	477.	379.	29803.	6.	86	1103.	526.	21607.	6.
37	490.	433.	27809.	4.	87	1114.	334.	18105.	5.
38	494.	463.	33673.	6.	88	1148.	525.	18718.	4.
39	509.	346.	27390.	4.	89	1149.	438.	19670.	4.
40	518.	172.	32458.	2.	90	1152.	412.	11171.	7.
41	534.	328.	25329.	3.	91	1177.	513.	19315.	3.
42	551.	439.	44988.	2.	92	1222.	365.	20390.	1.
43	585.	304.	50514.	3.	93	1239.	278.	13326.	3.
44	599.	366.	34509.	4.	94	1259.	567.	16406.	5.
45	612.	231.	41959.	4.	95	1271.	405.	6494.	3.
46	614.	397.	46125.	4.	96	1273.	396.	6798.	4.
47	615.	457.	57483.	3.	97	1310.	325.	28180.	4.
48	616.	484.	27071.	6.	98	1319.	289.	10349.	6.
49	669.	499.	6892.	4.	99	1337.	553.	14338.	4.
50	674.	582.	14408.	3.	100	1345.	284.	12639.	6.

The following summary statistics should be helpful to you in solving this problem.

$r^2 = .42$

Source	df	SS	MS	F
Regression	2	20,089,496,467.43	10044748233.7	35.5
Error	97	27,446,761,165.32	282956300.7	

$b_0 = 13115.43$
$b_1 = 73.47$ (slope for HCOST)
$b_2 = -4544.26$ (slope for PEPLE)
$s(b_1) = 9.83$
$s(b_2) = 1131.71$

	HCOST	INCOM	PEPLE
HCOST	1.00	.572	.017
INCOM		1.00	−.300
PEPLE			1.00

12–9 Ms. Arlene Davis, director of the Stat City Hospital, believes that there may be a relationship between the average weekly supermarket bill per person (EATPL is a measure of nutrition) and the average number of trips a family makes to the hospital per year (HOSP), as of January 1980. A random sample of 10 Stat City families was drawn and appears below.

HOSP	EATPL
1.	23.
1.	22.
3.	22.
1.	23.
4.	18.
2.	23.
1.	29.
6.	23.
4.	21.
1.	26.

a. Set up a scatter diagram.
b. Assuming a linear relationship exists between EATPL and HOSP, compute the slope and y-intercept of EATPL(y) on HOSP(x).
c. Forecast EATPL for families that make five trips to the hospital per year.
d. Compute the "90 percent confidence" interval for the mean value of EATPL given HOSP = 5.
e. Compute the "90 percent confidence" interval for an individual forecasted value of EATPL given HOSP = 5.
f. Compute the coefficient of determination.

12–10 The mayor of Stat City believes that there is a relationship between HCOST and HEAT, as of January 1980. A random sample of 10 Stat City families was drawn and appears below.

HCOST	HEAT
888.	1309.
386.	752.
546.	624.
327.	851.
620.	657.
462.	475.
403.	391.
467.	413.
268.	229.
350.	481.

a. Set up a scatter diagram.

b. Assuming that a linear relationship exists between HCOST and HEAT, compute the slope and y-intercept of HCOST(y) on HEAT(x).

c. Forecast HCOST for a family whose heating bill is $500 per year.

d. Compute a "90 percent confidence" interval for the mean value of HCOST given HEAT = $500.

e. Compute a "90 percent confidence" interval for an individual forecasted value of HCOST given HEAT = $500.

f. Compute the coefficient of determination.

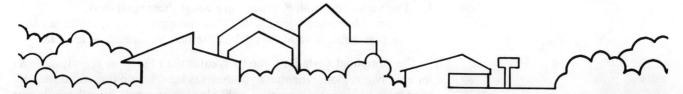

THE ANALYSIS OF VARIANCE

13

Hypotheses concerning the difference between two groups' means or proportions was introduced and explained in depth in Chapter 10. In this chapter you will be testing hypotheses concerning the differences among three or more groups' means. The name of the statistical technique used to test hypotheses concerning the differences among three or more groups' means is "analysis of variance."

The null and alternative hypotheses necessary to test if there are statistically significant differences among K groups' means are:

$$H_0: \mu_1 = \mu_2 = \ldots = \mu_K$$
$$H_A: \text{Not all the means are equal, where}$$
$$\mu_i = \text{the mean of group } i.$$

The statistics which must be computed to test the above null hypothesis are detailed in Table 13–1.

Table 13–1

Source	Degrees of freedom	Sum of squares	Mean squares	F
Between group variation	$K - 1$	SS_B	$SS_B/(K - 1)$	MS_B/MS_w
Within group variation	$n_T - K$	SS_w	$SS_w/(n_T - K)$	
Total variation	$n_T - 1$	SS_T		

where

K = the number of groups
n_T = the total sample size
n_j = the sample size for group j
y_{ij} = the ith observation in group j

$$SS_T = \Sigma_{j=1}^{K}\Sigma_{i=1}^{n_j}(y_{ij}^2) - \frac{(\Sigma_{j=1}^{K}\Sigma_{i=1}^{n_j}y_{ij})^2}{n_T}$$

$$SS_B = \Sigma_{j=1}^{K}\left[\frac{(\Sigma_{i=1}^{n_j}y_{ij})^2}{n_j}\right] - \frac{(\Sigma_{j=1}^{K}\Sigma_{i=1}^{n_j}y_{ij})^2}{n_T}$$

$$SS_w = \Sigma_{j=1}^{K}\Sigma_{i=1}^{n_j}(y_{ij}^2) - \Sigma_{j=1}^{K}\left[\frac{(\Sigma_{i=1}^{n_j}y_{ij})^2}{n_j}\right]$$

If the computed F-statistic is greater than $F(1 - \alpha, K - 1, n_T - K)$, reject the null hypothesis.

The computational formulas presented in your textbook may be different from the formulas presented above; either set of formulas will lead you to compute the same F-statistic.

The assumptions necessary to perform a one-way analysis of variance (fixed model) are:

1. The process being studied is in control; that is, it is repeatable.
2. The probability distribution being sampled is normal.
3. The variances in all K groups are equal (homogeneous).
4. The null hypothesis is tenable. This assumption is important because the F-statistic is valid only when the null hypothesis is tenable.

The computed F-statistic can be greater than $F(1 - \alpha, K - 1, n_T - K)$ for only two reasons; the null hypothesis is tenable and the sample results happened by chance (however unlikely), *or* the null hypothesis is not tenable. The above alternatives imply that the null hypothesis will be rejected 100α percent of the time when, in fact, it is tenable. This error (the type I error) represents the chance that the sample under study will lead to rejecting the null hypothesis when it is tenable.

Understanding hypothesis tests concerning three or more groups' means is more difficult than a person without formal training in statistics would suspect. Consequently, it is important that the footnote which appears in Exhibit 13–1 (or some version thereof) appears whenever a hypothesis test concerning three or more means is reported in a

Exhibit 13–1

Footnote

[5] The sample estimators permit us to perform a hypothesis test (for a given sample size).

To illustrate, if all possible samples were selected and each of these were surveyed under essentially the same conditions, then:

a. Approximately nine tenths of the hypothesis tests conducted at the "90 percent level of confidence" (or 10 percent level of significance) would fail to reject the null hypothesis when it was tenable (or 10 percent would reject the null hypothesis when it was tenable).

b. Approximately 19/20 of the hypothesis test conducted at the "95 percent level of confidence" (or 5 percent level of significance) would fail to reject the null hypothesis when it was tenable (or 5 percent would reject the null hypothesis when it was tenable).

Once the level of confidence (or significance) has been set, the chances of failing to reject the null hypothesis when it is false must be computed. If these probabilities are unacceptably high (type II errors), then corrective action must be taken.

memorandum. The footnote in Exhibit 13–1 serves as a partial disclaimer for the errors that can be encountered in hypothesis testing.

There are many different types of analysis of variance (ANOVA) procedures. Discussing them all is the subject for an entire course. The only ANOVA model that will be discussed in this chapter is the model already mentioned: the one-way ANOVA model with fixed factor levels.

A final point: do not be confused by the phrase "analysis of variance." Although the phrase mentions variance and not means, ANOVA is still a technique for testing if significant differences exist among group means.

All of the following examples and problems deal with testing for significant differences among three or more groups' means. Pay particular attention to the assumptions stated earlier when conducting your analysis.

Example 13–1 Redefining the tax assessment policy for Stat City

Mr. Lee Kaplowitz, the mayor of Stat City, is considering changing the property tax structure of Stat City from one tax assessment rate for all Stat City families to separate tax assessment rates for families in each zone in Stat City. The mayor will enact the above change in tax assessment rates if there is a statistically significant difference in the 1979 average family incomes among the four zones in Stat City at the 5 percent level of significance.

You have been retained by the mayor to conduct a survey and determine if there is a statistically significant difference in the 1979 average family incomes among the four zones in Stat City. Sufficient monies have been budgeted for you to draw a simple random sample of 130 Stat City dwelling units (see Table 13–2 for the random sample). Conduct the survey and type a memorandum to the mayor of Stat City reporting your findings (see Exhibit 13–2).

Table 13–2

CASE-N	ID	ZONE	INCOM	CASE-N	ID	ZONE	INCOM
1	2.	1.	23280.	27	316.	3.	28784.
2	7.	1.	19469.	28	321.	3.	142251.
3	22.	1.	56263.	29	349.	3.	39124.
4	27.	1.	52135.	30	350.	3.	23889.
5	55.	1.	33910.	31	355.	3.	23584.
6	58.	1.	30830.	32	374.	3.	25648.
7	61.	1.	47583.	33	375.	3.	31998.
8	81.	1.	33260.	34	379.	3.	31622.
9	91.	1.	44914.	35	386.	3.	32667.
10	94.	1.	36165.	36	394.	3.	29338.
11	144.	2.	58321.	37	437.	3.	15515.
12	152.	2.	83484.	38	440.	3.	21729.
13	175.	2.	76434.	39	446.	3.	30436.
14	178.	2.	91058.	40	457.	3.	26552.
15	199.	2.	95761.	41	459.	3.	42806.
16	248.	2.	83195.	42	462.	3.	42231.
17	256.	2.	121005.	43	474.	3.	30138.
18	257.	2.	64797.	44	495.	3.	34054.
19	260.	2.	78070.	45	510.	3.	32901.
20	262.	2.	52098.	46	515.	3.	33009.
21	263.	2.	70732.	47	520.	3.	39430.
22	268.	2.	86760.	48	539.	3.	49245.
23	274.	2.	57519.	49	542.	3.	40026.
24	284.	2.	68773.	50	569.	3.	39785.
25	297.	3.	33745.	51	575.	3.	48159.
26	306.	3.	31869.	52	582.	3.	33408.

Table 13–2 (*continued*)

CASE-N	ID	ZONE	INCOM	CASE-N	ID	ZONE	INCOM
53	590.	3.	44993.	92	992.	4.	33790.
54	592.	3.	54468.	93	998.	4.	21591.
55	593.	3.	49830.	94	1014.	4.	24059.
56	594.	3.	36794.	95	1029.	4.	12212.
57	599.	3.	34509.	96	1036.	4.	16847.
58	620.	3.	24156.	97	1045.	4.	9780.
59	626.	4.	14081.	98	1053.	4.	12661.
60	631.	4.	12037.	99	1095.	4.	27968.
61	634.	4.	17987.	100	1098.	4.	22100.
62	644.	4.	16730.	101	1107.	4.	31397.
63	647.	4.	23346.	102	1112.	4.	22603.
64	649.	4.	18425.	103	1127.	4.	18341.
65	666.	4.	27030.	104	1133.	4.	20791.
66	671.	4.	15948.	105	1148.	4.	18718.
67	679.	4.	15597.	106	1159.	4.	18141.
68	680.	4.	14055.	107	1163.	4.	19334.
69	720.	4.	16777.	108	1164.	4.	7369.
70	723.	4.	13377.	109	1168.	4.	8998.
71	725.	4.	15310.	110	1171.	4.	7406.
72	820.	4.	8473.	111	1176.	4.	15981.
73	826.	4.	23257.	112	1179.	4.	13622.
74	830.	4.	25456.	113	1181.	4.	18194.
75	835.	4.	13661.	114	1182.	4.	15918.
76	836.	4.	14572.	115	1191.	4.	7520.
77	842.	4.	15166.	116	1213.	4.	21847.
78	847.	4.	15130.	117	1214.	4.	13461.
79	851.	4.	12642.	118	1217.	4.	20175.
80	862.	4.	17522.	119	1224.	4.	25525.
81	873.	4.	11509.	120	1225.	4.	26071.
82	874.	4.	11410.	121	1234.	4.	17580.
83	895.	4.	11960.	122	1242.	4.	24743.
84	900.	4.	14657.	123	1245.	4.	22413.
85	911.	4.	5375.	124	1250.	4.	10145.
86	912.	4.	16601.	125	1265.	4.	23745.
87	922.	4.	9648.	126	1301.	4.	20073.
88	931.	4.	8987.	127	1306.	4.	29034.
89	955.	4.	8055.	128	1317.	4.	26335.
90	959.	4.	24312.	129	1365.	4.	9715.
91	980.	4.	22087.	130	1366.	4.	8967.

Solution　Before any data are collected the hypothesis under investigation must be stated. The hypothesis of concern and its alternative are listed below:

$$H_0: \ \mu_1 = \mu_2 = \mu_3 = \mu_4$$
$$H_A: \ \text{Not all the means are equal}$$

where

μ_1 = the average 1979 family income in Zone 1,
μ_2 = the average 1979 family income in Zone 2,
μ_3 = the average 1979 family income in Zone 3, and
μ_3 = the average 1979 family income in Zone 4.

The desired level of significance for the above hypothesis test is 5 percent. In layman's language, if the mayor had to revise the tax assessment 100 times he would be willing to change the tax assessment policy erroneously 5 times.

The computations required to test the null hypothesis are shown below and in Table 13–3.

Table 13–3

	Zone 1	Zone 2	Zone 3	Zone 4
n_j	10	14	34	72
$\sum_{i=1}^{n_j} y_{ij}$	377,089	1,088,007	1,278,693	1,236,350
$\sum_{i=1}^{n_j} (y_{ij}^2)$	15,600,504,576	88,814,668,800	61,887,816,704	24,117,963,520

$(\Sigma\Sigma y_{ij})^2 = (3,980,859)^2$

$\Sigma\Sigma(y_{ij}^2) = 190,420,951,040$

$$SS_T = (190,420,951,040) - \left(\frac{3,980,859^2}{130}\right) = 68,519,129,600.0000$$

$$SS_B = \left[\frac{377,809^2}{10} + \frac{1,088,007^2}{14} + \frac{1,278,693^2}{34} + \frac{1,236,350^2}{72}\right] - \left[\frac{3,980,859^2}{130}\right]$$
$$= 46,246,254,737.7861$$

$$SS_w = [190,420,951,040] - \left[\frac{377,809^2}{10} + \frac{1,088,007^2}{14} + \frac{1,278,693^2}{34} + \frac{1,236,350^2}{72}\right]$$
$$= 176,768,848.0000$$

The ANOVA table is shown in Table 13–4. The computed F-statistic (87.207) exceeds the critical F-value ($F_{(1-\alpha=.95,3,126)} = 2.67$). This indicates that a statistically significant proportion of the variation in income is explained by zone-to-zone variability, as opposed to variation inside the zones. The above statistics lead to rejecting the null hypothesis. Rejecting the null hypothesis indicates that there are important zone-by-zone differentials in income.

Table 13–4

Source	Degrees of freedom	Sum of squares	Mean squares	F-ratio
Between groups	3	46,246,254,737.7881	15,415,418,240.0000	87.207
Within groups	126	22,272,875,024.0000	176,768,848.0000	
Total	129	68,519,129,600.0000		

The memorandum appears in Exhibit 13–2.

Exhibit 13–2

HOWARD S. GITLOW, PH.D.
STATISTICAL CONSULTANT

MEMORANDUM

TO: Mr. Lee Kaplowitz
        Mayor, Stat City

FROM: Howard Gitlow

DATE: October 20, 1980

RE: Redefining the tax assessment policy for Stat City

In accordance with our consulting contract and using commonly accepted statistical techniques, I have computed several summary measures concerning the 1979 average family income in each zone of Stat City.[1] My research indicates that the average 1979 family income for each zone in Stat City is as follows:[2]

Zone	Average 1979 family income
1	$37,780.90
2	$77,714.79
3	$37,608.62
4	$17,171.53
Overall	$30,621.99

Allowing for the errors in survey estimates, you can have "95 percent confidence" that there is a statistically

Exhibit 13–2 (*continued*)

significant difference in the average 1979 family incomes among the four residential zones in Stat City.[3]

I hope the above information meets your needs. If you have any questions, please do not hesitate to call me at 305–999–9999.

Footnotes

[1] Insert standard footnote 1 (insert a sample of 130 drawn via simple random sampling).
[2] Insert standard footnote 2.
[3] Insert standard footnote 5.

Example 13–2   Promotion of the "eat at home" concept

The Stat City Supermarket Association is trying to decide if they should use the same point-of-purchase displays in all three supermarkets in Stat City to promote the concept of "eating at home." The association has decided that the critical factor in making the above decision is whether or not the average weekly supermarket bills per family (as of January 1980) are the same in the three supermarkets. If the average weekly supermarket bills per family are not the same in the three supermarkets, the association will set up different point-of-purchase displays (promoting the "eat at home" concept) in each store.

You have been retained by the association to conduct a survey to determine if the average weekly supermarket bills per family are significantly different among the three Stat City supermarkets, at the 5 percent level of significance. The association has advanced funds for you to draw a sample of 120 Stat City families. Conduct the necessary survey and type a memorandum to the Stat City Supermarket Association reporting your findings (see Exhibit 13–3).

Use Table 13–5 as a simple random sample of 120 Stat City families.

Table 13–5

CASE-N	ID	EAT	FEAT	CASE-N	ID	EAT	FEAT
1	5.	52.	3.	71	722.	77.	2.
2	11.	67.	3.	72	740.	75.	2.
3	12.	122.	3.	73	743.	85.	2.
4	18.	101.	3.	74	759.	73.	2.
5	39.	68.	2.	75	762.	97.	2.
6	65.	69.	1.	76	765.	154.	2.
7	85.	51.	3.	77	772.	71.	2.
8	37.	107.	2.	78	878.	50.	2.
9	96.	58.	3.	79	879.	101.	2.
10	99.	84.	3.	80	895.	161.	2.
11	110.	126.	1.	81	902.	106.	2.
12	139.	67.	3.	82	914.	70.	2.
13	172.	95.	3.	83	917.	69.	2.
14	178.	62.	1.	84	918.	70.	2.
15	187.	139.	1.	85	934.	58.	2.
16	190.	52.	3.	86	945.	29.	2.
17	199.	76.	3.	87	950.	89.	2.
18	202.	156.	1.	88	969.	44.	3.
19	214.	43.	3.	89	973.	74.	2.
20	219.	96.	3.	90	987.	83.	2.
21	222.	45.	1.	91	990.	73.	2.
22	225.	95.	2.	92	991.	88.	2.
23	231.	80.	3.	93	999.	127.	2.
24	236.	74.	1.	94	1017.	46.	2.
25	257.	58.	1.	95	1024.	59.	2.
26	273.	22.	3.	96	1036.	182.	2.
27	277.	70.	1.	97	1038.	149.	2.
28	290.	48.	2.	98	1066.	95.	2.
29	292.	73.	2.	99	1080.	132.	2.
30	301.	139.	1.	100	1088.	125.	2.
31	307.	100.	2.	101	1108.	43.	2.
32	310.	71.	1.	102	1121.	114.	2.
33	314.	153.	1.	103	1127.	73.	2.
34	321.	34.	1.	104	1128.	114.	2.
35	323.	74.	1.	105	1146.	25.	2.
36	326.	98.	1.	106	1168.	186.	2.
37	335.	102.	2.	107	1169.	25.	2.
38	374.	109.	2.	108	1196.	81.	2.
39	378.	63.	1.	109	1231.	68.	2.
40	389.	83.	2.	110	1237.	50.	3.
41	392.	88.	1.	111	1257.	80.	2.
42	395.	130.	1.	112	1262.	33.	2.
43	398.	146.	2.	113	1268.	80.	2.
44	412.	108.	1.	114	1275.	83.	1.
45	413.	76.	2.	115	1297.	109.	2.
46	421.	101.	2.	116	1324.	39.	2.
47	425.	53.	1.	117	1337.	101.	2.
48	428.	94.	1.	118	1340.	85.	2.
49	482.	87.	2.	119	1342.	132.	2.
50	484.	51.	3.	120	1361.	72.	2.
51	488.	120.	1.				
52	491.	51.	1.				
53	531.	97.	2.				
54	532.	127.	1.				
55	541.	93.	1.				
56	543.	118.	2.				
57	554.	61.	1.				
58	586.	110.	1.				
59	603.	76.	2.				
60	615.	62.	2.				
61	617.	75.	1.				
62	628.	93.	2.				
63	640.	67.	2.				
64	680.	83.	2.				
65	697.	90.	2.				
66	703.	65.	2.				
67	705.	74.	2.				
68	711.	143.	2.				
69	716.	128.	3.				
70	718.	130.	2.				

Solution   Remember, before any data are collected the hypothesis under investigation must be stated. The null and alternative hypotheses are listed below:

$$H_0: \quad \mu_1 = \mu_2 = \mu_3$$
$$H_A: \quad \text{Not all the means are equal}$$

where

$\mu_1 = $ the average weekly supermarket bill, as of January 1980, for families favoring Food Fair,

μ_2 = the average weekly supermarket bill, as of January 1980, for families favoring Grand Union, and

μ_3 = the average weekly supermarket bill, as of January 1980, for families favoring A&P.

The desired level of significance for the above hypothesis test is 5 percent. To restate, if the association had to make the point-of-promotion format decision 100 times, it would be willing to pay for the three-pronged point-of-sale promotion when it was unnecessary 5 times.

The computations required to test the null hypothesis are shown below and in Table 13–6.

Table 13–6

Statistics	Supermarket		
	Food Fair	Grand Union	A&P
n_j	29	72	19
$\sum\limits_{i=1}^{n_j} y_{ij}$	2629.	6439.	1349.
$\sum\limits_{i=1}^{n_j} (y_{ij}^2)$	270,501.	658,295.	109,783.

$$(\Sigma\Sigma y_{ij})^2 = 10,417^2$$
$$(\Sigma\Sigma y_{ij}^2) = 1,038,579.$$

$$SS_T = (1,038,579.) - \frac{10,417^2}{120} = 134,296.58$$

$$SS_B = \left[\frac{2,629^2}{29} + \frac{6,439^2}{72} + \frac{1,349^2}{19}\right] - \left[\frac{10,417^2}{120}\right] = 5,672.39$$

$$SS_W = (1,038,579) - \left[\frac{2,629^2}{29} + \frac{6,439^2}{72} + \frac{1,349^2}{19}\right] = 128,624.19$$

The ANOVA table is shown in Table 13–7.

Table 13–7

Source	SS	df	MS	F
Between	5,672.39	2	2836.20	2.58
Within	128,624.19	117	1099.35	
Total	134,296.58	119		

The computed F-statistic (2.58) is less than the critical F-value ($F_{[1-\alpha=.95,2,117]} = 3.08$) and the null hypothesis cannot be rejected. This indicates that a statistically insignificant proportion of the variation in average weekly supermarket bills is explained by supermarket-to-supermarket variability. Consequently, the association, in accordance with their decision criteria, should continue the standard "point-of-purchase" display for all supermarkets.

The memorandum appears in Exhibit 13–3.

Exhibit 13–3

HOWARD S. GITLOW, PH.D.
STATISTICAL CONSULTANT
———

MEMORANDUM

TO: Stat City Supermarket Association

FROM: Howard Gitlow

DATE: October 23, 1980

RE: Promoting the "eat at home" concept

As per your request and using commonly accepted
statistical techniques, I have computed several summary
measures concerning the average weekly supermarket bills
for Stat City families, as of January 1980.[1] My research
indicates that the average weekly supermarket bills per
family among Stat City's three supermarkets are as
follows.[2]

Supermarket	Average weekly supermarket bill per family as of January 1980
Food Fair	$90.66
Grand Union	$89.43
A&P	$71.00
Overall	$86.81

Allowing for the errors in survey estimates, you can have "95
percent confidence" that there is no statistically

SUMMARY

In this chapter you learned how to extend hypothesis tests concerning
means from two groups (populations) to three or more groups (populations). In this chapter the F-distribution was introduced for the first time
and you were made aware of the severe limitations as to the choices of α
when using an F-statistic. These limitations occur because F-statistics are
usually tabulated for alphas of 0.01, 0.05, and 0.10. However, computer
programs and hand calculators are now available that will compute
F-statistics for any value of α.

This chapter will not deal with computing the power curve or sample
size requirements for a one-way analysis of variance. These topics are
considered beyond the scope of this book.

ADDITIONAL PROBLEMS

**13–3 Special promotional campaign to boost Stat City gas station
retail sales**   Mr. Paul Sugrue, chairman of the Stat City Gas Station Association, needs to determine if the average number of trips to the gas
station per month are the same for families who perform their own automotive repairs, have their repairs performed by a gas station, or have
their repairs performed by a dealer (as of January 1980). If the average
number of trips to the gas station per month are significantly different

The analysis of variance

among the above types of families, the association will pay for a special advertising campaign to boost the average number of trips to the gas station per month for the category of family that is the lowest.

You have been retained by Mr. Sugrue to conduct the necessary survey at the 10 percent level of significance. He has advanced sufficient funds to draw a sample of 100 Stat City families. Perform the necessary survey and type a memorandum to Mr. Sugrue reporting your findings.

Use the accompanying simple random sample of 100 Stat City families.

Random sample for Problem 13–3

CASE-N	ID	GASTR	REPAR	CASE-N	ID	GASTR	REPAR
1	8.	3.	2.	51	766.	6.	0.
2	18.	5.	1.	52	768.	5.	0.
3	41.	8.	1.	53	775.	3.	1.
4	42.	4.	2.	54	777.	6.	1.
5	78.	4.	2.	55	779.	5.	1.
6	90.	3.	2.	56	784.	4.	1.
7	95.	6.	2.	57	791.	6.	0.
8	117.	5.	1.	58	848.	7.	0.
9	121.	4.	1.	59	857.	4.	0.
10	128.	4.	2.	60	873.	7.	1.
11	146.	4.	2.	61	872.	5.	0.
12	168.	4.	2.	62	916.	7.	1.
13	182.	6.	2.	63	930.	5.	1.
14	187.	4.	2.	64	957.	4.	2.
15	188.	6.	2.	65	965.	6.	1.
16	218.	6.	2.	66	985.	6.	1.
17	238.	5.	1.	67	1027.	4.	2.
18	253.	9.	2.	68	1030.	4.	0.
19	298.	5.	2.	69	1035.	12.	0.
20	312.	8.	1.	70	1039.	3.	0.
21	321.	2.	2.	71	1042.	6.	0.
22	326.	7.	2.	72	1043.	7.	1.
23	371.	6.	0.	73	1047.	6.	1.
24	373.	7.	0.	74	1055.	5.	1.
25	405.	6.	1.	75	1057.	3.	0.
26	432.	3.	1.	76	1059.	6.	0.
27	435.	4.	0.	77	1072.	2.	1.
28	448.	5.	2.	78	1086.	5.	1.
29	468.	6.	1.	79	1091.	9.	0.
30	475.	10.	2.	80	1124.	4.	0.
31	479.	6.	1.	81	1125.	9.	1.
32	499.	8.	0.	82	1147.	7.	0.
33	502.	7.	1.	83	1151.	4.	0.
34	519.	5.	1.	84	1167.	9.	1.
35	548.	4.	0.	85	1188.	3.	1.
36	573.	10.	2.	86	1217.	5.	2.
37	585.	5.	2.	87	1244.	2.	1.
38	596.	3.	1.	88	1262.	5.	1.
39	601.	5.	1.	89	1263.	6.	1.
40	637.	6.	1.	90	1272.	5.	0.
41	647.	8.	2.	91	1291.	5.	1.
42	652.	9.	2.	92	1328.	5.	1.
43	669.	7.	0.	93	1331.	3.	1.
44	671.	9.	0.	94	1334.	5.	0.
45	690.	6.	0.	95	1339.	5.	0.
46	711.	6.	0.	96	1340.	4.	1.
47	718.	6.	1.	97	1349.	4.	2.
48	721.	7.	1.	98	1358.	6.	1.
49	744.	3.	0.	99	1363.	7.	1.
50	748.	5.	0.	100	1367.	2.	0.

13–4 Survey to determine zoning policies Mr. Lee Kaplowitz, the mayor of Stat City, needs to determine if the average number of rooms per dwelling unit are the same in each residential housing zone in Stat City, as of January 1980. If the average number of rooms per dwelling unit differ among zones, the mayor will set up separate zoning policies controlling room size for each housing zone.

You have been retained by the mayor to conduct the necessary survey at the 5 percent level of significance. Funds are available to draw a sample of 100 Stat City families. Perform the necessary survey and report your findings to the mayor.

Use the accompanying simple random sample of Stat City families.

Random sample for Problem 13-4

CASE-N	ID	ZONE	ROOMS	CASE-N	ID	ZONE	ROOMS
1	35.	1.	10.	71	1066.	4.	7.
2	39.	1.	8.	72	1071.	4.	6.
3	66.	1.	9.	73	1077.	4.	5.
4	74.	1.	8.	74	1084.	4.	5.
5	85.	1.	7.	75	1097.	4.	4.
6	87.	1.	8.	76	1100.	4.	6.
7	102.	1.	7.	77	1104.	4.	7.
8	106.	1.	9.	78	1114.	4.	6.
9	119.	1.	6.	79	1131.	4.	5.
10	126.	1.	5.	80	1140.	4.	6.
11	149.	2.	8.	81	1155.	4.	7.
12	169.	2.	9.	82	1169.	4.	4.
13	187.	2.	10.	83	1177.	4.	7.
14	193.	2.	9.	84	1186.	4.	6.
15	217.	2.	5.	85	1210.	4.	8.
16	236.	2.	9.	86	1228.	4.	6.
17	237.	2.	8.	87	1247.	4.	6.
18	243.	2.	8.	88	1256.	4.	4.
19	247.	2.	7.	89	1259.	4.	7.
20	297.	3.	10.	90	1267.	4.	5.
21	307.	3.	10.	91	1292.	4.	6.
22	325.	3.	7.	92	1296.	4.	5.
23	330.	3.	7.	93	1303.	4.	6.
24	365.	3.	10.	94	1308.	4.	6.
25	375.	3.	10.	95	1311.	4.	7.
26	421.	3.	6.	96	1329.	4.	6.
27	440.	3.	8.	97	1334.	4.	5.
28	471.	3.	6.	98	1340.	4.	7.
29	481.	3.	7.	99	1342.	4.	6.
30	494.	3.	6.	100	1347.	4.	7.
31	498.	3.	6.				
32	499.	3.	6.				
33	510.	3.	6.				
34	528.	3.	8.				
35	530.	3.	7.				
36	533.	3.	10.				
37	571.	3.	6.				
38	574.	3.	7.				
39	598.	3.	9.				
40	606.	3.	8.				
41	610.	3.	7.				
42	613.	3.	8.				
43	620.	3.	4.				
44	646.	4.	9.				
45	676.	4.	7.				
46	691.	4.	4.				
47	706.	4.	7.				
48	715.	4.	3.				
49	718.	4.	6.				
50	720.	4.	5.				
51	742.	4.	6.				
52	759.	4.	8.				
53	760.	4.	5.				
54	765.	4.	5.				
55	770.	4.	7.				
56	771.	4.	6.				
57	795.	4.	5.				
58	838.	4.	5.				
59	869.	4.	6.				
60	870.	4.	6.				
61	882.	4.	6.				
62	890.	4.	5.				
63	891.	4.	4.				
64	920.	4.	3.				
65	953.	4.	6.				
66	955.	4.	5.				
67	979.	4.	5.				
68	1008.	4.	7.				
69	1020.	4.	8.				
70	1034.	4.	6.				

13-5 Survey concerning the number of people per dwelling unit Mr. Joseph Moder, division head of the Stat City Department of Public Works, needs to determine if there is a statistically significant difference in the average number of people per dwelling unit among the residential housing zones in Stat City, as of January 1980. You have been employed by Mr. Moder to conduct a survey of 90 Stat City Dwelling units to ascertain the required information. Conduct the survey and report your findings to Mr. Moder at the 5 percent level of significance.

Use the accompanying simple random sample of Stat City families.

Random sample for Problem 13–5

CASE-N	ID	ZONE	PEPLE	CASE-N	ID	ZONE	PEPLE
1	2.	1.	5.	71	1098.	4.	7.
2	15.	1.	4.	72	1101.	4.	5.
3	22.	1.	3.	73	1146.	4.	1.
4	34.	1.	7.	74	1149.	4.	4.
5	38.	1.	6.	75	1173.	4.	5.
6	95.	1.	1.	76	1175.	4.	5.
7	98.	1.	2.	77	1189.	4.	5.
8	125.	1.	4.	78	1196.	4.	4.
9	157.	2.	4.	79	1212.	4.	5.
10	165.	2.	6.	80	1222.	4.	1.
11	174.	2.	5.	81	1230.	4.	2.
12	185.	2.	2.	82	1234.	4.	3.
13	191.	2.	1.	83	1241.	4.	5.
14	211.	2.	3.	84	1274.	4.	4.
15	213.	2.	3.	85	1280.	4.	5.
16	222.	2.	2.	86	1294.	4.	9.
17	234.	2.	2.	87	1316.	4.	1.
18	241.	2.	3.	88	1325.	4.	4.
19	267.	2.	2.	89	1351.	4.	4.
20	292.	3.	4.	90	1372.	4.	5.
21	302.	3.	2.				
22	343.	3.	6.				
23	366.	3.	2.				
24	368.	3.	4.				
25	379.	3.	3.				
26	389.	3.	4.				
27	409.	3.	4.				
28	430.	3.	2.				
29	442.	3.	7.				
30	472.	3.	1.				
31	482.	3.	5.				
32	488.	3.	4.				
33	509.	3.	4.				
34	510.	3.	5.				
35	511.	3.	4.				
36	540.	3.	4.				
37	558.	3.	3.				
38	559.	3.	3.				
39	592.	3.	5.				
40	594.	3.	4.				
41	600.	3.	4.				
42	610.	3.	3.				
43	619.	3.	1.				
44	655.	4.	8.				
45	706.	4.	4.				
46	717.	4.	5.				
47	726.	4.	4.				
48	732.	4.	9.				
49	755.	4.	4.				
50	781.	4.	5.				
51	825.	4.	4.				
52	841.	4.	1.				
53	857.	4.	6.				
54	868.	4.	5.				
55	886.	4.	5.				
56	917.	4.	4.				
57	936.	4.	2.				
58	952.	4.	4.				
59	982.	4.	3.				
60	983.	4.	4.				
61	1012.	4.	5.				
62	1026.	4.	3.				
63	1028.	4.	1.				
64	1029.	4.	4.				
65	1037.	4.	5.				
66	1039.	4.	5.				
67	1064.	4.	5.				
68	1080.	4.	6.				
69	1091.	4.	4.				
70	1093.	4.	6.				

PART SIX

SOME NONPARAMETRIC STATISTICS

14

Nonparametric statistics is a term that describes statistical techniques: (1) that are not concerned with inferences about population parameters, and/or (2) whose mathematical justification for inferences is not based on the probability distribution of the population, and/or (3) that involve data which are inappropriate for classical (parametric) analysis.

You will recall that most classical (parametric) hypothesis tests are concerned with inferences about parameters (means, regression coefficients, variances, and so on). However, many nonparametric tests are not concerned with testing hypotheses about parameters (e.g., H_0: $\mu = 10$ or H_0: $\beta_1 = 0$); rather, they are concerned with testing hypotheses regarding the shape of a probability distribution or randomness in data. But, there are nonparametric procedures which are concerned with testing hypotheses about parameters (e.g., H_0: Median = 7).

Nonparametric tests generally make fewer and less stringent assumptions than do parametric tests. Consequently, nonparametric procedures are more widely applicable and afford conclusions that are more broad-based in scope than parametric procedures. However parametric procedures should be used as opposed to nonparametric procedures when all the assumptions upon which a parametric procedure is based are satisfied. For, parametric procedures generally become more powerful than nonparametric procedures when all their assumptions are met.

Finally, nonparametric procedures allow for the analysis of data that lack sufficient strength to perform meaningful arithmetic operations. For example, parametric procedures lack the facility to test hypotheses regarding the association between "favorite gas station" (FAVGA) and "housing zone" (ZONE): fortunately, nonparametric procedures have such facility.

The nonparametric procedures to be discussed in this chapter fall into three categories: procedures concerning inferences dealing with population distributions, procedures concerning inferences dealing with location or central tendency, and procedures concerning inference dealing with association. Procedures concerning inferences dealing with dispersion and randomness (or trend) will not be discussed. Unfortunately, there is no standard statistical notation for nonparametric tests. Consequently, formulas will be presented in this chapter. If the formulas in this book differ from your textbook, use the formulas presented in class. Both formulas will yield the same answer.

INTRODUCTION

Some nonparametric statistics

INFERENCES CONCERNING POPULATION DISTRIBUTIONS

In many statistical studies questions arise concerning the shape of one or more population probability distributions. For example, as of January 1980, is the distribution of average monthly electric bills in Stat City normal, or, are the distributions of average bimonthly automobile gasoline bills identical for families who favor Howie's Gulf Station and families who favor Paul's Texaco Station?

The first type of test is referred to as a one-sample, goodness-of-fit test. There are several nonparametric tests in this category; however, this text will only deal with the chi-square (χ^2) goodness-of-fit test. The test procedure is as follows:

STEP 1:   State the hypothesis to be tested.

H_0:   The probability distribution of the observed random variable is $F_0(x)$
H_A: The probability distribution of the observed random variable is different from $F_0(x)$

where $F_0(x)$ is a theoretical probability distribution (e.g., normal, uniform, binomial, Poisson, and so on).

STEP 2:   Set α.

STEP 3:   List r mutually exclusive categories in which to compare the observed and theoretical probability distributions.

STEP 4:   Select the test statistic.

$$Q = \sum_{i=1}^{r} \left[\frac{(f_i - e_i)^2}{e_i} \right]$$

where

r = the number of categories the data has been classified into,
f_i = the number of observations classified into the ith category, and
e_i = the number of observations expected to be classified into category i if the null hypothesis is tenable.

STEP 5:   Determine the sample size and draw the sample.

STEP 6:   Compute the expected frequencies (e_i) and Q.

STEP 7:   Determine if the null hypothesis should be rejected given the appropriate degrees of freedom ($r - w - 1$ for this test, where the w = the number of parameters that must be estimated from sample observations). The distribution of Q is χ^2_{r-w-1} only if the sample size is large (greater than 50) and all the expected cell frequencies are greater than 5.

Example 14–1   Distribution of the average monthly electric bills in Stat City

Mr. Saul Reisman, chief financial officer of the Stat City Electric Company, needs to know if the distribution of average monthly electric bills in Stat City, as of January 1980, is normal. He has provided you with sufficient funds to draw a random sample of 100 families. Conduct the necessary survey and report your findings to Mr. Reisman (see Exhibit 14–1).

Solution   STEP 1:   The hypothesis under investigation is:

H_0:   The probability distribution of average monthly electric bills in Stat City, as of January 1980, is normal
H_A: The probability distribution of average monthly electric bills in Stat City, as of January 1980, is not normal

STEP 2: $\alpha = 0.05$

STEP 3: The $r (= 6)$ categories are:

Under $20
$20 to less than $40
$40 to less than $60
$60 to less than $80
$80 to less than $100
$100 or more

STEP 4: The appropriate test statistic is Q.

STEP 5: $n = 100$ (We will not deal with sample size determination for χ^2 goodness-of-fit tests.) The sample is listed in Table 14–1. The sample mean is $55.85 and the sample standard deviation is $19.00.

Table 14–1

CASE-N	ID	ELEC	CASE-N	ID	ELEC
1	7.	30.	51	679.	54.
2	10.	80.	52	691.	32.
3	19.	84.	53	709.	35.
4	44.	88.	54	719.	20.
5	88.	42.	55	723.	69.
6	90.	40.	56	725.	45.
7	122.	86.	57	726.	59.
8	131.	57.	58	729.	49.
9	133.	67.	59	731.	47.
10	185.	40.	60	771.	44.
11	189.	67.	61	809.	63.
12	220.	100.	62	816.	57.
13	232.	89.	63	863.	36.
14	233.	64.	64	885.	39.
15	239.	45.	65	900.	50.
16	250.	82.	66	914.	40.
17	262.	40.	67	918.	27.
18	264.	90.	68	949.	39.
19	294.	64.	69	956.	45.
20	310.	56.	70	961.	63.
21	321.	54.	71	971.	70.
22	337.	109.	72	986.	57.
23	349.	90.	73	988.	50.
24	368.	78.	74	1023.	28.
25	409.	31.	75	1032.	40.
26	410.	51.	76	1035.	42.
27	412.	54.	77	1042.	46.
28	414.	75.	78	1049.	41.
29	417.	42.	79	1051.	63.
30	436.	32.	80	1084.	38.
31	464.	78.	81	1085.	37.
32	473.	72.	82	1092.	56.
33	494.	47.	83	1115.	52.
34	497.	42.	84	1128.	68.
35	520.	67.	85	1129.	57.
36	521.	80.	86	1136.	43.
37	528.	71.	87	1145.	59.
38	534.	56.	88	1153.	33.
39	573.	66.	89	1180.	71.
40	578.	92.	90	1211.	46.
41	587.	84.	91	1222.	36.
42	591.	62.	92	1239.	41.
43	602.	14.	93	1251.	46.
44	604.	65.	94	1283.	27.
45	611.	48.	95	1284.	52.
46	616.	84.	96	1295.	37.
47	624.	69.	97	1320.	46.
48	646.	82.	98	1322.	66.
49	653.	45.	99	1326.	44.
50	660.	68.	100	1344.	61.

STEP 6: Given the sample observations, the sample mean and standard deviation, and the r previously determined categories, the r expected frequencies can be computed. They are shown in Table 14–2. The computations for the expected probabilities of occurrence are shown below:

Table 14–2

Average monthly electric bills	Actual frequency	Expected probability of occurrence given H_0 is tenable	Expected frequency (probability $\times$ n)
Under $20	1	.0296	2.96
$20 to less than $40	17	.1725	17.25
$40 to less than $60	43	.3843	38.43
$60 to less than $80	24	.3117	31.17
$80 to less than $100	13	.0917	9.17
$100 or more	2	.0102	1.02
Total	100	1.0000	Sample size = n = 100

$$P(\text{ELEC} < \$20) \rightarrow Z = \frac{20 - 55.85}{19.00} = -1.89$$
$$P(Z < -1.89) = .0296$$

$P(\$20 \leq \text{ELEC} < \$40) \rightarrow$

$Z_1 = \dfrac{20 - 55.85}{19}$ $Z_1 = -1.89$ $P(Z_1 < -1.89) = .0296$	$Z_2 = \dfrac{40 - 55.85}{19}$ $Z_2 = -.83$ $P(Z_2 < -.83) = .2021$
$P(-1.89 \leq Z < -.83) = .2021 - .0296 = .1725$	

$P(\$40 \leq \text{ELEC} < \$60) \rightarrow$

$Z_1 = \dfrac{40 - 55.85}{19}$ $Z_1 = -.83$ $P(Z_1 < -.83) = .2021$	$Z_2 = \dfrac{60 - 55.85}{19}$ $Z_2 = +.22$ $P(Z_2 < +.22) = .5864$
$P(-.83 \leq Z < .22) = .5864 - .2021 = .3843$	

$P(\$60 \leq \text{ELEC} < \$80) \rightarrow$

$Z_1 = \dfrac{60 - 55.85}{19}$ $Z_1 = +.22$ $P(Z_1 < +.22) = .5864$	$Z_2 = \dfrac{80 - 55.85}{19}$ $Z_2 = +1.27$ $P(Z_2 < +1.27) = .8981$
$P(+.22 \leq Z < +1.27) = .8981 - .5864 = .3117$	

$P(\$80 \leq \text{ELEC} < \$100) \rightarrow$

$Z_1 = \dfrac{80 - 55.85}{19}$ $Z_1 = +1.27$ $P(Z_1 \leq +1.27) = .8981$	$Z_2 = \dfrac{100 - 55.85}{19}$ $Z_2 = 2.32$ $P(Z_2 < +2.32) = .9898$
$P(+1.27 \leq Z < +2.32) = .9898 - .8981 = .0917$	

$$P(\text{ELEC} \geq \$100) \rightarrow Z = \frac{100 - 55.85}{19} = 2.32$$
$$P(Z \geq 2.32) = .0102$$

Before Q can be computed, the categories with expected cell frequencies below five must be collapsed into other categories so that all remaining expected cell frequencies are five or more (see Table 14–3).

Table 14–3

Average monthly electric bill	Observed (f_i)	Expected (e_i)	$\dfrac{(f_i - e_i)^2}{e_i}$
Under $20 $20 to less than $40 under $40	1 17 }18	2.96 17.25 }20.21	0.24
$40 to less than $60	43	38.43	0.54
$60 to less than $80	24	31.17	1.65
$80 to less than $100 $80 or more .... $100 or more	13 2 }15	9.17 1.02 }10.19	2.27
Total	100	100	4.70

Now, Q can be computed.

$$Q = 4.70$$
$$df = 4 - 2 - 1 = 1$$

(where 2 refers to the fact that $\bar{x}$ and s had to be estimated to compute the expected probabilities).

STEP 7:

$$Q = 4.70$$
$$\chi^2_{(.95,1)} = 3.84$$

The null hypothesis is rejected at the 5 percent level of significance because Q is greater than $\chi^2_{(.95,1)} = 3.84$. Consequently, it is not reasonable to assume that the distribution of average monthly electric bills in Stat City, as of January 1980, is normal.

The memorandum appears in Exhibit 14–1.

Exhibit 14–1

HOWARD S. GITLOW, PH.D.
STATISTICAL CONSULTANT
_____

MEMORANDUM

TO: Mr. Saul Reisman, Chief Financial Officer
        Stat City Electric Company

FROM: Howard Gitlow

DATE: April 3, 1981

RE: Distribution of average monthly electric bills in
        Stat City, as of January 1980

   As per your request and using commonly accepted
statistical techniques, I have computed several summary
statistics concerning the distribution of average monthly
electric bills in Stat City, as of January 1980.[1] Allowing
for the errors in survey estimates, you can have "95 percent
confidence" that the distribution of average monthly
electric bills in Stat City is not normal (bell-shaped).[2]
   If you have any questions, please call me at
305-999-9999.

Exhibit 14–1 (*continued*)

Footnotes

[1] Insert standard footnote 1 (insert a sample of 100 Stat City families selected via simple random sampling).

[2] The particular sample used in this survey is one of a large number of all possible samples of the same size that could have been selected using the same sample design. The sample estimates derived from the different samples would differ from each other. The difference between the sample estimates is called sampling error.

The notion of sampling error permits a researcher to say that only 5 percent of the possible samples could produce sample statistics that indicate a normal distribution does not exist when, in fact, a normal distribution does exist. Consequently, if the sample statistic strongly indicates that the distribution is not normal, a researcher must reject the notion that the distribution is normal with the understanding that he/she may be wrong 5 percent of the time.

The second type of test is referred to as a two-sample, goodness-of-fit test. The test procedure is given here.

STEP 1:   State the hypothesis to be tested.

H_0:   The population distributions are identical

$$[F_1(x) = F_2(x) \text{ for all } x]$$

H_A: The population distributions differ in some way

$$[F_1(x) \neq F_2(x) \text{ for some } x]$$

where

$F_1(x)$ = the probability distribution of the first population
$F_2(x)$ = the probability distribution of the second population.

STEP 2:   Set α.

STEP 3:   List r mutually exclusive categories in which to compare the two probability distributions.

STEP 4:   Select the test statistic

$$Q = \sum^{r} \frac{(f_{i1} - e_{i1})^2}{e_{i1}} + \sum^{r} \frac{(f_{i2} - e_{i2})^2}{e_{i2}}$$

where

f_{i1} = the number of observations classified into category i from population 1,

f_{i2} = the number of observations classified into category i from population 2,

e_{i1} = the number of observations expected to be classified into category i from population 1 if the null hypothesis is tenable, and

e_{i2} = the number of observations expected to be classified into category i from population 2 if the null hypothesis is tenable.

STEP 5: Determine the sample size and draw the sample.

STEP 6: Compute the expected frequencies and Q.

STEP 7: Determine if the null hypothesis should be rejected given the appropriate degrees of freedom ($r - 1$ for this test).

Example 14–2   Survey of average bimonthly gasoline bills for Stat City service stations

Ms. Marsha Lubitz, manager of the Stat City Chapter of the American Automobile Association (AAA), needs to determine if the distributions of average bimonthly automobile gas bills, as of January 1980, are the same for Howie's Gulf Station and Paul's Texaco Station. She has provided you with sufficient funds to draw a random sample of 130 Stat City dwelling units. Conduct the necessary survey and report your findings to Ms. Lubitz at the 5 percent level of significance (see Exhibit 14–2).

STEP 1: The hypothesis under investigation is: **Solution**

H_0: The distributions of average bimonthly automobile gas bills, as of January 1980, are the same for families who favor Howie's Gulf Station and families who favor Paul's Texaco Station

H_A: The distributions of average bimonthly automobile gas bills, as of January 1980, are not the same for families who favor the two stations in question

STEP 2: $\alpha = 0.05$

STEP 3: The r categories are:

$50 to less than $100
$100 to less than $150
150 to less than $200
$200 or more

STEP 4: The appropriate test statistic is Q.

STEP 5: $n = 130$ (We will not deal with sample size determination for χ^2 goodness-of-fit tests.) The sample is shown in Table 14–4.

CASE-N	ID	GAS	FAVGA	CASE-N	ID	GAS	FAVGA	
1	9.	177.	1.	12	159.	203.	1.	**Table 14–4**
2	17.	149.	2.	13	163.	151.	1.	
3	25.	201.	2.	14	167.	211.	1.	
4	66.	188.	2.	15	169.	190.	2.	
5	77.	204.	2.	16	183.	144.	2.	
6	81.	193.	2.	17	188.	136.	2.	
7	93.	231.	2.	18	190.	125.	1.	
8	121.	194.	1.	19	197.	170.	1.	
9	136.	190.	2.	20	209.	148.	2.	
10	148.	156.	1.	21	215.	163.	2.	
11	154.	161.	2.	22	232.	206.	1.	

Table 14–4 (*continued*)

CASE-N	ID	GAS	FAVGA	CASE-N	ID	GAS	FAVGA
23	245.	89.	1.	77	753.	65.	1.
24	246.	121.	2.	78	779.	120.	2.
25	269.	65.	2.	79	790.	120.	2.
26	278.	209.	2.	80	793.	65.	2.
27	292.	203.	2.	81	838.	120.	2.
28	310.	141.	2.	82	872.	65.	1.
29	311.	155.	2.	83	876.	120.	2.
30	329.	221.	2.	84	877.	65.	2.
31	330.	175.	2.	85	885.	120.	1.
32	332.	151.	2.	86	890.	120.	1.
33	335.	199.	1.	87	905.	120.	2.
34	337.	120.	1.	88	915.	120.	2.
35	345.	122.	2.	89	925.	120.	2.
36	359.	65.	2.	90	927.	120.	2.
37	363.	186.	1.	91	950.	65.	2.
38	365.	198.	2.	92	952.	120.	1.
39	369.	177.	2.	93	976.	120.	2.
40	386.	160.	1.	94	984.	120.	2.
41	390.	220.	2.	95	1016.	185.	1.
42	404.	219.	2.	96	1030.	120.	2.
43	411.	158.	1.	97	1037.	120.	2.
44	412.	189.	2.	98	1071.	65.	2.
45	416.	174.	2.	99	1087.	120.	1.
46	432.	157.	2.	100	1088.	120.	1.
47	445.	148.	2.	101	1107.	158.	2.
48	448.	209.	2.	102	1111.	197.	1.
49	451.	172.	2.	103	1116.	165.	2.
50	466.	179.	1.	104	1138.	65.	2.
51	471.	234.	2.	105	1139.	120.	2.
52	478.	219.	2.	106	1148.	120.	1.
53	482.	200.	1.	107	1151.	120.	1.
54	489.	142.	2.	108	1159.	120.	1.
55	499.	120.	1.	109	1167.	120.	1.
56	511.	157.	1.	110	1168.	120.	2.
57	531.	176.	2.	111	1176.	65.	2.
58	541.	166.	1.	112	1182.	120.	1.
59	542.	230.	2.	113	1192.	65.	2.
60	551.	135.	2.	114	1209.	120.	2.
61	573.	145.	2.	115	1231.	65.	2.
62	574.	166.	2.	116	1239.	120.	1.
63	589.	165.	1.	117	1266.	140.	1.
64	592.	214.	2.	118	1281.	65.	2.
65	614.	181.	2.	119	1292.	65.	2.
66	622.	148.	1.	120	1298.	120.	2.
67	641.	191.	1.	121	1303.	120.	1.
68	643.	120.	1.	122	1306.	190.	1.
69	645.	65.	1.	123	1311.	65.	1.
70	660.	194.	1.	124	1329.	137.	2.
71	681.	120.	1.	125	1342.	120.	1.
72	682.	120.	2.	126	1345.	65.	2.
73	689.	65.	1.	127	1346.	165.	1.
74	703.	122.	1.	128	1357.	120.	1.
75	712.	120.	2.	129	1358.	65.	2.
76	722.	65.	2.	130	1371.	120.	2.

STEP 6:   Given the sample observations and the r previously defined categories, the expected frequencies are computed in Table 14–5.

Table 14–5

Average bimonthly automobile gas bills	Observed frequencies			Expected frequencies	
	Howie's Gulf f_1	Paul's Texaco f_2	Total	Howie's Gulf e_1	Paul's Texaco e_2
\$50 to less than \$100	$f_{11} = 15$	$f_{12} = 6$	$f_1 = 21$	$\frac{n_1 f_1}{n} = 12.76$	$\frac{n_2 f_1}{n} = 8.24$
\$100 to less than \$150	$f_{21} = 31$	$f_{22} = 22$	$f_2 = 53$	$\frac{n_1 f_2}{n} = 32.21$	$\frac{n_2 f_2}{n} = 20.79$
\$150 to less than \$200	$f_{31} = 20$	$f_{32} = 19$	$f_3 = 39$	$\frac{n_1 f_3}{n} = 23.70$	$\frac{n_2 f_3}{n} = 15.30$
\$200 or more	$f_{41} = 13$	$f_{42} = 4$	$f_4 = 17$	$\frac{n_1 f_4}{n} = 10.33$	$\frac{n_2 f_4}{n} = 6.67$
Total	$n_1 = 79$	$n_2 = 51$	$n = 130$	$n_1 = 79$	$n_2 = 51$

Now, Q can be computed.

$$Q = \sum_{i=1}^{r} \frac{(f_{i1} - e_{i1})^2}{e_{i1}} + \sum_{i=1}^{r} \frac{(f_{i2} - e_{i2})^2}{e_{i2}}$$

$$Q = \left[\frac{(15 - 12.76)^2}{12.76} + \frac{(31 - 32.21)^2}{32.21} + \frac{(20 - 23.70)^2}{23.70} + \frac{(13 - 10.33)^2}{10.33} \right]$$

$$+ \left[\frac{(6 - 8.24)^2}{8.24} + \frac{(22 - 20.79)^2}{20.79} + \frac{(19 - 15.3)^2}{15.3} + \frac{(4 - 6.67)^2}{6.67} \right]$$

$$= 1.71 + 2.64 = 4.35$$

$Q = 4.35$

$df = r - 1 = 4 - 1 = 3$.

STEP 7:

$$Q = 4.35$$
$$\chi^2_{(1-\alpha=.95,3)} = 7.815$$

The null hypothesis should not be rejected because Q is less than $\chi^2_{(.95,3)}$. Consequently, it is unreasonable to assume that there is a statistically significant difference between the distributions of average bimonthly automobile gas bills (as of January 1980) for families favoring Howie's Gulf Station and families favoring Paul's Texaco Station, at the 5 percent level of significance.

The memorandum appears in Exhibit 14–2.

Exhibit 14–2

HOWARD S. GITLOW, PH.D.
STATISTICAL CONSULTANT
———

MEMORANDUM

TO: Ms. Marsha Lubitz, Manager
        Stat City Chapter of The American
        Automobile Association

FROM: Howard Gitlow

DATE: April 3, 1981

RE: Distribution of average bimonthly automobile gas
        bills for Howie's Gulf Station and Paul's Texaco
        Station, as of January 1980.

In accordance with our consulting contract and commonly
accepted statistical techniques, I have computed several
summary statistics concerning the distributions of average
bimonthly automobile gas bills for Howie's Gulf Station and
Paul's Texaco Station, as of January 1980.[1] Allowing for the
errors in survey estimates, you can have "95 percent
confidence" that there is not a statistically significant
difference between the distributions of average bimonthly
automobile gas bills for Howie's Gulf Station and Paul's
Texaco Station.[2]

If you have any questions, please call me at
305–999–9999.

Exhibit 14-2 (*continued*)

Footnotes

[1] Insert standard footnote 1 (insert a sample of 130 Stat City families drawn via simple random sampling).

[2] The particular samples used in this survey are two of a large number of all possible pairs of samples of the same size that could have been selected using the same sample design. The sample estimates derived from the different pairs of samples would differ from each other. The difference between the pairs of sample estimates is called sampling error.

The notion of sampling error permits a researcher to say that only 5 percent of the possible pairs of samples could produce sample statistics that indicate the distributions are different when, in fact, the distributions are not different. Consequently, if the sample statistic gives a strong indication that the distributions are different, a researcher must reject the notion that the distributions are the same with the understanding that he/she may be wrong 5 percent of the time.

INFERENCES CONCERNING LOCATION OR CENTRAL TENDENCY

Many situations arise in which a test concerning the location of a distribution is called for; however, the assumptions required to perform Z- or t-tests are not satisfied. Consequently, analysts must rely on nonparametric testing procedures. There are three nonparametric tests of location (or central tendency) that will be discussed in this chapter.

The first nonparametric location test to be discussed is the Wilcoxon one-sample signed rank test. This test uses a set of observations from a single sample to make an inference about a population median. This procedure is the nonparametric counterpart to the classical one-sample t-test on means. The assumptions for the Wilcoxon one-sample signed rank test are:

1. The variable under study must be continuous and at least ordinal in scale.
2. The data must be drawn by simple random sampling.
3. The distribution of the difference between the observed data and the hypothesized median should be symmetric (not necessarily normal).

The test procedure is as follows.

STEP 1: State the hypothesis to be tested (considering if the test is one- or two-tailed).

	Two-tailed test	One-tailed test	One-tailed test

H_0: Median $= M_0$ OR H_0: Median $\geq M_0$ OR H_0: Median $\leq M_0$

H_A: Median $\neq M_0$ H_A: Median $< M_0$ H_A: Median $> M_0$

where

M_0 = the hypothesized value of the median.

STEP 2: Set α.

STEP 3: Select the test statistic and the procedure required for its computation.

a. Compute the difference score (D_i) between each sample observation (X_i) and the hypothesized median (M_0) for each of the n observations in the sample.

$$D_i = X_i - M_0 \text{ where } i = 1 \text{ to } n$$

b. Convert the n difference scores into absolute values (neglect the signs of the D_i^s).

$$|D_i| = |X_i - M_0|$$

c. Drop from analysis all $D_i = 0$. This will produce a reduced sample size, called m ($m \leq n$).

d. Assign ranks (R_i) to each of the m remaining $|D_i|$s such that the smallest $|D_i|$ receives a rank of 1 and the largest $|D_i|$ receives a rank of m. If two or more $|D_i|$s are equal, assign each of the tied $|D_i|$s the average rank of the ranks they otherwise would have individually received had this not occurred.

e. Assign a $+$ or $-$ to each of the m nonzero absolute difference scores ($|D_i|$), depending on whether $X_i - M_0$ was positive or negative.

f. The Wilcoxon test statistic (W) is the sum of the R_i whose corresponding $|D_i|$ was assigned a$+$.

$$W = \sum_{i=1}^{m} R_i^{(+)}$$

The above statistic can be computed using the following chart.

X_i	$D_i = X_i - M_0$	D_i	R_i	*Sign of* D_i

STEP 4: Determine the sample size and draw the sample.

STEP 5: Compute W.

STEP 6: Determine if the null hypothesis should be rejected. This determination depends upon whether $m \leq 20$ or $m > 20$.

If $m \leq 20$, Table B–7 can be used to determine the critical values for W for both one- and two-tailed tests at several significance levels. Given that a two-tailed test is being performed at the α level of significance, reject the null hypothesis if W lies either below the lower critical value or above the upper critical value. If a one-tailed test is being performed at the α level of significance in which the null hypothesis is H_0: Median $\geq M_0$, reject the null hypothesis if W lies below the lower critical value. Finally, if a one-tailed test is being performed at the α level of significance in which the null hypothesis is H_0: Median $\leq M_0$, reject the null hypothesis if W lies above the upper critical value.

If $m > 20$, W is approximately normal and the test statistic is

$$Z \cong \left[\frac{W - \dfrac{m(m+1)}{4}}{\sqrt{\dfrac{m(m+1)(2m+1)}{24}}} \right]$$

The rejection region(s) for the above statistic depend on α and whether the test is one- or two-tailed, as with any Z-test.

Example 14–3 Median telephone bill in Stat City

Mr. Jack Davis, chairman of the Stat City Telephone Company, needs to determine if the median monthly telephone bill in Stat City is less than $60 per month. If the median telephone bill is less than $60 per month, Mr. Davis will institute a promotional campaign to increase telephone usage. The chairman is far more concerned about instituting the promotional campaign when it is not called for than not instituting the campaign when it is called for. Mr. Davis is extremely sensitive about unnecessary expenditures of stockholders' money. Mr. Davis is willing to accept a 5 percent risk of instituting an unnecessary campaign.

You have been retained by Mr. Davis to conduct the necessary survey. He has provided you with enough funds to sample 25 Stat City families. Conduct the survey and report your findings to Mr. Davis (see Exhibit 14–3)

Solution

STEP 1: State the hypothesis to be tested.

Recall, to set up the null and alternative hypotheses, the states of nature, the alternative courses of action, and the error the decision maker considers more severe must be known. Table 14–6 summarizes the information required to determine the null and alternative hypotheses for this example.

Table 14–6

Alternative courses of action	States of nature	
	Median < $60	Median ≥ $60
Institute promotional campaign	Correct decision given state of nature	More severe incorrect decision given state of nature
Do not institute promotional campaign	Less severe incorrect decision given state of nature	Correct decision given state of nature

It is customary when working with one-tail tests to designate as the null hypothesis that hypothesis which, if rejected when tenable, leads to the type I error (the more severe error). Consequently, in this problem the null hypothesis is

$$H_0: \text{Median} \geq \$60$$

and the alternative hypothesis is

$$H_A: \text{Median} < \$60$$

STEP 2: Set $\alpha = 0.05$. Mr. Davis is willing to accept a 5 percent risk of instituting an unnecessary promotional campaign. In other words, if this decision were being made 100 times, Mr. Davis would tolerate being wrong (and institute an unnecessary promotional campaign) 5 times. In fact, the chairman only makes this decision once.

STEP 3: The appropriate test statistic is W and its computation can be accomplished through the procedure stated in Step 3 on page 238 of this chapter.

$$W = \sum_{i=1}^{m} R_i^{(+)}$$

STEP 4: The sample size has been set at 25. (This chapter will not cover the procedure for determining the sample size for a Wilcoxon one-sample signed rank test.) The sample appears in Table 14–7.

Table 14–7

Random identification number	Average monthly telephone bill, as of January 1980 (variable 11)
88	42
185	40
239	45
310	56
409	31
436	32
520	67
578	92
611	48
660	68
723	69
745	39
771	44
900	50
961	63
1032	40
1081	53
1084	38
1109	68
1194	19
1211	46
1320	46
1322	66
1326	44
1344	61

STEP 5: The computation of W is presented in Table 14–8.

Table 14–8

X_i	$D_i = X_i - \$60$	$\lvert D_i \rvert$	R_i	Sign of D_i
42	−18	18	17	−
40	−20	20	18.5	−
45	−15	15	14	−
56	− 4	4	3	−
31	−29	29	23	−
32	−28	28	22	−
67	+ 7	7	5.5	+
92	+32	32	24	+
48	−12	12	11	−
68	+ 8	8	7.5	+
69	+ 9	9	9	+
39	−21	21	20	−
44	−16	16	15.5	−
50	−10	10	10	−
63	+ 3	3	2	+
40	−20	20	18.5	−
53	− 7	7	5.5	−
38	−22	22	21	−
68	+ 8	8	7.5	+
19	−41	41	25	−
46	−14	14	12.5	−
46	−14	14	12.5	−
66	+ 6	6	4	+
44	−16	16	15.5	−
61	+ 1	1	1	+

$$W = 5.5 + 24 + 7.5 + 9 + 2 + 7.5 + 4 + 1 = 60.5$$

STEP 6: The computed $W = 60.5$. Since $m > 20$, Z must be computed ($m = n = 25$ in this example).

$$Z \cong \left[\frac{W - \dfrac{m(m+1)}{4}}{\sqrt{\dfrac{m(m+1)(2m+1)}{24}}}\right] = \left[\frac{60.5 - \dfrac{25(26)}{4}}{\sqrt{\dfrac{25(26)(51)}{24}}}\right]$$

$$= \frac{60.5 - 162.5}{37.17} = \frac{-102}{37.17} = -2.74$$

Since $-2.74 < -1.64$, the null hypothesis is rejected at the 5 percent level of significance. This finding will lead Mr. Davis to assume that the median telephone bill in Stat City, as of January 1980, was less than $60.00. Consequently, Mr. Davis will institute the promotional campaign.

The memorandum appears in Exhibit 14–3.

The second nonparametric location test to be discussed is the Wilcoxon rank sum test. This test is used to determine if there is a statistically significant difference between two population medians. The Wilcoxon rank sum test is the nonparametric counterpart to the two-sample (independent) t-test. The assumptions required to perform the Wilcoxon rank sum test are:

1. Both samples are selected independently and randomly from their respective populations.
2. The underlying variable is continuous and at least at the ordinal level of measurement.
3. No differences exist between the two populations other than differences in medians.

Exhibit 14–3

HOWARD S. GITLOW, PH.D.
STATISTICAL CONSULTANT

MEMORANDUM

TO: Mr. Jack Davis, Chairman
        Stat City Telephone Company

FROM: Howard Gitlow

DATE: April 1, 1981

RE: Median telephone bill in Stat City, as of January
        1980

In accordance with our consulting contract I have
computed several statistics concerning the "average
monthly telephone bills" of Stat City families, as of
January 1980.[1] My research indicates that the median
"average monthly telephone bill" in Stat City, as of January
1980, was $46.00.[2] Allowing for the errors in survey
estimates, you can have "95 percent confidence" that the
median telephone bill in Stat City was below $60.00.[3]

The above survey was conducted using commonly accepted
statistical techniques. If you have any questions, please
call me at 305–999–9999.

Footnotes

[1] Insert standard footnote 1 (insert a simple random sample of 25
Stat City families).
[2] Insert standard footnote 2.
[3] Insert standard footnote 4.

The differences between the Wilcoxon rank sum test and the t-test are
summarized in Table 14–9. From this table, it is clear that the Wilcoxon
rank sum test has less stringent assumptions than the t-test.

Table 14–9

Characteristics	Wilcoxon rank sum test	Two independent samples t-test
Sample selection	Random and independent samples	Random and independent samples
Variability in population	Variances are equal in both populations	Variances are equal in both populations
Population distribution	Continuous and symmetric populations	Normal populations
Level of measurement	Ordinal, interval or ratio	Interval or ratio

The steps involved in performing the Wilcoxon rank sum test are listed
below.

STEP 1:   State the hypothesis to be tested (considering if the test is
one- or two-tailed).

$$\begin{array}{ccccc}
\textit{Two-tailed test} & & \textit{One-tailed test} & & \textit{One-tailed test} \\
H_0\colon M_1 = M_2 & & H_0\colon M_1 \le M_2 & & H_0\colon M_1 \ge M_2 \\
& \text{OR} & & \text{OR} & \\
H_A\colon M_1 \ne M_2 & & H_A\colon M_1 > M_2 & & H_A\colon M_1 < M_2
\end{array}$$

where

M_1 = population median for the group having n_1 sample observations

M_2 = population median for the group having n_2 sample observations, and n_1 is assumed to be less than or equal to n_2 $(n_1 \le n_2)$

(Define population 1 to be the population from which the smaller sample is selected.)

STEP 2: Set α.

STEP 3: Select the test statistic and the procedure required for its computations.

a. Combine the observations from both samples and rank order the combined sample ($n = n_1 + n_2$) from lowest (1) to highest (n). In the case of ties, assign each value the average of the ranks that would otherwise have been assigned.

b. Compute T_{n_1}.

$$T_{n_1} = \sum_{i=1}^{n_1} R_i^1$$

where

R_i^1 = the ranking of the ith sample data point drawn from population 1, and

n_1 = the size of the sample drawn from population 1.

If both n_1 and n_2 are less than or equal to 10, use Table B–8 to obtain the critical values of T_{n_1}. For a two-tailed test and a given level of significance (α), reject the null hypothesis if T_{n_1} is below the lower critical limit or above the upper critical limit. For a one-tailed test in which the alternative hypothesis is $H_A\colon M_1 > M_2$ and a given level of significance (α), reject the null hypothesis if T_{n_1} is greater than the upper critical value. For a one-tailed test in which the alternative hypothesis is $H_A\colon M_1 < M_2$ and a given level of significance (α), reject the null hypothesis if T_{n_1} is less than the lower critical value.

If either (or both) n_1 or n_2 are greater than 10, the test statistic is approximately normal:

$$Z = \frac{T_{n_1} - \dfrac{n_1(n+1)}{2}}{\sqrt{[n_1 n_2(n+1)]/12}}$$

The rejection region(s) for the above statistic depended on α and whether the test is one- or two-tailed, as with any Z-test.

STEP 4: Determine the sample size and draw the sample.

STEP 5: Compute T_{n_1} and/or Z.

STEP 6: Determine if the null hypothesis should be rejected.

Example 14–4 Comparison of median electric bills between homeowners and apartment dwellers

Mr. Saul Reisman, CFO of the Stat City Electric Company, must find out if the median "average monthly electric bills" for homeowners and apartment dwellers were equal as of January 1980. He needs this information for his quarterly report. Mr. Reisman is willing to accept a 1 percent risk of claiming that there is a difference between homeowners' and apartment dwellers' "average monthly electric bills" when, in fact, there is no statistically significant difference.

You have been hired by Mr. Reisman to conduct the necessary survey. He has provided you with enough funds to sample 30 Stat City families. Conduct the survey and report your findings to Mr. Reisman (see Exhibit 14–4).

STEP 1:   State the hypothesis being tested.   **Solution**

$$H_0: M_1 = M_2$$
$$H_A: M_1 \neq M_2$$

where

$M_1 =$ the population median "average monthly electric bill" for homeowners as of January 1980,

$M_2 =$ the population median "average monthly electric bill" for apartment dwellers as of January 1980.

STEP 2: $\alpha = 0.01$

STEP 3:   Select the test statistic and the procedure required for its computation.

$$Z = \frac{T_{n_1} - \dfrac{n_1(n + 1)}{2}}{\sqrt{\dfrac{n_1 n_2(n + 1)}{12}}}$$

where

$$T_{n_1} = \sum_{i=1}^{n_1} R_i^1$$

STEP 4:   Determine the sample size and draw the random sample. (This example will not deal with sample size determinations.) A simple random sample of 30 Stat City families appears in Table 14–10.

Table 14–10

CASE-N	ID	DWELL	ELEC	CASE-N	ID	DWELL	ELEC
1	33.	1.	84.	16	781.	0.	43.
2	98.	1.	53.	17	848.	0.	50.
3	120.	1.	54.	18	871.	0.	37.
4	126.	1.	40.	19	879.	0.	47.
5	141.	1.	62.	20	907.	0.	52.
6	268.	1.	64.	21	1071.	0.	44.
7	320.	1.	81.	22	1141.	0.	63.
8	399.	0.	48.	23	1198.	0.	31.
9	420.	0.	37.	24	1217.	0.	39.
10	444.	0.	45.	25	1221.	0.	45.
11	507.	0.	48.	26	1247.	0.	45.
12	525.	1.	45.	27	1271.	0.	33.
13	561.	1.	90.	28	1318.	0.	38.
14	611.	1.	48.	29	1329.	0.	42.
15	618.	1.	81.	30	1330.	0.	55.

STEP 5: The computation of T_{n_1} and Z are shown below and in Table 14–11.

Table 14–11

Homeowners = 1			Apartment dwellers = 0	
ELEC	Combined ranks		ELEC	Combined ranks
84	29		48	17
53	21		37	3.5
54	22		45	12.5
40	7		48	17
62	24		43	9
64	26		50	19
81	27.5		37	3.5
45	12.5		47	15
90	30		52	20
48	17		44	10
81	27.5		63	25
$n_1 = 11$			31	1
median = 62			39	6
			45	12.5
			45	12.5
			33	2
			38	5
			42	8
			55	23
			$n_2 = 19$	
			median = 45	

$$T_{n_1} = 29 + 21 + 22 + 7 + 24 + 26 + 27.5 + 12.5 + 30 + 17 + 27.5$$
$$= 243.5$$

$$Z \cong \frac{243.5 - \dfrac{11(30+1)}{2}}{\sqrt{\dfrac{11(19)(30+1)}{12}}} = \frac{243.5 - 170.5}{23.24}$$

$$\cong \frac{73}{23.24} = 3.14$$

STEP 6: Since $3.14 > |2.58|$, the null hypothesis is rejected at the 1 percent level of significance. This finding will lead Mr. Reisman to assume that there is a statistically significant difference in "average monthly electric bills" (as of January 1980) between homeowners and apartment dwellers.

The memorandum appears in Exhibit 14–4.

The last nonparametric location test to be discussed is the Kruskal-Wallis H-test. The H-test is used to test for differences in location among K (more than two) populations. The measure of location used by the H-test is the median. The Kruskal-Wallis H-test is the nonparametric counterpart to the one-way analysis of variance F-test. The assumptions required to perform the H-test are:

1. All K samples are selected randomly and independently.
2. The underlying variable is continuous and at least at the ordinal level of measurement.
3. No differences exist among the populations other than difference in location.

As you can see, the H-test does not require interval/ratio scale data or normality.

The steps involved in performing the H-test are summarized below.

STEP 1: State the hypothesis to be tested.

$$H_0: M_1 = M_2 = \ldots M_K$$
$$H_A: \text{Not all } M_i \text{s are equal } (i = 1 \text{ to } K)$$

Exhibit 14-4

```
┌─────────────────────────────────────────────────────────┐
│  HOWARD S. GITLOW, PH.D.                                  │
│     STATISTICAL CONSULTANT                                │
│     ─────────                                             │
│                                                           │
│                    MEMORANDUM                             │
│                                                           │
│  TO:    Mr. Saul Reisman, CFO                             │
│         Stat City Electric Company                        │
│                                                           │
│  FROM:  Howard Gitlow                                     │
│                                                           │
│  DATE:  April 2, 1981                                     │
│                                                           │
│  RE:    Comparison of median "average monthly electric    │
│         bills" between homeowners and apartment dwellers  │
│         in Stat City, as of January 1980.                 │
│                                                           │
│     Per your request I have computed several summary      │
│  statistics concerning the "average monthly electric bills"│
│  of homeowners and apartment dwellers in Stat City, as of │
│  January 1980.¹ My research indicates that the median     │
│  "average monthly electric bill" for homeowners was $62 and│
│  $45 for apartment dwellers, as of January 1980.² Allowing│
│  for the errors in survey estimates, you can have 99 percent│
│  confidence that there was a statistically significant    │
│  difference between the median "average monthly electric  │
│  bills" for homeowners and apartment dwellers, as of January│
│  1980.³                                                    │
│     The above statistics were computed using commonly     │
│  accepted statistical techniques. If you have any questions,│
│  please call me at 305-999-9999.                          │
│  ─────────────                                            │
│                    Footnotes                              │
│                                                           │
│    ¹ Insert standard footnote 1 (insert a simple random sample of 30│
│  Stat City families).                                     │
│    ² Insert standard footnote 2.                          │
│    ³ Insert standard footnote 5.                          │
└─────────────────────────────────────────────────────────┘
```

where

M_i = the population median for population (group) i

STEP 2:   Set α.

STEP 3:   Select the test statistic and the procedure required for its computation.

a. Combine the observations from all K samples and rank order the combined sample ($n = n_1 + n_2 + \cdots + n_K$) from lowest (1) to highest (n). In the case of ties, assign each value the average of the ranks that would otherwise have been assigned.

b. Compute H.

$$H = \left[\frac{12}{n(n+1)} \sum_{i=1}^{K} \frac{T_{n_i}^2}{n_i} \right] - 3(n+1)$$

where

n_i = the number of observations in the sample from the ith population

n = the total number of observations in all K samples ($n = n_1 + n_2 + \cdots + n_K$)

T_{n_i} = the sum of all the ranks assigned to the sample from the ith population

Some nonparametric statistics

245

c. If all of the n_i's are greater than 5, the H-statistic can be approximated by the χ^2 distribution with $K - 1$ degrees of freedom. For a given level of significance (α) and K, reject the null hypothesis if H exceeds the critical χ^2 value.

STEP 4: Determine the sample size and draw the random sample.

STEP 5: Compute H.

STEP 6: Determine if the null hypothesis should be rejected.

Example 14–5 Median incomes among Stat City residential zones

Mr. Lee Kaplowitz, the mayor of Stat City, has to prepare a report for the federal government for which he needs to know if the median incomes in all four housing zones of Stat City were significantly different, as of January 1980. In this report he is willing to accept responsibility for a 5 percent error of stating the median incomes were significantly different when, in fact, they were not significantly different.

You have been asked by the mayor to conduct the necessary survey. He has provided you with funds to draw a simple random sample of 50 families. Conduct the survey and report your findings to Mr. Kaplowitz (see Exhibit 14–5).

Solution   STEP 1: State the hypothesis to be tested.

$$H_0: \quad M_1 = M_2 = M_3 = M_4$$
$$H_A: \quad \text{Not all medians are equal}$$

STEP 2: $\alpha = 0.05$

STEP 3: Select the test statistic and the procedure required for its computation.

$$H = \left[\frac{12}{n(n + 1)} \sum_{i=1}^{K} \frac{T_{n_i}^2}{n_i} \right] - 3(n + 1)$$

STEP 4: A simple random sample of 50 Stat City families appears in Table 14–12.

CASE-N	ID	ZONE	INCOM	Table 14–12
1	49.	1.	19626.	
2	74.	1.	43249.	
3	89.	1.	34680.	
4	97.	1.	57553.	
5	105.	1.	49068.	
6	126.	1.	26316.	
7	150.	2.	69898.	
8	201.	2.	96448.	
9	272.	2.	66364.	
10	280.	2.	134443.	
11	286.	2.	50602.	
12	287.	2.	43982.	
13	404.	3.	33909.	
14	425.	3.	28988.	
15	454.	3.	37265.	
16	499.	3.	31801.	
17	509.	3.	27390.	
18	543.	3.	36756.	
19	588.	3.	44824.	
20	603.	3.	24343.	
21	734.	4.	20553.	
22	738.	4.	12821.	
23	740.	4.	11186.	
24	748.	4.	10380.	
25	763.	4.	8571.	
26	793.	4.	18811.	
27	796.	4.	16200.	
28	801.	4.	19261.	
29	806.	4.	22557.	
30	815.	4.	13161.	
31	863.	4.	16916.	
32	903.	4.	17922.	
33	929.	4.	11873.	
34	977.	4.	21897.	
35	980.	4.	22087.	
36	990.	4.	19681.	
37	1005.	4.	20075.	
38	1030.	4.	15273.	
39	1060.	4.	7288.	
40	1073.	4.	7306.	
41	1091.	4.	35832.	
42	1110.	4.	23902.	
43	1117.	4.	30937.	
44	1182.	4.	15918.	
45	1267.	4.	8466.	
46	1286.	4.	8521.	
47	1291.	4.	14120.	
48	1293.	4.	13723.	
49	1302.	4.	24761.	
50	1341.	4.	24789.	

STEP 5:   Compute H; see Table 14–13.

Table 14–13

	Zones						
	1		*2*		*3*		*4*
Income	*Combined rank*	*Income*	*Combined rank*	*Income*	*Combined rank*	*Income*	*Combined rank*
19626.	20	69898.	48	33909.	36	20553.	23
43249.	41	96448.	49	28988.	33	12821.	9
34680.	37	66364.	47	37265.	40	11186.	7
57553.	46	134443.	50	31801.	35	10380.	6
49068.	44	50602.	45	27390.	32	8571.	5
26316.	31	43982.	42	36756.	39	18811.	18
				44824.	43	16200.	15
				24343.	28	19261.	19
						22557.	26
						13161.	10
						16916.	16
						17922.	17
						11873.	8
						21897.	24
						22087.	25
						19681.	21
						20075.	22
						15273.	13
						7288.	1
						7306.	2
						35832.	38
						23902.	27
						30937.	34
						15918.	14
						8466.	3
						8521.	4
						14120.	12
						13723.	11
						24761.	29
						24789.	30

$$T^2_{n_1} = (20 + 41 + 37 + 46 + 44 + 31)^2 = 219^2 = 47{,}961$$
$$T^2_{n_2} = (48 + 49 + 47 + 50 + 45 + 42)^2 = 281^2 = 78{,}961$$
$$T^2_{n_3} = (36 + 33 + 40 + 35 + 32 + 39 + 43 + 28)^2 = 286^2 = 81{,}796$$
$$T^2_{n_4} = (23 + 9 + 7 + 6 + 5 + 18 + 15 + 19 + 26 + 10 + 16 + 17$$
$$+ 8 + 24 + 25 + 21 + 22 + 13 + 1 + 2 + 38 + 27 + 34$$
$$+ 14 + 3 + 4 + 12 + 11 + 29 + 30)^2 = 489^2 = 239{,}121$$

$$H = \left[\frac{12}{50(50 + 1)} \left(\frac{47{,}961}{6} + \frac{78{,}961}{6} + \frac{81{,}796}{8} + \frac{239{,}121}{30} \right) \right] - 3(50 + 1)$$

$$H = \left[\frac{12}{2{,}550} (7{,}993.5 + 13{,}160.17 + 10{,}224.5 + 7{,}970.7) \right] - 153$$

$$H = \left[\frac{12}{2{,}550} (39{,}348.87) \right] - 153 = 185.17 - 153 = 32.17$$

STEP 6:   Since $32.17 > (\chi^2_{(1-\alpha = .95, K-1 = 3)} = 7.815)$, the null hypothesis is rejected at the 5 percent level of significance. This finding will lead Mr. Kaplowitz to report that the median 1979 incomes among the four housing zones are not equal.

The memorandum appears in Exhibit 14–5.

Exhibit 14–5

HOWARD S. GITLOW, PH.D.
STATISTICAL CONSULTANT

MEMORANDUM

TO: Mr. Lee Kaplowitz, Mayor
        Stat City

FROM: Howard Gitlow

DATE: April 3, 1981

RE: Median incomes among Stat City housing zones.

   In accordance with our consulting contract I have
computed several summary statistics concerning the 1979
total family incomes of Stat City families.[1] The chart below
displays the median incomes in each residential housing
zone, and for all of Stat City.[2]

	Median 1979 total family income
Zone 1	$38,964.50
Zone 2	$68,131.00
Zone 3	$32,855.00
Zone 4	$16,558.00
All of Stat City	$22,322.00

Allowing for the errors in survey estimates, you can have "95
percent confidence" that there was a statistically
significant difference among the median 1979 total family
incomes for the housing zones in Stat City.[3]

Exhibit 14–5 (*continued*)

The above statistics were computed using commonly
accepted statistical techniques. If you have any questions,
please call me at 305-999-9999.

Footnotes

[1] Insert standard footnote 1 (insert a sample of 50 families drawn via simple random sampling).
[2] Insert standard footnote 2.
[3] Insert standard footnote 5.

INFERENCES CONCERNING ASSOCIATION

Frequently decision makers are not concerned with inferences concerning distributions or location; rather, they are concerned with inferences dealing with the association between variables. For example, is there a statistically significant relationship between favorite supermarket (FEAT) and residential housing zone (ZONE) in Stat City? (You will recall the problem dealing with territorial shopping behavior in Stat City.) This section of Chapter 14 deals with determining if such relationships are statistically significant. There are two nonparametric procedures that will be discussed in this section: χ^2-tests of independence and Spearman rank correlation tests.

χ^2-tests of independence are used to determine if there is a statistically significant relationship (association) between two (or more) variables presented in a contingency table format (count data). The steps involved in performing a χ^2-test of independence follow.

STEP 1: State the hypothesis to be tested.

H_0: There is no relationship between variable A and variable B

H_A: There is a relationship between variable A and variable B

or

$$H_0: \quad \pi_i\pi_j = \pi_{ij}$$
$$H_A: \quad \pi_i\pi_j \neq \pi_{ij} \text{ for any } i \text{ and } j$$

where

π_i = the population proportion of observations in category i of variable A

π_j = the population proportion of observations in category j of variable B

π_{ij} = the population proportion of observations in category i of variable A and category j of variable B

Recall from the discussion of independent events in Chapter 5 that if $P(A \text{ and } B) = P(A) \cdot P(B)$, then events A and B are independent. This concept carries over to the analysis of contingency table data. If the product of marginal probability π_i from variable A and marginal probability π_j from variable B equals the joint probability π_{ij} for all i and j, then variables A and B are independent.

STEP 2:   Set α.

STEP 3:   Set up R mutually exclusive categories for variable A and C mutually exclusive categories for variable B.

STEP 4:   Select the test statistic.

$$Q = \sum_{i=1}^{R} \sum_{j=1}^{C} \left(\frac{[0_{ij} - E_{ij}]^2}{E_{ij}} \right)$$

where

R = the number of categories in variable A

C = the number of categories in variable B

0_{ij} = the number of observations in category i of variable A and category j of variable B

$0_i = \sum\limits_{j=1}^{C} 0_{ij}$ = the number of observations in category i of variable A

$0_j = \sum\limits_{i=1}^{R} 0_{ij}$ = the number of observations in category j of variable B

n = the total number of observations

$\quad = \sum\limits_{i=1}^{R} \sum\limits_{j=1}^{C} 0_{ij} = n$

E_{ij} = the expected number of observations in category i of variable A and category j of variable B assuming the null hypothesis is tenable (variables A and B are independent)

$\quad = \dfrac{(0_i)(0_j)}{n}$

STEP 5:   Determine the sample size and draw the sample. (This chapter will not cover the formulas for computing the sample size for an $R \times C$ contingency table test of independence.)

STEP 6:   Compute the E_{ij}'s and Q.

STEP 7:   Determine if the null hypothesis should be rejected given the appropriate degrees of freedom [$(R - 1)(C - 1)$ for this test]. The distribution of Q can be approximated by the χ^2 distribution with $(R - 1)(C - 1)$ degrees of freedom if the sample size is large (greater than or equal to 50), none of the E_{ij} are less than 1, and at least 80 percent of the E_{ij} are greater than 10 for a 2×2 table and greater than 5 for an $R \times C$ table. If Q is greater than $\chi^2_{(R-1)(C-1)}$, reject the null hypothesis because the 0_{ij} are too far from the E_{ij} to assume that variables A and B are independent.

Example 14–6 Relationship between beer and soda purchases

The Stat City Supermarket Association would like to find out if there is a statistically significant relationship between a family's weekly purchases of beer and soda (all types). Specifically, the association is interested in the following purchase classifications:

Average number of six-packs purchased per week, as of January 1980

$$\text{Beer} = \begin{cases} 0 & \text{(nonpurchasers)} \\ 1 \text{ or more} & \text{(purchasers)} \end{cases}$$

$$\text{Soda} = \begin{cases} 0 & \text{(nonpurchasers)} \\ 1 \text{ or more} & \text{(purchasers)} \end{cases}$$

If there is a significant relationship, the association will capitalize on this relationship in constructing their advertising campaign to promote soda and beer purchases.

You have been retained by the Stat City Supermarket Association to determine if there is a statistically significant relationship between a family's soda and beer purchases at the 5 percent level of significance (see Exhibit 14–6). The association has provided you with enough funds to draw a random sample of 100 Stat City families. Use the randomly selected ID numbers in Table 14–14.

Table 14–14	CASE-N	ID	LSODA	HSODA	BEER
	1	16.	0.	1.	10.
	2	48.	0.	0.	0.
	3	52.	0.	0.	0.
	4	67.	2.	0.	8.
	5	68.	1.	0.	2.
	6	73.	0.	0.	0.
	7	77.	0.	0.	0.
	8	92.	0.	1.	0.
	9	101.	0.	0.	0.
	10	142.	1.	0.	3.
	11	146.	2.	0.	0.
	12	164.	0.	1.	0.
	13	177.	2.	0.	1.
	14	182.	0.	0.	0.
	15	183.	0.	0.	0.
	16	186.	0.	3.	3.
	17	196.	0.	1.	3.
	18	216.	0.	6.	0.
	19	241.	2.	0.	0.
	20	244.	0.	0.	0.
	21	258.	0.	0.	10.
	22	268.	0.	5.	6.
	23	270.	0.	4.	0.
	24	281.	0.	0.	7.
	25	287.	2.	0.	9.
	26	319.	2.	0.	3.
	27	321.	2.	0.	0.
	28	322.	0.	0.	0.
	29	341.	0.	0.	0.
	30	346.	0.	1.	3.
	31	386.	2.	0.	1.
	32	420.	0.	0.	0.
	33	422.	1.	0.	6.
	34	433.	0.	5.	2.
	35	443.	0.	0.	6.
	36	448.	1.	0.	0.
	37	449.	0.	0.	0.
	38	476.	0.	1.	2.
	39	486.	0.	0.	0.
	40	507.	0.	0.	0.
	41	524.	0.	1.	3.
	42	532.	0.	5.	1.

CASE-N	ID	LSODA	HSODA	BEER	Table 14–14 (*continued*)
43	553.	0.	3.	0.	
44	566.	2.	0.	2.	
45	595.	2.	0.	0.	
46	597.	0.	4.	5.	
47	604.	0.	0.	0.	
48	621.	0.	0.	0.	
49	634.	0.	0.	0.	
50	657.	0.	0.	9.	
51	674.	0.	0.	0.	
52	715.	0.	2.	0.	
53	727.	0.	0.	0.	
54	741.	0.	0.	0.	
55	751.	0.	0.	0.	
56	755.	2.	0.	0.	
57	763.	0.	6.	5.	
58	767.	0.	0.	0.	
59	785.	0.	4.	0.	
60	802.	0.	0.	2.	
61	819.	0.	0.	0.	
62	827.	0.	1.	3.	
63	844.	0.	0.	9.	
64	878.	1.	0.	0.	
65	890.	0.	3.	1.	
66	896.	0.	3.	2.	
67	916.	1.	0.	3.	
68	941.	1.	0.	5.	
69	944.	0.	0.	0.	
70	950.	0.	0.	0.	
71	952.	0.	0.	0.	
72	955.	0.	5.	3.	
73	960.	2.	0.	0.	
74	971.	0.	2.	1.	
75	975.	1.	0.	0.	
76	997.	0.	1.	0.	
77	1005.	0.	0.	1.	
78	1048.	0.	4.	0.	
79	1075.	0.	1.	10.	
80	1100.	2.	0.	0.	
81	1124.	0.	3.	0.	
82	1152.	2.	0.	0.	
83	1158.	0.	4.	0.	
84	1164.	2.	0.	2.	
85	1166.	0.	0.	2.	
86	1174.	0.	0.	4.	
87	1188.	2.	0.	5.	
88	1194.	2.	0.	0.	
89	1205.	0.	6.	3.	
90	1225.	1.	0.	3.	
91	1232.	0.	0.	0.	
92	1233.	1.	0.	3.	
93	1256.	2.	0.	7.	
94	1266.	0.	0.	0.	
95	1274.	0.	0.	0.	
96	1281.	2.	0.	2.	
97	1294.	0.	0.	0.	
98	1305.	0.	0.	10.	
99	1326.	1.	0.	0.	
100	1367.	0.	1.	3.	

STEP 1: BEFORE any data are collected the hypothesis of interest **Solution**
must be specified. In this example the relevant hypothesis is:

H_0: There is no relationship between weekly purchases of
beer and soda

H_A: There is a relationship between weekly purchases of
beer and soda

STEP 2: $\alpha = 0.05$.

STEPS 3, 4, 5, and 6: The sample contingency table and the expected
cell frequencies (in circles) are in Table 14–15.

Table 14–15

	Beer				
Soda	*Nonpurchasers (0)*		*Purchasers (1 or more)*		*Total*
Nonpurchasers (0)	30	$\boxed{21.6}$	10	$\boxed{18.4}$	40
Purchasers (1 or more)	24	$\boxed{32.4}$	36	$\boxed{27.6}$	60
Total	54		46		100

$$E_{11} = \frac{54(40)}{100} = \frac{2{,}160}{100} = 21.6$$

$$E_{12} = \frac{46(40)}{100} = \frac{1{,}840}{100} = 18.4$$

$$E_{21} = \frac{54(60)}{100} = \frac{3{,}240}{100} = 32.4$$

$$E_{22} = \frac{46(60)}{100} = \frac{2{,}760}{100} = 27.6$$

The Q statistic is computed as follows:

$$Q = \frac{(30 - 21.6)^2}{21.6} + \frac{(10 - 18.4)^2}{18.4} + \frac{(24 - 32.4)^2}{32.4} + \frac{(36 - 27.6)^2}{27.6}$$
$$= 3.27 + 3.83 + 2.18 + 2.56$$
$$= 11.84$$

STEP 7: The critical χ^2 value for one degree of freedom [(R − 1)(C − 1) = 1] and the 5 percent level of significance is 3.841. As you can see, 11.84 is greater than 3.841; hence, the null hypothesis is rejected and you can assume that there is a statistically significant relationship between weekly beer and soda purchases.

The memorandum appears in Exhibit 14–6.

Exhibit 14–6

HOWARD S. GITLOW, PH.D.
STATISTICAL CONSULTANT
———

MEMORANDUM

TO: Stat City Supermarket Association

FROM: Howard Gitlow

DATE: November 6, 1980

RE: Relationship between beer and soda purchases in Stat
       City, as of January 1980.

   As per your request and using commonly accepted
statistical techniques, I have investigated the
relationship between "a families' weekly purchase of beer
as of January 1980 (BEER)" and "a families' weekly purchase
of all types of soda as of January 1980 (SODA)."[1] Allowing for
the errors in survey estimates, you can have 95 percent
confidence that there is a statistically significant
relationship between BEER and SODA purchases.[2]
   If you have any further questions, please do not hesitate
to call me at 305–999–9999.

Exhibit 14–6 (*continued*)

Footnotes

[1] Insert standard footnote 1 (insert a sample of 100 families drawn via simple random sampling).

[2] The particular sample used in this survey is one of a large number of all possible samples of the same size that could have been selected using the same sample design. The sample estimates derived from the different samples would differ from each other. The difference between the sample estimates is called sampling error.

The concept of sampling error permits a researcher to say that only 5 percent of the possible samples could produce sample statistics that indicate a relationship exists when, in fact, no relationship exists. Consequently, if the sample statistic gives a strong indication that a relationship exists, a researcher must reject the notion of no relationship with the understanding that he/she may be wrong 5 percent of the time.

Example 14–7 Survey to investigate the relationship between the number of cars in household and favorite place for car repairs

Mr. Rose, the owner of the Rose Auto Repair Store, must find out if there is any relationship between a family's favorite place to have their automotive repairs performed and the number of cars in a family, as of January 1980. Mr. Rose is interested in the following classification of cars: family owns one car, family owns two cars, or family owns three or more cars. If there is a significant relationship, Mr. Rose will utilize this fact when constructing his promotional campaign.

You have been retained by Mr. Rose to obtain the necessary information at the 5 percent level of significance (see Exhibit 14–7). He has provided you with enough funds to draw a random sample of 100 Stat City families. Use the random sample in Table 14–16.

Table 14–16

CASE-N	ID	CARS	REPAR	CASE-N	ID	CARS	REPAR
1	24.	3.	0.	51	646.	3.	2.
2	30.	2.	1.	52	683.	1.	0.
3	31.	3.	2.	53	690.	2.	0.
4	39.	3.	2.	54	702.	1.	0.
5	49.	2.	0.	55	725.	2.	1.
6	58.	4.	1.	56	728.	2.	1.
7	64.	3.	0.	57	735.	1.	0.
8	67.	4.	2.	58	769.	2.	1.
9	87.	4.	2.	59	772.	2.	0.
10	94.	2.	1.	60	780.	2.	0.
11	96.	2.	1.	61	783.	2.	0.
12	119.	1.	0.	62	791.	1.	0.
13	185.	1.	2.	63	793.	1.	1.
14	187.	3.	2.	64	799.	2.	0.
15	194.	2.	1.	65	812.	3.	0.
16	200.	3.	2.	66	813.	4.	2.
17	219.	3.	1.	67	860.	2.	1.
18	226.	2.	1.	68	885.	2.	1.
19	259.	3.	2.	69	901.	2.	1.
20	262.	1.	0.	70	905.	2.	1.
21	289.	2.	0.	71	966.	2.	0.
22	293.	1.	0.	72	973.	3.	2.
23	324.	3.	2.	73	974.	2.	1.
24	333.	2.	2.	74	977.	1.	2.
25	345.	2.	1.	75	989.	3.	0.
26	360.	2.	2.	76	1012.	4.	2.
27	384.	3.	2.	77	1027.	2.	2.
28	390.	4.	2.	78	1055.	1.	1.
29	392.	3.	1.	79	1061.	2.	0.
30	397.	2.	1.	80	1105.	3.	1.
31	434.	3.	1.	81	1136.	2.	2.
32	438.	2.	1.	82	1151.	2.	0.
33	448.	4.	2.	83	1167.	2.	1.
34	450.	2.	1.	84	1171.	2.	1.
35	459.	2.	0.	85	1184.	2.	0.
36	470.	3.	1.	86	1199.	2.	1.
37	489.	2.	1.	87	1201.	1.	0.
38	492.	2.	2.	88	1209.	2.	1.
39	508.	4.	2.	89	1220.	2.	1.
40	511.	2.	1.	90	1231.	1.	0.
41	523.	3.	2.	91	1241.	2.	0.
42	528.	3.	1.	92	1242.	3.	2.
43	553.	2.	0.	93	1262.	2.	1.
44	576.	2.	1.	94	1277.	1.	1.
45	577.	2.	1.	95	1279.	2.	1.
46	583.	3.	1.	96	1285.	2.	1.
47	600.	2.	1.	97	1342.	2.	1.
48	618.	3.	1.	98	1343.	2.	2.
49	623.	4.	2.	99	1354.	1.	0.
50	642.	3.	2.	100	1372.	1.	0.

Solution

STEP 1: BEFORE any data are collected the hypothesis of interest must be specified. In this example the relevant hypothesis is:

H_0: There is no relationship between the number of cars in a family (CARS) and a family's favorite place to have their automotive repairs performed (REPAR), as of January 1980

H_A: There is a relationship between the number of cars in a family and a family's favorite place to have their automotive repairs performed, as of January 1980,

where

$$
\text{REPAR} = \begin{cases} 0 \text{ performs own repairs} \\ 1 \text{ if repairs done by gas station} \\ 2 \text{ if repairs done by dealer} \end{cases}
$$

$$
\text{CARS} = \begin{cases} 1 \text{ if one car} \\ 2 \text{ if two cars} \\ 3 \text{ if three or more cars} \end{cases}
$$

STEP 2: $\alpha = 0.05$.

STEP 3: The sample contingency table and the expected cell frequencies (in circles) are shown in Table 14–17.

Table 14–17

Cars	Repair						Total
	Performs own repairs (0)		Repairs done by station (1)		Repairs done by dealer (2)		
1	11	(4.64)	3	(6.88)	2	(4.48)	16
2	14	(14.79)	31	(21.93)	6	(14.28)	51
3 or more	4	(9.57)	9	(14.19)	20	(9.24)	33
Total	29		43		28		100

$$E_{11} = \frac{29(16)}{100} = 4.64 \qquad E_{12} = \frac{43(16)}{100} = 6.88 \qquad E_{13} = \frac{28(16)}{100} = 4.48$$

$$E_{21} = \frac{29(51)}{100} = 14.79 \qquad E_{22} = \frac{43(51)}{100} = 21.93 \qquad E_{23} = \frac{28(51)}{100} = 14.28$$

$$E_{31} = \frac{29(33)}{100} = 9.57 \qquad E_{32} = \frac{43(33)}{100} = 14.19 \qquad E_{33} = \frac{28(33)}{100} = 9.24$$

The chart reveals that two expected cell frequencies are less than five (more than 20 percent of the expected cell frequencies are less than 5.) Consequently, either the χ^2 test must be terminated or the first two categories of the car variable must be collapsed to increase the expected cell frequencies above five. Assuming that the collapsed variable contains satisfactory information to test the relationship between CARS and REPAR, you can continue. The revised test is shown below and in Table 14–18.

H_0: There is no relationship between CARS and REPAR

H_A: There is a relationship between CARS and REPAR

where

$$REPAR = \begin{cases} 0 \text{ if perform own repairs} \\ 1 \text{ if repairs performed by gas station} \\ 2 \text{ if repairs performed by dealer} \end{cases}$$

$$CARS = \begin{cases} 1 \text{ or } 2 \text{ cars} \\ 3 \text{ or more cars} \end{cases}$$

Table 14–18

Cars	Repair						Total
	Performs own repairs (0)		Repairs done by station (1)		Repairs done by dealer (2)		
1 or 2	25	(19.43)	34	(28.81)	8	(18.76)	67
3 or more	4	(9.57)	9	(14.19)	20	(9.24)	33
Total	29		43		28		100

Notice that 19.43 is 4.64 + 14.79, 28.81 is 6.88 + 21.93, and 18.76 is 4.48 + 14.28. In other words, it is not necessary to recompute the expected cell frequencies.

The Q statistic is computed as follows:

$$Q = \frac{(25 - 19.43)^2}{19.43} + \frac{(34 - 28.81)^2}{28.81} + \frac{(8 - 18.76)^2}{18.76}$$

$$+ \frac{(4 - 9.57)^2}{9.57} + \frac{(9 - 14.19)^2}{14.19} + \frac{(20 - 9.24)^2}{9.24}$$

$$= 1.60 + 0.93 + 6.17 + 3.24 + 1.90 + 12.53$$

$$= 26.37$$

The critical χ^2 value for two degrees of freedom $[(2-1)(3-1)=2]$ and the 5 percent level of significance is 5.99. Since 26.37 is greater than 5.99, the null hypothesis is rejected and you can say there is a statistically significant relationship between REPAR and CARS.

The memorandum appears in Exhibit 14–7.

Exhibit 14–7

HOWARD S. GITLOW, PH.D.
STATISTICAL CONSULTANT

MEMORANDUM

TO: Mr. Rose, Owner
        Rose Auto Repair Store

FROM: Howard Gitlow

DATE: November 11, 1980

RE: Relationship between the number of cars in a family
        and a family's favorite place to have their
        automotive repairs performed, as of January 1980.

   As per your request and using commonly accepted statistical techniques, I have investigated the relationship between "the number of cars in a family (CARS)" and "a family's favorite place to have their automotive repairs performed (REPAR)," as of January 1980.[1] The following definitions were used for the variables under investigation:

$$REPAR = \begin{cases} 0 \text{ if the family performs own repairs} \\ 1 \text{ if repairs were performed by gas station} \\ 2 \text{ if repairs were performed by a dealer} \end{cases}$$

$$CARS = \begin{cases} 1 \text{ or } 2 \text{ cars} \\ 3 \text{ or more cars} \end{cases}$$

Allowing for the errors in survey estimates, you can have "95 percent confidence" that there is a statistically significant relationship between CARS and REPAR.[2]

Exhibit 14-7 (*continued*)

If you have any further questions, please do not hesitate
to call me at 305-999-9999.

Footnotes

[1] Insert standard footnote 1 (insert a sample of 100 Stat City
families drawn via simple random sampling).

[2] The particular sample used in this survey is one of a large
number of all possible samples of the same size that could have been
selected using the same sample design. The sample estimates derived
from the different samples would differ from each other. The
difference between the sample estimates is called sampling error.

The notion of sampling error permits a researcher to say that
only 5 percent of the possible samples could produce sample
statistics that indicate a relationship exists when, in fact, no
relationship exists. Consequently, if the sample statistic gives a
strong indication that a relationship exists, a researcher must
reject the notion of no relationship with the understanding that
he/she may be wrong 5 percent of the time.

The Spearman rank correlation test is used to determine if a statistically significant correlation (association) exists between two variables. Recall that the product moment correlation (r) that was discussed in Chapters 5 and 12 assumed that the underlying population was bivariate normal and that the data was interval or ratio scaled.

The Spearman rank correlation requires data that is ordinal (or better) and does not require bivariate normality. The steps involved in performing a Spearman rank correlation test follow.

STEP 1: State the hypothesis to be tested.

Two-tailed test		*One-tailed test*		*One-tailed test*
H_0: $\rho_s = 0$		H_0: $\rho_s \leq 0$		H_0: $\rho_s \geq 0$
H_A: $\rho_s \neq 0$	OR	H_A: $\rho_s > 0$	OR	H_A: $\rho_s < 0$

where

$$\rho_s = \text{the population Spearman rank correlation coefficient}$$

STEP 2: Set α.

STEP 3: Select the test statistic and the procedure required for its computation.

a.

$$r_s = 1 - \left[\frac{6 \sum\limits_{i=1}^{n} d_i^2}{n(n^2 - 1)} \right]$$

where

r_s = the sample Spearman rank correlation
n = the number of paired observations
x_i = rank assigned to the ith observation on variable x
y_i = rank assigned to the ith observation on variable y
$d_i = x_i - y_i$ = the difference between the x_i and y_i rankings associated with the ith paired observation.

If ties occur, assign each value the average of the ranks that would otherwise have been assigned.

b. Given $n \geq 10$, the statistic

$$t = r_s \sqrt{\frac{(n-2)}{(1-r_s^2)}}$$

approximates a t-distribution with $n - 2$ degrees of freedom. For a given level of significance, reject the null hypothesis if t exceeds the critical t-statistic.

STEP 4: Determine the sample size and draw the random sample.
STEP 5: Compute t.
STEP 6: Determine if the null hypothesis should be rejected.

Example 14-8   Energy consumption study

Ms. Donna Nelson, a reporter for the *Stat City Beacon,* is writing a story about energy consumption habits in Stat City. She would like to determine if there is a statistically significant correlation between "average total yearly heating bill" and "average monthly electric bill" for Stat City families, as of January 1980.

You have been retained by Ms. Nelson to conduct the necessary survey. She has provided you with funds to sample 15 families. Conduct your survey and report your findings to Ms. Nelson at the 5 percent level of significance (see Exhibit 14-8).

Note: You can not assume that the bivariate distribution of electric bills and heating bills is normal.

Table 14-19

ID	$X_i = ELEC_i$	$y_i = HEAT_i$
29	68	778
120	54	908
130	86	1067
189	67	966
225	69	747
265	64	957
276	95	1480
316	102	1009
527	66	726
700	33	288
921	43	462
980	77	748
1195	35	384
1344	61	646
1366	44	533

Solution

STEP 1: State the hypothesis to be tested.

H_0: $\rho_s = 0$ (there is no association between HEAT and ELEC)
H_A: $\rho_s \neq 0$ (there is an association between HEAT and ELEC)

STEP 2: $\alpha = 0.05$.
STEP 3: The test statistic is

$$t = r_s \sqrt{\frac{(n-2)}{(1-r_s^2)}}$$

STEP 4: The simple random sample of 15 Stat City families appears in Table 14-19. (Sample size determination will not be discussed in conjunction with the Spearman rank correlation test.)

STEP 5: Compute t; see Table 14–20 and the following computations.

Table 14–20

ID	ELEC	Ranked $ELEC_i = x_i$	$HEAT_i$	Ranked $HEAT_i = y_i$	$d_i = x_i - y_i$	d_i^2
29	68	10	778	9	+1	1
120	54	5	908	10	−5	25
130	86	13	1,067	14	−1	1
189	67	9	966	12	−3	9
225	69	11	747	7	+4	16
265	64	7	957	11	−4	16
276	95	14	1,480	15	−1	1
316	102	15	1,009	13	+2	4
527	66	8	726	6	+2	4
700	33	1	288	1	0	0
921	43	3	462	3	0	0
980	77	12	748	8	+4	16
1195	35	2	384	2	0	0
1344	61	6	646	5	+1	1
1366	44	4	533	4	0	0

$$\Sigma[d_i^2] = 94$$

$$r_s = 1 - \frac{6(94)}{15(225 - 1)} = 1 - \frac{6(94)}{3,360}$$
$$= 1 - .168 = .832$$

$$t = r_s \sqrt{\frac{(n-2)}{(1-r_s^2)}} = .832 \sqrt{\frac{(15-2)}{(1-.832^2)}}$$

$$= .832 \sqrt{\frac{13}{.308}} = .832(6.499) = 5.407$$

STEP 6: Determine if the null hypothesis should be rejected. Since 5.407 is greater than $|2.160|$, $[t_{(1-\alpha/2=.975,n-2=13)} = 2.160]$, reject the null hypothesis. Consequently, Ms. Nelson will assume that there is a statistically significant correlation between HEAT and ELEC.

The memorandum appears in Exhibit 14–8.

Exhibit 14–8

```
HOWARD S. GITLOW, PH.D.
    STATISTICAL CONSULTANT
    _____


                          MEMORANDUM


TO:     Ms. Donna Nelson, Reporter
        Stat City Beacon

FROM:   Howard Gitlow

DATE:   April 6, 1981

RE:     Energy consumption study.


   Per your request and using commonly accepted statistical
techniques I have investigated the relationship between a
family's "average total yearly heating bill" (HEAT) and a
family's "average monthly electric bill" (ELEC), as of
January 1980.[1,2] Allowing for the errors in survey estimates,
you can have "95 percent confidence" that there is a
statistically significant relationship between HEAT and
ELEC.[3]
   If you have any questions, please do not hesitate to call
me at 305-999-9999.


        _____


                          Footnotes


   [1] Insert standard footnote 1 (insert a random sample of 15 Stat
City families drawn via simple random sampling).
   [2] Insert standard footnote 2.
   [3] Insert standard footnote 4.
```

SUMMARY

In this chapter you became acquainted with nonparametric statistics. The major advantages of nonparametric techniques are as follows: (1) require fewer and less stringent assumptions than parametric techniques, (2) are "generally easy" to compute (given small sample sizes), (3) are appropriate with small sample sizes, and (4) produce more general results than their parametric counterparts. The disadvantages of nonparametric procedures are: (1) difficult to compute given large sample sizes, and (2) usually less powerful than their parametric counterparts when the assumptions required for a parametric test are met.

ADDITIONAL PROBLEMS

14–9 Investigation of the distribution of weekly supermarket bills Mr. Marc Cooper, manager of the A&P, needs to determine if the distribution of weekly supermarket bills in Stat City is normal. He is interested in the following breakdown of weekly supermarket bills: under $30, $30 to less than $60, $60 to less than $90, $90 to less than $120, $120 to less than $150, $150 to less than $180, $180 or more. Mr. Cooper has provided you with sufficient funds to draw a random sample of 80 families. Conduct the necessary survey and report your findings to Mr. Cooper at the 10 percent level of significance (one-sample χ^2 goodness-of-fit test).

Use the accompanying random sample of 80 Stat City families.

Random sample for Problem 14–9

CASE-N	ID	EAT	CASE-N	ID	EAT
1	12.	122.	41	618.	93.
2	30.	79.	42	681.	63.
3	44.	76.	43	709.	160.
4	64.	86.	44	750.	124.
5	69.	108.	45	753.	100.
6	77.	91.	46	771.	109.
7	86.	145.	47	778.	138.
8	87.	107.	48	810.	90.
9	95.	24.	49	816.	97.
10	97.	28.	50	839.	140.
11	102.	64.	51	852.	93.
12	108.	82.	52	890.	69.
13	122.	105.	53	896.	36.
14	141.	68.	54	936.	43.
15	145.	109.	55	968.	109.
16	150.	43.	56	986.	89.
17	220.	115.	57	989.	100.
18	231.	80.	58	1001.	151.
19	232.	97.	59	1018.	78.
20	233.	65.	60	1033.	138.
21	240.	62.	61	1062.	101.
22	275.	71.	62	1076.	112.
23	296.	68.	63	1098.	132.
24	314.	153.	64	1105.	106.
25	338.	91.	65	1112.	144.
26	377.	86.	66	1146.	26.
27	385.	112.	67	1160.	81.
28	389.	88.	68	1170.	84.
29	407.	84.	69	1171.	112.
30	440.	63.	70	1174.	132.
31	449.	130.	71	1203.	122.
32	450.	48.	72	1227.	109.
33	457.	109.	73	1230.	36.
34	494.	140.	74	1258.	46.
35	514.	39.	75	1266.	75.
36	524.	103.	76	1301.	88.
37	529.	100.	77	1323.	145.
38	537.	95.	78	1344.	51.
39	547.	134.	79	1350.	74.
40	567.	120.	80	1368.	128.

Note: $\Sigma x = 7,514$
$\Sigma(x^2) = 790,986$

14–10 Survey concerning the distribution of housing costs between homeowners and apartment dwellers Ms. Sharon Vigil, chairperson of the Stat City Real Estate Board, needs to determine if there is a statistically significant difference in the distribution of housing costs (as of January 1980) between homeowners and apartment dwellers. Ms. Vigil is interested in the following breakdown of monthly housing costs: under $200, $200 to less than $400, $400 to less than $600, $600 to less than $800, $800 to less than $1,000, $1,000 to less than $1,200, and $1,200 or more. She has provided you with funds to draw a random sample of 120 Stat City families. Conduct the survey and report your findings to Ms. Vigil at the 1 percent level of significance (two-sample goodness-of-fit test).

Use the accompanying random sample of 120 homes and apartments.

CASE-N	ID	DWELL	HCOST	CASE-N	ID	DWELL	HCOST
1	3.	1.	644.	61	739.	0.	514.
2	12.	1.	1022.	62	742.	0.	484.
3	17.	1.	683.	63	747.	0.	315.
4	27.	1.	511.	64	768.	0.	276.
5	48.	1.	645.	65	781.	0.	272.
6	54.	1.	566.	66	788.	0.	286.
7	77.	1.	781.	67	790.	0.	416.
8	80.	1.	509.	68	792.	0.	528.
9	100.	1.	819.	69	819.	0.	652.
10	101.	1.	580.	70	820.	0.	711.
11	106.	1.	560.	71	823.	0.	587.
12	114.	1.	658.	72	837.	0.	564.
13	122.	1.	537.	73	841.	0.	238.
14	142.	1.	702.	74	846.	0.	418.
15	148.	1.	956.	75	849.	0.	373.
16	166.	1.	740.	76	868.	0.	450.
17	172.	1.	654.	77	869.	0.	505.
18	178.	1.	1101.	78	881.	0.	689.
19	197.	1.	600.	79	887.	0.	372.
20	234.	1.	528.	80	892.	0.	571.
21	244.	1.	867.	81	954.	0.	477.
22	250.	1.	797.	82	958.	0.	363.
23	263.	1.	766.	83	965.	0.	453.
24	271.	1.	738.	84	1009.	0.	270.
25	272.	1.	891.	85	1024.	0.	458.
26	279.	1.	837.	86	1037.	0.	380.
27	281.	1.	599.	87	1043.	0.	531.
28	297.	1.	275.	88	1053.	0.	355.
29	317.	1.	360.	89	1086.	0.	222.
30	322.	1.	225.	90	1088.	0.	426.
31	353.	1.	640.	91	1100.	0.	432.
32	359.	1.	249.	92	1104.	0.	468.
33	366.	1.	55.	93	1108.	0.	552.
34	378.	1.	183.	94	1112.	0.	469.
35	390.	1.	472.	95	1115.	0.	358.
36	431.	0.	440.	96	1153.	0.	323.
37	439.	0.	615.	97	1169.	0.	230.
38	455.	0.	354.	98	1171.	0.	270.
39	457.	0.	437.	99	1177.	0.	513.
40	459.	0.	424.	100	1191.	0.	337.
41	463.	0.	479.	101	1192.	0.	405.
42	488.	0.	427.	102	1200.	0.	386.
43	521.	0.	609.	103	1205.	0.	264.
44	562.	1.	564.	104	1206.	0.	466.
45	571.	1.	208.	105	1208.	0.	504.
46	580.	1.	562.	106	1215.	0.	538.
47	589.	1.	520.	107	1217.	0.	267.
48	594.	1.	470.	108	1224.	0.	446.
49	612.	1.	231.	109	1235.	0.	282.
50	615.	1.	457.	110	1237.	0.	576.
51	617.	1.	611.	111	1240.	0.	498.
52	622.	1.	300.	112	1245.	0.	485.
53	629.	0.	328.	113	1262.	0.	291.
54	638.	0.	438.	114	1284.	0.	311.
55	677.	0.	616.	115	1287.	0.	394.
56	683.	0.	434.	116	1294.	0.	432.
57	689.	0.	433.	117	1316.	0.	261.
58	697.	0.	376.	118	1319.	0.	289.
59	717.	0.	505.	119	1332.	0.	156.
60	729.	0.	498.	120	1364.	0.	414.

14–11 Study of the assessed values of Stat City houses Mr. Lee Kaplowitz, mayor of Stat City, must determine if the median assessed value of houses in Stat City is greater than $70,000, as of January 1980. If it is, the mayor will not apply for special housing subsidies offered by the federal government. Mr. Kaplowitz is far more concerned about missing the opportunity of obtaining justified subsidies than he is about wasting city money in applying for unjustified subsidies.

You have been hired by the mayor to conduct a survey to obtain the required information. He has provided you with funds to sample 25 Stat City families. Conduct the survey and report your findings to the mayor at the 5 percent level of significance. (Wilcoxon one-sample signed rank test.)

Use the following random sample of Stat City houses.

CASE-N	ID	ASST
1	12.	94549.
2	27.	63601.
3	36.	79019.
4	53.	60875.
5	85.	73351.
6	88.	50844.
7	110.	57964.
8	167.	121338.
9	180.	70408.
10	183.	88804.
11	205.	94743.
12	235.	57906.
13	242.	100303.
14	257.	80570.
15	272.	110652.
16	334.	57954.
17	522.	49317.
18	537.	44747.
19	547.	56719.
20	553.	20833.
21	567.	51210.
22	575.	49758.
23	581.	50188.
24	590.	29101.
25	624.	36624.

14–12 Survey concerning the median telephone bills of homeowners and apartment dwellers Mr. Jack Davis, chairman of the Stat City Telephone Company, must find out if there was a statistically significant difference between the median monthly telephone bills of homeowners and apartment dwellers as of January 1980. He is willing to accept a 5 percent chance of stating that there was a difference between homeowners' and apartment dwellers' median monthly telephone bills when, in fact, there was no statistically significant difference.

Mr. Davis has retained you to conduct a survey to obtain the desired information. He has provided you with enough funds to sample 25 Stat City dwelling units. Conduct the study and report your findings to Mr. Davis (Wilcoxon Rank Sum Test).

Use the following random sample of 25 Stat City families.

ID Variable 4	DWELL Variable 6	PHONE Variable 12
933.	0.	64.
481.	0.	39.
576.	0.	17.
1318.	0.	20.
1348.	0.	19.
966.	0.	18.
1215.	0.	20.
573.	0.	15.
997.	0.	16.
1144.	0.	63.
130.	1.	11.
773.	0.	26.
1022.	0.	18.
317.	1.	30.
303.	1.	39.
1160.	0.	21.
117.	1.	37.
1221.	0.	72.
264.	1.	24.
404.	0.	57.
741.	0.	86.
884.	0.	139.
866.	0.	15.
371.	1.	23.
1276.	0.	37.

14–13 Supermarket bill survey   The Stat City Supermarket Association would like to find out if the median "average weekly supermarket bills," as of January 1980, were the same for the Food Fair, Grand Union, and A&P. The association is willing to accept a 10 percent chance of assuming that there was a statistically significant difference among the median "average weekly supermarket bills" when, in fact, there was none.

The association has retained you to conduct the required survey and provided you with enough funds to draw a sample of 70 families. Conduct the survey and report your findings to the association (Kruskal-Wallis H-test due to the inability to assume the normality of EAT).

Use the accompanying random sample of 70 Stat City families.

CASE-N	ID	EAT	FEAT
1	2.	81.	2.
2	6.	50.	3.
3	19.	101.	3.
4	25.	90.	3.
5	31.	99.	3.
6	69.	108.	3.
7	85.	51.	3.
8	97.	28.	3.
9	135.	86.	3.
10	138.	78.	3.
11	192.	80.	3.
12	219.	96.	3.
13	220.	115.	3.
14	225.	95.	2.
15	244.	86.	3.
16	260.	105.	3.
17	283.	71.	1.
18	289.	76.	1.
19	312.	88.	1.
20	315.	116.	2.
21	344.	78.	1.
22	349.	153.	3.
23	353.	111.	1.
24	399.	86.	1.
25	421.	101.	2.
26	422.	67.	1.
27	425.	58.	1.
28	493.	58.	1.
29	501.	125.	2.
30	504.	70.	1.
31	532.	127.	1.
32	543.	118.	2.
33	618.	93.	2.
34	620.	29.	2.
35	662.	33.	2.
36	667.	164.	2.
37	704.	189.	2.
38	717.	111.	2.
39	722.	77.	2.
40	753.	100.	2.
41	756.	155.	3.
42	762.	97.	2.
43	765.	154.	2.
44	807.	71.	2.
45	818.	56.	2.
46	822.	118.	2.
47	823.	56.	2.
48	842.	109.	2.
49	862.	56.	2.
50	867.	133.	2.
51	886.	111.	2.
52	898.	101.	2.
53	920.	96.	2.
54	945.	29.	2.
55	959.	70.	2.
56	970.	126.	3.
57	984.	110.	3.
58	985.	63.	2.
59	988.	129.	1.
60	1008.	95.	2.
61	1012.	109.	2.
62	1037.	99.	2.
63	1062.	101.	2.
64	1199.	93.	2.
65	1202.	85.	2.
66	1249.	84.	2.
67	1280.	110.	2.
68	1305.	107.	2.
69	1307.	127.	2.
70	1353.	82.	2.

14–14 Testing the income-alcohol theory Dr. Marsha Cox, chief psychiatrist for the Stat City Hospital, would like you to rewrite the memorandum you prepared for her. (See Example 6–2.) She would like you to include a statement about the statistical significance of the income-alcohol theory at the 5 percent level of significance.

You may recall that the psychiatrist believes that poorer families (families earning under $20,000 per year) consume more alcohol than wealthier families (families earning $20,000 or more per year). She believes that understanding the relationship between alcohol consumption and income will enhance her staff's ability to detect families with drinking problems.

If her theory has validity the Stat City Hospital psychological counselors should be alerted for alcohol-related problems when working with poorer families. Before she can issue a directive to her staff regarding this matter, she would like to sample Stat City families and investigate her theory. Basically she needs comparative statistics depicting the chances of poor and wealthy families drinking or not drinking alcohol.

She has decided that the most practical measure of alcohol consumption is whether a family purchases at least one six-pack of beer per week.

The sample survey data that was collected is listed for your convenience.

Sample survey data for Problem 14-14

ID	INCOM	BEER	ID	INCOM	BEER
5.	40603.	1.	767.	7809.	0.
82.	17046.	4.	778.	6476.	0.
94.	36165.	3.	793.	18811.	10.
122.	62205.	3.	851.	12642.	0.
123.	56479.	0.	862.	17522.	0.
134.	63330.	0.	884.	11445.	0.
140.	23576.	4.	897.	11969.	0.
211.	96618.	5.	912.	16601.	3.
292.	48054.	0.	931.	8987.	5.
316.	28784.	0.	942.	3938.	1.
320.	31665.	4.	967.	26820.	0.
328.	22232.	4.	972.	21765.	0.
408.	33300.	0.	992.	33790.	0.
439.	25971.	5.	1018.	14985.	5.
487.	23527.	9.	1024.	12822.	5.
501.	26669.	0.	1032.	18651.	0.
568.	51522.	3.	1078.	6894.	0.
581.	41974.	0.	1090.	24361.	2.
637.	16917.	0.	1127.	18341.	1.
655.	23675.	0.	1155.	14018.	0.
689.	14369.	7.	1242.	24743.	2.
712.	13527.	0.	1247.	21735.	1.
718.	19359.	3.	1299.	29004.	0.
723.	13377.	3.	1358.	9207.	0.
743.	16473.	0.	1368.	8603.	4.

14–15 Energy consumption article Ms. Donna Nelson, a reporter for the *Stat City Beacon,* is writing another story about energy consumption in Stat City. This time she would like to determine if there is a statistically significant relationship between the average bimonthly automobile gasoline bills and the average total yearly heating bills for Stat City families, as of January 1980.

You have been hired by Ms. Nelson to conduct a survey to obtain the desired information. She has provided you with funds to draw a sample of 20 families. Conduct your survey and report your findings to Ms. Nelson at the 5 percent level of significance.

Note: You cannot assume that the bivariate distribution of GAS and HEAT is bivariate normal.

Use the accompanying random sample of 20 Stat City families.

Random sample for Problem 14–15

CASE-N	ID	HEAT	GAS
1	38.	1204.	217.
2	158.	1443.	120.
3	235.	632.	134.
4	282.	1313.	201.
5	294.	735.	138.
6	354.	544.	65.
7	379.	698.	159.
8	487.	671.	162.
9	531.	655.	176.
10	730.	247.	120.
11	772.	261.	120.
12	793.	428.	65.
13	797.	392.	65.
14	872.	265.	65.
15	942.	475.	120.
16	1165.	229.	120.
17	1178.	545.	65.
18	1180.	666.	120.
19	1195.	384.	120.
20	1270.	412.	65.

PART SEVEN

STATISTICS AND THE COMPUTER

15

You have spent this semester becoming acquainted with and computing statistics. By this time, you must realize that statistical computations can be very time-consuming. You also know the frustration of getting the wrong answer to a problem because you made a careless math error or pushed the wrong button on your calculator. Well, take heart, because help with statistical computations is available with computers.

Computers are your friends, not your enemies. Once you become familiar with computers, you will achieve a whole new level of statistical proficiency and awareness. Statistical computations with a computer are no longer difficult or time-consuming. Computers virtually erase the difficulties encountered in statistical computations (clients always want accurate answers "yesterday").

SPSS (*S*tatistical *P*ackage for the *S*ocial *S*ciences) is a set of prewritten computer instructions which can calculate most of the statistics you will ever need. It is easy to use SPSS. All SPSS programs have exactly the same structure. As a matter of fact, the statements in an SPSS program vary very little from program to program.

SPSS programs are made up of control cards. Each control card has a specific command that the computer must perform. Control cards have two parts, or fields: a control field and a specification field. The control field occupies the first 15 columns of a control card and contains one or more words indicating the task the computer is being instructed to perform. The specification field occupies card columns 16 through 80 of the control card. The specification field enumerates the specifics of the operation to be performed that is stated in the control field.

The cards in Exhibit 15–1 are a complete listing of an SPSS program designed to analyze a sample survey of 10 Stat City dwelling units. The control cards exemplify what control and specification fields of control cards look like.

A brief discussion of the program in Exhibit 15–1 follows. The RUN NAME card simply gives the SPSS program a name so that it can be easily recognized at some future date. The VARIABLE LIST card names the variables that are to be analyzed. The INPUT FORMAT card tells the computer where each piece of data to be analyzed is located so the computer can properly read the data. In this example, the number of rooms in a dwelling unit is located in data card columns 1 and 2, the number of

Exhibit 15-1

IBM   FORTRAN CODING FORM   X28-7327-6 U/M 050
                                                                  Printed in U.S.A.

| PROGRAM | STAT CITY SURVEY | | GRAPHIC | | | | | | | PAGE 1 OF 1 |
| PROGRAMMER | HOWARD GITLOW | DATE DEC. 3, 1980 | PUNCHING INSTRUCTIONS | PUNCH | | | | | | CARD ELECTRO NUMBER* |

CONTROL FIELD	SPECIFICATION FIELD — FORTRAN STATEMENT	IDENTIFICATION SEQUENCE
RUN NAME	STAT CITY SURVEY OF THE # OF ROOMS & PEOPLE PER DWELLING UNIT	
VARIABLE LIST	ROOMS, PEOPLE, DWELL	
INPUT FORMAT	FIXED(F2.0,F2.0,F1.0)	
N OF CASES	10	
INPUT MEDIUM	CARD	
VAR LABELS	ROOMS, # OF ROOMS IN A HOUSEHOLD AS OF 1-80/	
	PEOPLE, # OF PEOPLE IN A HOUSEHOLD AS OF 1-80/	
	DWELL, TYPE OF DWELLING UNIT	
VALUE LABELS	DWELL (0)APARTMENT (1)HOUSE	
CONDESCRIPTIVE	ROOMS, PEOPLE	
STATISTICS	1,2,5,9,10,11	
READ INPUT DATA		
09041		
09051		
08031		
05040		
07031		
10061		
05050		
03020		
08030		
05070		
FREQUENCIES	GENERAL=DWELL	
STATISTICS	4	
FINISH		

people in a dwelling unit is located in data card columns 3 and 4, and the type of dwelling unit is located in data card column 5. Further, this tells us that there is only one data card per dwelling unit. The N of CASES card indicates the number of data points (questionnaires) to be analyzed. The INPUT MEDIUM card indicates that the data is located on computer cards. The VAR LABELS card gives additional details on the exact definition of each variable under analysis. The VALUE LABELS card verbally defines the numeric codes assigned to nominal or ordinal variables. The CONDESCRIPTIVE and STATISTICS cards indicate that the mean, standard error, standard deviation, range, minimum, and maximum should be computed for ROOMS and PEPLE. The READ INPUT DATA card tells the computer that the data previously defined are about to be read into the computer. The first three lines of data are interpreted as follows:

09041   The first family selected in the sample lives in a *house* (1) with *9 rooms* and is comprised of *4 people*

09051   The second family selected in the sample lives in a *house* (*1*) with *9 rooms* and is comprised of *5 people*

08031   The third family selected in the sample lives in a *house* (*1*) with *8 rooms* and is comprised of *3 people*

Please note that the READ INPUT DATA card and the data must immediately follow the first statistical task to be performed. The FREQUENCIES and STATISTICS cards indicate that a frequency distribution and mode should be constructed for the type of dwelling unit. The FINISH card signifies the end of the SPSS program.

The exact meanings of the control cards are explained in the following section of this chapter. Of course, the *SPSS* manual should be consulted for any detailed information that might be needed. The *SPSS* manual can be obtained from your college bookstore or the publisher, McGraw-Hill Book Company (New York, St. Louis, or San Francisco). The authors of the *SPSS* manual are: Norman Nie, C. Hadlai Hull, Jean Jenkins, Karen Steinbrunner, and Dale Brent.

All SPSS programs are composed of three basic parts: (1) a data definition section, (2) a data modification section, and (3) a task definition section.

The data definition section is comprised of SPSS control cards that explain the data being studied. The data definition section of an SPSS program:

1. Names the analysis being performed.
2. Names the variables under investigation.
3. Defines the location of each variable on a data card.
4. Enumerates how many cases (dwelling units) are going to be analyzed.
5. Defines how the data will be read into the computer (cards, tapes, disks, and so on).
6. Affords expanded definitions for the variables under investigation.
7. Gives verbal descriptions of all the categories of nominal or ordinal variables.

The data definition section of an SPSS program used to analyze Stat City data can be seen in Exhibit 15–2.

The data modification section of an SPSS program is comprised of control cards that modify the coding scheme for one or more variables in the data file. There are several types of data modification cards in SPSS.

1. The RECODE card. The RECODE card allows a researcher to replace (recode) any value or set of values for a variable with a new value of his/her choice. For example, the data modification card required to recode average monthly electric bills into the following categories: $10 to less than $40, $40 to less than $70, $70 to less than $100, and $100 to less than $130, is:

```
Card        Card
column 1    column 16
   ↓           ↓
   RECODE    ELEC (10 THRU 39=0) (40 THRU 69=1) (70 THRU 99=2)
                  (100 THRU 130=3)
```

A researcher may want to perform the above transformation to obtain the percentage of Stat City families with average monthly electric bills in the above categories. See Example 3–3.

2. The COMPUTE card.   The COMPUTE card allows a researcher to assign a variable or value according to some mathematical formula for every case in his/her data file. For example, the following modification card is required if a researcher wanted to transform all Stat City incomes into Z-scores $[Z = (X - \mu)/\sigma]$:

```
Card        Card
column 1    column 16
   ↓           ↓
   COMPUTE    ZINCOM = (INCOM − 29394.572)/22847.118
```

The above transformation requires knowing that $\mu = \$29,394.572$ and $\sigma = \$22,847.118$. The above type of transformation is useful in avoiding rounding errors when working with SPSS.

Exhibit 15–2

Assigns a name to the ─────► 1.   RUN NAME   STAT CITY SPSS DRIVER PROGRAM
analysis being performed. ─► 2.   VARIABLE LIST NAME1,NAME2,NAME3,NAME4,ADDR1,ADDR2,ADDR3,ADDR4,
                  3.   BLOCK,ID,ZONE,DWELL,HCOST,ASSES,ROOMS,HEAT,ELEC,
Lists the variables ─────── 4.   PHONE,INCOM,PEPLE,CARS,GAS,GASCA,GASTR,REPAR,
under study.   5.   FAVGA,HOSP,EAT,EATPL,FEAT,LSODA,HSODA,BEER
                  6.   INPUT FORMAT FIXED (1X,4A3,1X,4A4,F3.0,2X,F4.0,1X,F1.0,1X,
Defines the location ───── 7.   F1.0,1X,F4.0,1X,F6.0,1X,F2.0,1X,F4.0,1X,
of each variable on   8.   F4.0,1X,F3.0,1X,F6.0,1X/F2.0,1X,F1.0,1X,F3.0,1X,
a computer card.   9.   F3.0,1X,F2.0,1X,F1.0,1X,F1.0,1X,
               10.   F2.0,1X,F3.0,1X,F2.0,1X,F1.0,1X,F2.0,
               11.   1X,F2.0,1X,F2.0)

ACCORDING TO YOUR INPUT FORMAT, VARIABLES ARE TO BE READ AS FOLLOWS

VARIABLE	FORMAT	RECORD	COLUMNS
NAME1	A 3	1	2- 4
NAME2	A 3	1	5- 7
NAME3	A 3	1	8- 10
NAME4	A 3	1	11- 13
ADDR1	A 4	1	15- 18
ADDR2	A 4	1	19- 22
ADDR3	A 4	1	23- 26
ADDR4	A 4	1	27- 30
BLOCK	F 3. 0	1	31- 33
ID	F 4. 0	1	36- 39
ZONE	F 1. 0	1	41- 41
DWELL	F 1. 0	1	43- 43
HCOST	F 4. 0	1	45- 48
ASSES	F 6. 0	1	50- 55
ROOMS	F 2. 0	1	57- 58
HEAT	F 4. 0	1	60- 63
ELEC	F 4. 0	1	65- 68
PHONE	F 3. 0	1	70- 72
INCOM	F 6. 0	1	74- 79
PEPLE	F 2. 0	2	1- 2
CARS	F 1. 0	2	4- 4
GAS	F 3. 0	2	6- 8
GASCA	F 3. 0	2	10- 12
GASTR	F 2. 0	2	14- 15
REPAR	F 1. 0	2	17- 17
FAVGA	F 1. 0	2	19- 19
HOSP	F 2. 0	2	21- 22
EAT	F 3. 0	2	24- 26
EATPL	F 2. 0	2	28- 29
FEAT	F 1. 0	2	31- 31
LSODA	F 2. 0	2	33- 34
HSODA	F 2. 0	2	36- 37
BEER	F 2. 0	2	39- 40

THE INPUT FORMAT PROVIDES FOR 33 VARIABLES.   33 WILL BE READ
IT PROVIDES FOR 2 RECORDS ('CARDS') PER CASE. A MAXIMUM OF 80 'COLUMNS' ARE USED ON A RECORD.
```

States how many ───────► 12.   N OF CASES     1373
cases will be analyzed. ─► 13.   INPUT MEDIUM   CARD
               14.   COMMENT       THE ENTIRE STAT CITY DATA BASE WAS COLLECTED VIA
               15.   COMMENT       QUESTIONNAIRE SURVEY DURING FEBRUARY AND MARCH OF
               16.   COMMENT       1980.  THE DATA BASE REFLECTS THE CHARACTERISTICS
Defines how the ─────────17.   COMMENT       OF STAT CITY DWELLING UNITS AS OF JANUARY 1980.
computer will         18.   COMMENT
read the data.        19.   COMMENT
               20.   COMMENT
               21.   COMMENT       THE ELEMENTARY UNIT FOR ANALYSIS IN STAT CITY IS THE
               22.   COMMENT       DWELLING UNIT, NOT THE INDIVIDUAL.  ALL PROBLEMS WILL
               23.   COMMENT       CONSIDER THE DWELLING UNIT (FAMILY OR HOME) AS THE
               24.   COMMENT       BASIC UNIT FOR ANALYSIS.
               25.   COMMENT
               26.   COMMENT
Affords expanded      27.   COMMENT
definitions of the ─────► 28.   VAR LABELS    NAME1,FIRST 4 LETTERS OF LAST NAME/
variables under study.
```

Exhibit 15-2 (*continued*)

```
STAT CITY SPSS DRIVER PROGRAM
           29.                    NAME2,SECOND 4 LETTERS OF LAST NAME/
           30.                    NAME3,THIRD 4 LETTERS OF LAST NAME/
           31.                    NAME4,FOURTH 4 LETTERS OF LAST NAME/
           32.                    ADDR1,FIRST 4 LETTERS OF ADDRESS/
           33.                    ADDR2,SECOND 4 LETTERS OF ADDRESS/
           34.                    ADDR3,THIRD 4 LETTERS OF ADDRESS/
           35.                    ADDR4,FOURTH 4 LETTERS OF ADDRESS/
           36.                    BLOCK,BLOCK DWELLING UNIT IS LOCATED ON/
           37.                    ID,IDENTIFICATION NUMBER/
           38.                    ZONE,RESIDENTIAL HOUSING ZONE/
           39.                    DWELL,DWELLING TYPE/
           40.                    HCOST,HOUSING COST AS OF 1-80/
           41.                    ASSES,ASSESSED VALUE OF HOUSE AS OF 1-80/
           42.                    ROOMS,# OF ROOMS IN DWELLING UNIT AS OF 1-80/
           43.                    HEAT,AVERAGE YEARLY HEATING BILL AS OF 1-80/
           44.                    ELEC,AVERAGE MONTHLY ELECTRIC BILL AS OF 1-80/
           45.                    PHONE,AVERAGE MONTHLY PHONE BILL AS OF 1-80/
           46.                    INCOM,TOTAL FAMILY INCOME IN 1979/
           47.                    PEPLE,NUMBER OF PEOPLE IN HOUSEHOLD AS OF 1-80/
           48.                    CARS,NUMBER OF CARS IN HOUSEHOLD AS OF 1-80/
           49.                    GAS,AVERAGE BIMONTHLY GAS BILL AS OF 1-80/
           50.                    GASCA,AVE. BIMONTHLY GAS PER CAR AS OF 1-80/
           51.                    GASTR,AVE. MONTHLY TRIPS FOR GAS AS OF 1-80/
           52.                    REPAR,FAVORITE PLACE FOR CAR REPAIRS AS OF 1-80/
           53.                    FAVGA,FAVORITE GAS STATION AS OF 1-80/
           54.                    HOSP,AVE YRLY HOSP. TRIPS PER HOME AS OF 1-80/
           55.                    EAT,AVERAGE WEEKLY FOOD BILL AS OF 1-80/
           56.                    EATPL,AVE WKLY FOOD BILL PER PERSON AS OF 1-80/
           57.                    FEAT,FAVORITE SUPERMARKET AS OF 1-80/
           58.                    LSODA,AVE WKLY DIET SODA PURCHASE AS OF 1-80/
           59.                    HSODA,AVE WKLY REG. SODA PURCHASE AS OF 1-80/
           60.                    BEER,AVE WKLY BEER PURCHASE AS OF 1-80
           61.   COMMENT          LSODA,HSODA, AND BEER PURCHASES ARE IN SIX PACK UNITS.
           62.   VALUE LABELS     ZONE (1) ZONE 1 (2) ZONE 2 (3) ZONE 3 (4) ZONE 4/
           63.                    DWELL (0)APARTMENT (1)HOUSE/
           64.                    REPAR (0)PERFORMS OWN REPAIRS (1)REPAIRS DONE BY STATION
           65.                          (2)REPAIRS DONE BY DEALER/
           66.                    FAVGA (1)PAUL'S TEXACO STATION (2)HOWIE'S GULF STATION/
           67.                    FEAT (1)FOODFAIR (2)GRAND UNION (3)A & P
```

Gives verbal descriptions of all the categories of nominal and ordinal variables.

3. The IF card. The IF card allows a researcher to perform logical operations. The IF card is especially useful in constructing new variables. For example, if a researcher needs a variable to indicate whether a Stat City family purchased beer or not, the following control card sequence could be used:

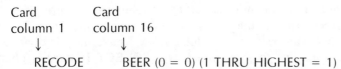

```
Card        Card
column 1    column 16
   ↓           ↓
   IF       (BEER EQ 0) BEER1 = 0
   IF       (BEER GT 0) BEER1 = 1
```

The above control card sequence has created a new variable, BEER1, which indicates if a family purchase beer (BEER1 = 1) or not (BEER1 = 0). The above data modification could also have been done with the following RECODE card:

```
Card         Card
column 1     column 16
   ↓            ↓
  RECODE     BEER (0 = 0) (1 THRU HIGHEST = 1)
```

Please note, the RECODE card transforms the variable BEER, while the IF card sequence creates a new variable, BEER1, and leaves the variable BEER intact.

4. The SELECT IF card. The SELECT IF card allows a researcher to include only those cases for analysis which pass a test. For example, if a researcher wanted to perform some analysis on only Zone 1 residents, the following card would be used:

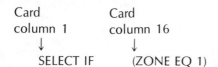

```
Card           Card
column 1       column 16
   ↓              ↓
 SELECT IF     (ZONE EQ 1)
```

The task definition section is comprised of control cards that define the statistical procedures to be performed. There are many statistical procedures that can be easily used via SPSS. However, only six statistical procedures will be discussed in this chapter.

1. FREQUENCIES. The FREQUENCIES control card initiates a statistical program that computes and presents frequency distributions for one or more variables (usually nominal or ordinal variables). The FREQUENCIES card is capable of presenting frequency distributions, means, medians, modes, standard deviations, variances, standard errors, minimums, maximums, and ranges. It is important to remember that not all of the above statistics are meaningful for the types of variables that are commonly analyzed with FREQUENCIES (nominal or ordinal data).

An example of the format of the FREQUENCIES card is:

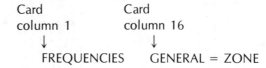

The above card will cause a frequency distribution to be printed indicating the numbers and percent of Stat City families in each residential housing zone.

An SPSS printout that computes and presents the above frequency distribution can be seen in Exhibit 15–3.

2. CONDESCRIPTIVE. The CONDESCRIPTIVE control card initiates a statistical program that computes a set of statistics for one or more variables (usually interval or ratio variables). The statistics that are available can be selectively called via the following numbering scheme:

1	Mean	8	Skewness
2	Standard error	9	Range
5	Standard deviation	10	Minimum
6	Variance	11	Maximum
7	Kurtosis		

An example of the format of the CONDESCRIPTIVE card is:

```
Card                  Card
column 1              column 16
   ↓                     ↓
   CONDESCRIPTIVE        HOSP
   STATISTICS           1, 5, 9, 10, 11
```

The above card sequence will cause the mean, standard deviation, range, minimum and maximum of HOSP to be computed and printed.

An SPSS printout that computes the above parameters for HOSP, *IN ZONE 1*, can be seen in Exhibit 15–4. I have included only Zone 1 dwelling units just to show you how a SELECT IF card is utilized in conjunction with a statistical routine.

3. CROSSTABS. The CROSSTABS control card initiates a statistical program that computes and prints a bivariate frequency distribution (contingency table) for one or more variables (usually nominal or ordinal). The CROSSTABS card is capable of presenting cross-tabulations, marginal probabilities, joint probabilities, conditional probabilities, χ^2 statistics, plus other information.

Exhibit 15–3

SPSS BATCH SYSTEM

SPSS FOR SPERRY UNIVAC 1100 EXEC 8, VERSION H, RELEASE 8.0-UW1.0, JUNE 1979

```
SPACE ALLOCATION..              ALLOWS FOR..      20 TRANSFORMATIONS
WORKSPACE    4375 WORDS                           82 RECODE VALUES + LAG VARIABLES
TRANSPACE     625 WORDS                          164 IF/COMPUTE OPERATIONS
              1.  RUN NAME        STAT CITY SPSS DRIVER PROGRAM
              2.  VARIABLE LIST   NAME1,NAME2,NAME3,NAME4,ADDR1,ADDR2,ADDR3,ADDR4,
              3.                  BLOCK,ID,ZONE,DWELL,HCOST,ASSES,ROOMS,HEAT,ELEC,
              4.                  PHONE,INCOM,PEPLE,CARS,GAS,GASCA,GASTR,REPAR,
              5.                  FAVGA,HOSP,EAT,EATPL,FEAT,LSODA,HSODA,BEER
              6.  INPUT FORMAT    FIXED (1X,4A3,1X,4A4,F3.0,2X,F4.0,1X,F1.0,1X,
              7.                  F1.0,1X,F4.0,1X,F6.0,1X,F2.0,1X,F4.0,1X,
              8.                  F4.0,1X,F3.0,1X,F6.0,1X,F2.0,1X,F1.0,1X,F3.0,1X,
              9.                  F3.0,1X,F2.0,1X,F1.0,1X,F1.0,1X,
             10.                  F2.0,1X,F3.0,1X,F2.0,1X,F1.0,1X,F2.0,
             11.                  1X,F2.0,1X,F2.0)
```

ACCORDING TO YOUR INPUT FORMAT, VARIABLES ARE TO BE READ AS FOLLOWS

```
          VARIABLE   FORMAT   RECORD      COLUMNS
          NAME1       A  3      1         2-     4
          NAME2       A  3      1         5-     7
          NAME3       A  3      1         8-    10
          NAME4       A  3      1        11-    13
          ADDR1       A  4      1        15-    18
          ADDR2       A  4      1        19-    22
          ADDR3       A  4      1        23-    26
          ADDR4       A. 4      1        27-    30
          BLOCK       F  3. 0   1        31-    33
          ID          F  4. 0   1        36-    39
          ZONE        F  1. 0   1        41-    41
          DWELL       F  1. 0   1        43-    43
          HCOST       F  4. 0   1        45-    48
          ASSES       F  6. 0   1        50-    55
          ROOMS       F  2. 0   1        57-    58
          HEAT        F  4. 0   1        60-    63
          ELEC        F  4. 0   1        65-    68
          PHONE       F  3. 0   1        70-    72
          INCOM       F  6. 0   1        74-    79
          PEPLE       F  2. 0   2         1-     2
          CARS        F  1. 0   2         4-     4
          GAS         F  3. 0   2         6-     8
          GASCA       F  3. 0   2        10-    12
          GASTR       F  2. 0   2        14-    15
          REPAR       F  1. 0   2        17-    17
          FAVGA       F  1. 0   2        19-    19
          HOSP        F  2. 0   2        21-    22
          EAT         F  3. 0   2        24-    26
          EATPL       F  2. 0   2        28-    29
          FEAT        F  1. 0   2        31-    31
          LSODA       F  2. 0   2        33-    34
          HSODA       F  2. 0   2        36-    37
          BEER        F  2. 0   2        39-    40
```

THE INPUT FORMAT PROVIDES FOR 33 VARIABLES.   33 WILL BE READ
IT PROVIDES FOR 2 RECORDS ('CARDS') PER CASE. A MAXIMUM OF   80 'COLUMNS' ARE USED ON A RECORD.

```
             12.  N OF CASES      1373
             13.  INPUT MEDIUM    CARD
             14.  COMMENT         THE ENTIRE STAT CITY DATA BASE WAS COLLECTED VIA
             15.  COMMENT         QUESTIONNAIRE SURVEY DURING FEBRUARY AND MARCH OF
             16.  COMMENT         1980.  THE DATA BASE REFLECTS THE CHARACTERISTICS
             17.  COMMENT         OF STAT CITY DWELLING UNITS AS OF JANUARY 1980.
             18.  COMMENT
             19.  COMMENT
             20.  COMMENT
             21.  COMMENT         THE ELEMENTARY UNIT FOR ANALYSIS IN STAT CITY IS THE
             22.  COMMENT         DWELLING UNIT, NOT THE INDIVIDUAL.  ALL PROBLEMS WILL
             23.  COMMENT         CONSIDER THE DWELLING UNIT (FAMILY OR HOME) AS THE
             24.  COMMENT         BASIC UNIT FOR ANALYSIS.
             25.  COMMENT
             26.  COMMENT
             27.  COMMENT
             28.  VAR LABELS      NAME1,FIRST 4 LETTERS OF LAST NAME/
             29.                  NAME2,SECOND 4 LETTERS OF LAST NAME/
             30.                  NAME3,THIRD 4 LETTERS OF LAST NAME/
             31.                  NAME4,FOURTH 4 LETTERS OF LAST NAME/
             32.                  ADDR1,FIRST 4 LETTERS OF ADDRESS/
             33.                  ADDR2,SECOND 4 LETTERS OF ADDRESS/
             34.                  ADDR3,THIRD 4 LETTERS OF ADDRESS/
             35.                  ADDR4,FOURTH 4 LETTERS OF ADDRESS/
             36.                  BLOCK,BLOCK DWELLING UNIT IS LOCATED ON/
             37.                  ID,IDENTIFICATION NUMBER/
             38.                  ZONE,RESIDENTIAL HOUSING ZONE/
             39.                  DWELL,DWELLING TYPE/
             40.                  HCOST,HOUSING COST AS OF 1-80/
             41.                  ASSES,ASSESSED VALUE OF HOUSE AS OF 1-80/
             42.                  ROOMS,# OF ROOMS IN DWELLING UNIT AS OF 1-80/
             43.                  HEAT,AVERAGE YEARLY HEATING BILL AS OF 1-80/
             44.                  ELEC,AVERAGE MONTHLY ELECTRIC BILL AS OF 1-80/
             45.                  PHONE,AVERAGE MONTHLY PHONE BILL AS OF 1-80/
             46.                  INCOM,TOTAL FAMILY INCOME IN 1979/
             47.                  PEPLE,NUMBER OF PEOPLE IN HOUSEHOLD AS OF 1-80/
             48.                  CARS,NUMBER OF CARS IN HOUSEHOLD AS OF 1-80/
             49.                  GAS,AVERAGE BIMONTHLY GAS BILL AS OF 1-80/
             50.                  GASCA,AVE. BIMONTHLY GAS PER CAR AS OF 1-80/
```

Exhibit 15–3 (*continued*)

STAT CITY SPSS DRIVER PROGRAM

```
        51.                     GASTR,AVE. MONTHLY TRIPS FOR GAS AS OF 1-80/
        52.                     REPAR,FAVORITE PLACE FOR CAR REPAIRS AS OF 1-80/
        53.                     FAVGA,FAVORITE GAS STATION AS OF 1-80/
        54.                     HOSP,AVE YRLY HOSP. TRIPS PER HOME AS OF 1-80/
        55.                     EAT,AVERAGE WEEKLY FOOD BILL AS OF 1-80/
        56.                     EATPL,AVE WKLY FOOD BILL PER PERSON AS OF 1-80/
        57.                     FEAT,FAVORITE SUPERMARKET AS OF 1-80/
        58.                     LSODA,AVE WKLY DIET SODA PURCHASE AS OF 1-80/
        59.                     HSODA,AVE WKLY REG. SODA PURCHASE AS OF 1-80/
        60.                     BEER,AVE WKLY BEER PURCHASE AS OF 1-80
        61.     COMMENT         LSODA,HSODA, AND BEER PURCHASES ARE IN SIX PACK UNITS.
        62.     VALUE LABELS    ZONE (1) ZONE 1 (2) ZONE 2 (3) ZONE 3 (4) ZONE 4/
        63.                     DWELL (0)APARTMENT (1)HOUSE/
        64.                     REPAR (0)PERFORMS OWN REPAIRS (1)REPAIRS DONE BY STATION
        65.                           (2)REPAIRS DONE BY DEALER/
        66.                     FAVGA (1)PAUL'S TEXACO STATION (2)HOWIE'S GULF STATION/
        67.                     FEAT  (1)FOODFAIR (2)GRAND UNION (3)A & P
        68.     FREQUENCIES     GENERAL=ZONE
```

GIVEN WORKSPACE ALLOWS FOR   2187 VARIABLES WITH   1093 VALUES AND   437 LABELS PER VARIABLE FOR 'FREQUENCIES'

```
        69.   READ INPUT DATA
```

FILE   NONAME   (CREATION DATE = 05/18/81)

ZONE   RESIDENTIAL HOUSING ZONE

CATEGORY LABEL	CODE	ABSOLUTE FREQ	RELATIVE FREQ (PCT)	ADJUSTED FREQ (PCT)	CUM FREQ (PCT)
ZONE 1	1.	130	9.5	9.5	9.5
ZONE 2	2.	157	11.4	11.4	20.9
ZONE 3	3.	338	24.6	24.6	45.5
ZONE 4	4.	748	54.5	54.5	100.0
	TOTAL	1373	100.0	100.0	

VALID CASES   1373   MISSING CASES   0

CPU TIME REQUIRED..   1.84 SECONDS

```
        70.   FINISH
        NORMAL END OF JOB.
          70 CONTROL CARDS WERE PROCESSED.
           0 ERRORS WERE DETECTED.
```

Exhibit 15–4

SPSS BATCH SYSTEM

SPSS FOR SPERRY UNIVAC 1100 EXEC 8, VERSION H, RELEASE 8.0-UW1.0, JUNE 1979

```
SPACE ALLOCATION..              ALLOWS FOR..    20 TRANSFORMATIONS
WORKSPACE      4375 WORDS                       82 RECODE VALUES + LAG VARIABLES
TRANSPACE       625 WORDS                      164 IF/COMPUTE OPERATIONS

         1.  RUN NAME        STAT CITY SPSS DRIVER PROGRAM
         2.  VARIABLE LIST   NAME1,NAME2,NAME3,NAME4,ADDR1,ADDR2,ADDR3,ADDR4,
         3.                  BLOCK,ID,ZONE,DWELL,HCOST,ASSES,ROOMS,HEAT,ELEC,
         4.                  PHONE,INCOM,PEPLE,CARS,GAS,GASCA,GASTR,REPAR,
         5.                  FAVGA,HOSP,EAT,EATPL,FEAT,LSODA,HSODA,BEER
         6.  INPUT FORMAT    FIXED (1X,4A3,1X,4A4,F3.0,2X,F4.0,1X,F1.0,1X,
         7.                  F1.0,1X,F4.0,1X,F6.0,1X,F2.0,1X,F4.0,1X,
         8.                  F4.0,1X,F3.0,1X,F6.0,1X/F2.0,1X,F1.0,1X,F3.0,1X,
         9.                  F3.0,1X,F2.0,1X,F1.0,1X,F1.0,1X,
        10.                  F2.0,1X,F3.0,1X,F2.0,1X,F1.0,1X,F2.0,
        11.                  1X,F2.0,1X,F2.0)
```

ACCORDING TO YOUR INPUT FORMAT, VARIABLES ARE TO BE READ AS FOLLOWS

VARIABLE	FORMAT	RECORD	COLUMNS
NAME1	A 3	1	2- 4
NAME2	A 3	1	5- 7
NAME3	A 3	1	8- 10
NAME4	A 3	1	11- 13
ADDR1	A 4	1	15- 18
ADDR2	A 4	1	19- 22
ADDR3	A 4	1	23- 26
ADDR4	A 4	1	27- 30
BLOCK	F 3. 0	1	31- 33
ID	F 4. 0	1	36- 39
ZONE	F 1. 0	1	41- 41
DWELL	F 1. 0	1	43- 43
HCOST	F 4. 0	1	45- 48
ASSES	F 6. 0	1	50- 55
ROOMS	F 2. 0	1	57- 58
HEAT	F 4. 0	1	60- 63
ELEC	F 4. 0	1	65- 68
PHONE	F 3. 0	1	70- 72
INCOM	F 6. 0	1	74- 79
PEPLE	F 2. 0	2	1- 2
CARS	F 1. 0	2	4- 4
GAS	F 3. 0	2	6- 8
GASCA	F 3. 0	2	10- 12
GASTR	F 2. 0	2	14- 15
REPAR	F 1. 0	2	17- 17
FAVGA	F 1. 0	2	19- 19
HOSP	F 2. 0	2	21- 22
EAT	F 3. 0	2	24- 26
EATPL	F 2. 0	2	28- 29
FEAT	F 1. 0	2	31- 31
LSODA	F 2. 0	2	33- 34
HSODA	F 2. 0	2	36- 37
BEER	F 2. 0	2	39- 40

THE INPUT FORMAT PROVIDES FOR 33 VARIABLES.   33 WILL BE READ
IT PROVIDES FOR 2 RECORDS ('CARDS') PER CASE. A MAXIMUM OF 80 'COLUMNS' ARE USED ON A RECORD.

```
        12.  N OF CASES      1373
        13.  INPUT MEDIUM    CARD
        14.  COMMENT         THE ENTIRE STAT CITY DATA BASE WAS COLLECTED VIA
        15.  COMMENT         QUESTIONNAIRE SURVEY DURING FEBRUARY AND MARCH OF
        16.  COMMENT         1980.  THE DATA BASE REFLECTS THE CHARACTERISTICS
        17.  COMMENT         OF STAT CITY DWELLING UNITS AS OF JANUARY 1980.
        18.  COMMENT
        19.  COMMENT
        20.  COMMENT
        21.  COMMENT         THE ELEMENTARY UNIT FOR ANALYSIS IN STAT CITY IS THE
        22.  COMMENT         DWELLING UNIT, NOT THE INDIVIDUAL.  ALL PROBLEMS WILL
        23.  COMMENT         CONSIDER THE DWELLING UNIT (FAMILY OR HOME) AS THE
        24.  COMMENT         BASIC UNIT FOR ANALYSIS.
        25.  COMMENT
        26.  COMMENT
        27.  COMMENT
        28.  VAR LABELS      NAME1,FIRST 4 LETTERS OF LAST NAME/
        29.                  NAME2,SECOND 4 LETTERS OF LAST NAME/
        30.                  NAME3,THIRD 4 LETTERS OF LAST NAME/
        31.                  NAME4,FOURTH 4 LETTERS OF LAST NAME/
        32.                  ADDR1,FIRST 4 LETTERS OF ADDRESS/
        33.                  ADDR2,SECOND 4 LETTERS OF ADDRESS/
        34.                  ADDR3,THIRD 4 LETTERS OF ADDRESS/
        35.                  ADDR4,FOURTH 4 LETTERS OF ADDRESS/
        36.                  BLOCK,BLOCK DWELLING UNIT IS LOCATED ON/
        37.                  ID,IDENTIFICATION NUMBER/
        38.                  ZONE,RESIDENTIAL HOUSING ZONE/
        39.                  DWELL,DWELLING TYPE/
        40.                  HCOST,HOUSING COST AS OF 1-80/
        41.                  ASSES,ASSESSED VALUE OF HOUSE AS OF 1-80/
        42.                  ROOMS,# OF ROOMS IN DWELLING UNIT AS OF 1-80/
        43.                  HEAT,AVERAGE YEARLY HEATING BILL AS OF 1-80/
        44.                  ELEC,AVERAGE MONTHLY ELECTRIC BILL AS OF 1-80/
        45.                  PHONE,AVERAGE MONTHLY PHONE BILL AS OF 1-80/
        46.                  INCOM,TOTAL FAMILY INCOME IN 1979/
        47.                  PEPLE,NUMBER OF PEOPLE IN HOUSEHOLD AS OF 1-80/
        48.                  CARS,NUMBER OF CARS IN HOUSEHOLD AS OF 1-80/
        49.                  GAS,AVERAGE BIMONTHLY GAS BILL AS OF 1-80/
        50.                  GASCA,AVE. BIMONTHLY GAS PER CAR AS OF 1-80/
```

Exhibit 15–4 (*continued*)

STAT CITY SPSS DRIVER PROGRAM

```
51.                    GASTR,AVE. MONTHLY TRIPS FOR GAS AS OF 1-80/
52.                    REPAR,FAVORITE PLACE FOR CAR REPAIRS AS OF 1-80/
53.                    FAVGA,FAVORITE GAS STATION AS OF 1-80/
54.                    HOSP,AVE YRLY HOSP. TRIPS PER HOME AS OF 1-80/
55.                    EAT,AVERAGE WEEKLY FOOD BILL AS OF 1-80/
56.                    EATPL,AVE WKLY FOOD BILL PER PERSON AS OF 1-80/
57.                    FEAT,FAVORITE SUPERMARKET AS OF 1-80/
58.                    LSODA,AVE WKLY DIET SODA PURCHASE AS OF 1-80/
59.                    HSODA,AVE WKLY REG. SODA PURCHASE AS OF 1-80/
60.                    BEER,AVE WKLY BEER PURCHASE AS OF 1-80
61.   COMMENT          LSODA,HSODA, AND BEER PURCHASES ARE IN SIX PACK UNITS.
62.   VALUE LABELS     ZONE (1) ZONE 1 (2) ZONE 2 (3) ZONE 3 (4) ZONE 4/
63.                    DWELL (0)APARTMENT (1)HOUSE/
64.                    REPAR (0)PERFORMS OWN REPAIRS (1)REPAIRS DONE BY STATION
65.                          (2)REPAIRS DONE BY DEALER/
66.                    FAVGA (1)PAUL'S TEXACO STATION (2)HOWIE'S GULF STATION/
67.                    FEAT  (1)FOODFAIR (2)GRAND UNION (3)A & P
68.   SELECT IF        (ZONE EQ 1)
69.   CONDESCRIPTIVE   HOSP
```

***** GIVEN WORKSPACE ALLOWS FOR   336 VARIABLES FOR CONDESCRIPTIVE PROBLEM *****

```
70.   STATISTICS       1,5,9,10,11
71.   READ INPUT DATA
```

FILE   NONAME   (CREATION DATE = 05/18/81)

VARIABLE HOSP   AVE YRLY HOSP. TRIPS PER HOME AS OF 1-80

MEAN	3.185	STD DEV	2.141	RANGE	10.000
MINIMUM	.000	MAXIMUM	10.000		

VALID OBSERVATIONS -   130   MISSING OBSERVATIONS -   0

```
TRANSPACE REQUIRED..      30 WORDS
          1 TRANSFORMATIONS
          0 RECODE VALUES + LAG VARIABLES
          3 IF/COMPUTE OPERATIONS

CPU TIME REQUIRED..    1.74 SECONDS


          72.  FINISH

    NORMAL END OF JOB.
       72 CONTROL CARDS WERE PROCESSED.
        0 ERRORS WERE DETECTED.
```

An example of the format of the CROSSTABS card is:

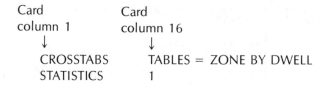

Card column 1	Card column 16
↓	↓
CROSSTABS	TABLES = ZONE BY DWELL
STATISTICS	1

The above card sequence will produce a cross-tabulation of zone by type of dwelling unit and a χ^2 statistic.

An SPSS printout that computes and presents the above cross-tabulation and χ^2 statistics can be seen in Exhibit 15–5.

4. REGRESSION. The REGRESSION control card initiates a statistical program that computes and presents repression coefficients, correlation coefficients, coefficients of determination, standard errors of regression coefficients, plus other statistics.

Exhibit 15-5

SPSS BATCH SYSTEM

SPSS FOR SPERRY UNIVAC 1100 EXEC 8, VERSION H, RELEASE 8.0-UWI.0, JUNE 1979

```
SPACE ALLOCATION..            ALLOWS FOR..    20 TRANSFORMATIONS
WORKSPACE      4375 WORDS                      82 RECODE VALUES + LAG VARIABLES
TRANSPACE       625 WORDS                     164 IF/COMPUTE OPERATIONS
        1.  RUN NAME        STAT CITY SPSS DRIVER PROGRAM
        2.  VARIABLE LIST   NAME1,NAME2,NAME3,NAME4,ADDR1,ADDR2,ADDR3,ADDR4,
        3.                  BLOCK,ID,ZONE,DWELL,HCOST,ASSES,ROOMS,HEAT,ELEC,
        4.                  PHONE,INCOM,PEPLE,CARS,GAS,GASCA,GASTR,REPAR,
        5.                  FAVGA,HOSP,EAT,EATPL,FEAT,LSODA,HSODA,BEER
        6.  INPUT FORMAT    FIXED (1X,4A3,1X,4A4,F3.0,2X,F4.0,1X,F1.0,1X,
        7.                  F1.0,1X,F4.0,1X,F6.0,1X,F2.0,1X,F4.0,1X,
        8.                  F4.0,1X,F3.0,1X,F6.0,1X/F2.0,1X,F1.0,1X,F3.0,1X,
        9.                  F3.0,1X,F2.0,1X,F1.0,1X,F1.0,1X,
       10.                  F2.0,1X,F3.0,1X,F2.0,1X,F1.0,1X,F2.0,
       11.                  1X,F2.0,1X,F2.0)
```

ACCORDING TO YOUR INPUT FORMAT, VARIABLES ARE TO BE READ AS FOLLOWS

```
        VARIABLE   FORMAT  RECORD     COLUMNS
        NAME1       A 3      1          2-  4
        NAME2       A 3      1          5-  7
        NAME3       A 3      1          8- 10
        NAME4       A 3      1         11- 13
        ADDR1       A 4      1         15- 18
        ADDR2       A 4      1         19- 22
        ADDR3       A 4      1         23- 26
        ADDR4       A 4      1         27- 30
        BLOCK       F 3.0    1         31- 33
        ID          F 4.0    1         36- 39
        ZONE        F 1.0    1         41- 41
        DWELL       F 1.0    1         43- 43
        HCOST       F 4.0    1         45- 48
        ASSES       F 6.0    1         50- 55
        ROOMS       F 2.0    1         57- 58
        HEAT        F 4.0    1         60- 63
        ELEC        F 4.0    1         65- 68
        PHONE       F 3.0    1         70- 72
        INCOM       F 6.0    1         74- 79
        PEPLE       F 2.0    2          1-  2
        CARS        F 1.0    2          4-  4
        GAS         F 3.0    2          6-  8
        GASCA       F 3.0    2         10- 12
        GASTR       F 2.0    2         14- 15
        REPAR       F 1.0    2         17- 17
        FAVGA       F 1.0    2         19- 19
        HOSP        F 2.0    2         21- 22
        EAT         F 3.0    2         24- 26
        EATPL       F 2.0    2         28- 29
        FEAT        F 1.0    2         31- 31
        LSODA       F 2.0    2         33- 34
        HSODA       F 2.0    2         36- 37
        BEER        F 2.0    2         39- 40
```

THE INPUT FORMAT PROVIDES FOR 33 VARIABLES.   33 WILL BE READ
IT PROVIDES FOR 2 RECORDS ("CARDS") PER CASE. A MAXIMUM OF   80 "COLUMNS" ARE USED ON A RECORD.

```
       12.  N OF CASES       1373
       13.  INPUT MEDIUM     CARD
       14.  COMMENT          THE ENTIRE STAT CITY DATA BASE WAS COLLECTED VIA
       15.  COMMENT          QUESTIONNAIRE SURVEY DURING FEBRUARY AND MARCH OF
       16.  COMMENT          1980.  THE DATA BASE REFLECTS THE CHARACTERISTICS
       17.  COMMENT          OF STAT CITY DWELLING UNITS AS OF JANUARY 1980.
       18.  COMMENT
       19.  COMMENT
       20.  COMMENT
       21.  COMMENT          THE ELEMENTARY UNIT FOR ANALYSIS IN STAT CITY IS THE
       22.  COMMENT          DWELLING UNIT, NOT THE INDIVIDUAL.  ALL PROBLEMS WILL
       23.  COMMENT          CONSIDER THE DWELLING UNIT (FAMILY OR HOME) AS THE
       24.  COMMENT          BASIC UNIT FOR ANALYSIS.
       25.  COMMENT
       26.  COMMENT
       27.  COMMENT
       28.  VAR LABELS       NAME1,FIRST 4 LETTERS OF LAST NAME/
       29.                   NAME2,SECOND 4 LETTERS OF LAST NAME/
       30.                   NAME3,THIRD 4 LETTERS OF LAST NAME/
       31.                   NAME4,FOURTH 4 LETTERS OF LAST NAME/
       32.                   ADDR1,FIRST 4 LETTERS OF ADDRESS/
       33.                   ADDR2,SECOND 4 LETTERS OF ADDRESS/
       34.                   ADDR3,THIRD 4 LETTERS OF ADDRESS/
       35.                   ADDR4,FOURTH 4 LETTERS OF ADDRESS/
       36.                   BLOCK,BLOCK DWELLING UNIT IS LOCATED ON/
       37.                   ID,IDENTIFICATION NUMBER/
       38.                   ZONE,RESIDENTIAL HOUSING ZONE/
       39.                   DWELL,DWELLING TYPE/
       40.                   HCOST,HOUSING COST AS OF 1-80/
       41.                   ASSES,ASSESSED VALUE OF HOUSE AS OF 1-80/
       42.                   ROOMS,# OF ROOMS IN DWELLING UNIT AS OF 1-80/
       43.                   HEAT,AVERAGE YEARLY HEATING BILL AS OF 1-80/
       44.                   ELEC,AVERAGE MONTHLY ELECTRIC BILL AS OF 1-80/
       45.                   PHONE,AVERAGE MONTHLY PHONE BILL AS OF 1-80/
       46.                   INCOM,TOTAL FAMILY INCOME IN 1979/
       47.                   PEPLE,NUMBER OF PEOPLE IN HOUSEHOLD AS OF 1-80/
       48.                   CARS,NUMBER OF CARS IN HOUSEHOLD AS OF 1-80/
       49.                   GAS,AVERAGE BIMONTHLY GAS BILL AS OF 1-80/
       50.                   GASCA,AVE. BIMONTHLY GAS PER CAR AS OF 1-80/
```

Exhibit 15–5 (*continued*)

```
STAT CITY SPSS DRIVER PROGRAM
         51.                     GASTR,AVE. MONTHLY TRIPS FOR GAS AS OF 1-80/
         52.                     REPAR,FAVORITE PLACE FOR CAR REPAIRS AS OF 1-80/
         53.                     FAVGA,FAVORITE GAS STATION AS OF 1-80/
         54.                     HOSP,AVE YRLY HOSP. TRIPS PER HOME AS OF 1-80/
         55.                     EAT,AVERAGE WEEKLY FOOD BILL AS OF 1-80/
         56.                     EATPL,AVE WKLY FOOD BILL PER PERSON AS OF 1-80/
         57.                     FEAT,FAVORITE SUPERMARKET AS OF 1-80/
         58.                     LSODA,AVE WKLY DIET SODA PURCHASE AS OF 1-80/
         59.                     HSODA,AVE WKLY REG. SODA PURCHASE AS OF 1-80/
         60.                     BEER,AVE WKLY BEER PURCHASE AS OF 1-80
         61.    COMMENT          LSODA,HSODA, AND BEER PURCHASES ARE IN SIX PACK UNITS.
         62.    VALUE LABELS     ZONE (1) ZONE 1 (2) ZONE 2 (3) ZONE 3 (4) ZONE 4/
         63.                     DWELL (0)APARTMENT (1)HOUSE/
         64.                     REPAR (0)PERFORMS OWN REPAIRS (1)REPAIRS DONE BY STATION
         65.                     (2)REPAIRS DONE BY DEALER/
         66.                     FAVGA (1)PAUL'S TEXACO STATION (2)HOWIE'S GULF STATION/
         67.                     FEAT (1)FOODFAIR (2)GRAND UNION (3)A & P
         68.    CROSSTABS        TABLES=ZONE BY DWELL
         69.    STATISTICS       1

***** GIVEN WORKSPACE ALLOWS FOR   729 CELLS,   485 TABLES WITH   2 DIMENSIONS FOR CROSSTAB PROBLEM *****

         70.    READ INPUT DATA


FILE   NONAME   (CREATION DATE = 05/18/81)
* * * * * * * * * * * * * * * *  C R O S S T A B U L A T I O N   O F  * * * * * *
    *  ZONE  *  RESIDENTIAL HOUSING ZONE                       BY  DWELL   DWELLING TYPE
* * * * * * * * * * * * * * * * * * * * * * * * * * * * * * * * * * * * *

                    DWELL
           COUNT  I
           ROW PCT IAPARTMEN HOUSE
           COL PCT IT
           TOT PCT I       0.I       1.I       ROW
                   I--------I--------I        TOTAL
ZONE       -------I--------I--------I
              1.   I      0 I    130 I        130
  ZONE 1         I     .0 I  100.0 I        9.5
                I     .0 I   26.1 I
                I     .0 I    9.5 I
              -I--------I--------I
              2.   I      0 I    157 I        157
  ZONE 2         I     .0 I  100.0 I        11.4
                I     .0 I   31.5 I
                I     .0 I   11.4 I
              -I--------I--------I
              3.   I    126 I    212 I        338
  ZONE 3         I   37.3 I   62.7 I        24.6
                I   14.4 I   42.5 I
                I    9.2 I   15.4 I
              -I--------I--------I
              4.   I    748 I      0 I        748
  ZONE 4         I  100.0 I     .0 I        54.5
                I   85.6 I     .0 I
                I   54.5 I     .0 I
              -I--------I--------I
           COLUMN      874      499       1373
           TOTAL      63.7     36.3      100.0

CHI SQUARE =  1031.39923 WITH   3 DEGREES OF FREEDOM   SIGNIFICANCE =   .0000

CPU TIME REQUIRED..    1.98 SECONDS


         71.  FINISH
       NORMAL END OF JOB.
         71 CONTROL CARDS WERE PROCESSED.
          0 ERRORS WERE DETECTED.
```

An example of the format of the REGRESSION card is:

Card column 1 ↓	Card column 16 ↓
SAMPLE	100 FROM 1373
LIST CASES	CASES = 1373/VARIABLES = ID, CARS, PEPLE
REGRESSION	VARIABLES = CARS, PEPLE/REGRESSION = CARS WITH PEPLE (2)
STATISTICS	ALL

The above card sequence will draw and print a simple random sample of 100 Stat City dwelling units. Further, a regression model predicting CARS as a function of PEPLE will be computed; a myriad of supporting statistics will also be computed and printed.

An SPSS printout for the above REGRESSION procedure can be seen in Exhibit 15–6.

Exhibit 15-6

SPSS BATCH SYSTEM

SPSS FOR SPERRY UNIVAC 1100 EXEC 8, VERSION H, RELEASE 8.0-UW1.0, JUNE 1979

```
SPACE ALLOCATION..              ALLOWS FOR..    20 TRANSFORMATIONS
WORKSPACE      4375 WORDS                        82 RECODE VALUES + LAG VARIABLES
TRANSPACE       625 WORDS                       164 IF/COMPUTE OPERATIONS
        1.    RUN NAME          STAT CITY SPSS DRIVER PROGRAM
        2.    VARIABLE LIST     NAME1,NAME2,NAME3,NAME4,ADDR1,ADDR2,ADDR3,ADDR4,
        3.                      BLOCK,ID,ZONE,DWELL,HCOST,ASSES,ROOMS,HEAT,ELEC,
        4.                      PHONE,INCOM,PEPLE,CARS,GAS,GASCA,GASTR,REPAR,
        5.                      FAVGA,HOSP,EAT,EATPL,FEAT,LSODA,HSODA,BEER
        6.    INPUT FORMAT      FIXED (1X,4A3,1X,4A4,F3.0,2X,F4.0,1X,F1.0,1X,
        7.                      F1.0,1X,F4.0,1X,F6.0,1X,F2.0,1X,F4.0,1X,
        8.                      F4.0,1X,F3.0,1X,F6.0,1X/F2.0,1X,F1.0,1X,F3.0,1X,
        9.                      F3.0,1X,F2.0,1X,F1.0,1X,F1.0,1X,
       10.                      F2.0,1X,F3.0,1X,F2.0,1X,F1.0,1X,F2.0,
       11.                      1X,F2.0,1X,F2.0)
```

ACCORDING TO YOUR INPUT FORMAT, VARIABLES ARE TO BE READ AS FOLLOWS

```
VARIABLE    FORMAT    RECORD    COLUMNS

NAME1        A 3        1         2-    4
NAME2        A 3        1         5-    7
NAME3        A 3        1         8-   10
NAME4        A 3        1        11-   13
ADDR1        A 4        1        15-   18
ADDR2        A 4        1        19-   22
ADDR3        A 4        1        23-   26
ADDR4        A 4        1        27-   30
BLOCK      F 3. 0       1        31-   33
ID         F 4. 0       1        36-   39
ZONE       F 1. 0       1        41-   41
DWELL      F 1. 0       1        43-   43
HCOST      F 4. 0       1        45-   48
ASSES      F 6. 0       1        50-   55
ROOMS      F 2. 0       1        57-   58
HEAT       F 4. 0       1        60-   63
ELEC       F 4. 0       1        65-   68
PHONE      F 3. 0       1        70-   72
INCOM      F 6. 0       1        74-   79
PEPLE      F 2. 0       2         1-    2
CARS       F 1. 0       2         4-    4
GAS        F 3. 0       2         6-    8
GASCA      F 3. 0       2        10-   12
GASTR      F 2. 0       2        14-   15
REPAR      F 1. 0       2        17-   17
FAVGA      F 1. 0       2        19-   19
HOSP       F 2. 0       2        21-   22
EAT        F 3. 0       2        24-   26
EATPL      F 2. 0       2        28-   29
FEAT       F 1. 0       2        31-   31
LSODA      F 2. 0       2        33-   34
HSODA      F 2. 0       2        36-   37
BEER       F 2. 0       2        39-   40
```

THE INPUT FORMAT PROVIDES FOR 33 VARIABLES.   33 WILL BE READ
IT PROVIDES FOR 2 RECORDS ('CARDS') PER CASE. A MAXIMUM OF   80 'COLUMNS' ARE USED ON A RECORD.

```
       12.    N OF CASES        1373
       13.    INPUT MEDIUM      CARD
       14.    COMMENT           THE ENTIRE STAT CITY DATA BASE WAS COLLECTED VIA
       15.    COMMENT           QUESTIONNAIRE SURVEY DURING FEBRUARY AND MARCH OF
       16.    COMMENT           1980.   THE DATA BASE REFLECTS THE CHARACTERISTICS
       17.    COMMENT           OF STAT CITY DWELLING UNITS AS OF JANUARY 1980.
       18.    COMMENT
       19.    COMMENT
       20.    COMMENT
       21.    COMMENT           THE ELEMENTARY UNIT FOR ANALYSIS IN STAT CITY IS THE
       22.    COMMENT           DWELLING UNIT, NOT THE INDIVIDUAL.  ALL PROBLEMS WILL
       23.    COMMENT           CONSIDER THE DWELLING UNIT (FAMILY OR HOME) AS THE
       24.    COMMENT           BASIC UNIT FOR ANALYSIS.
       25.    COMMENT
       26.    COMMENT
       27.    COMMENT
       28.    VAR LABELS        NAME1,FIRST 4 LETTERS OF LAST NAME/
       29.                      NAME2,SECOND 4 LETTERS OF LAST NAME/
       30.                      NAME3,THIRD 4 LETTERS OF LAST NAME/
       31.                      NAME4,FOURTH 4 LETTERS OF LAST NAME/
       32.                      ADDR1,FIRST 4 LETTERS OF ADDRESS/
       33.                      ADDR2,SECOND 4 LETTERS OF ADDRESS/
       34.                      ADDR3,THIRD 4 LETTERS OF ADDRESS/
       35.                      ADDR4,FOURTH 4 LETTERS OF ADDRESS/
       36.                      BLOCK,BLOCK DWELLING UNIT IS LOCATED ON/
       37.                      ID,IDENTIFICATION NUMBER/
       38.                      ZONE,RESIDENTIAL HOUSING ZONE/
       39.                      DWELL,DWELLING TYPE/
       40.                      HCOST,HOUSING COST AS OF 1-80/
       41.                      ASSES,ASSESSED VALUE OF HOUSE AS OF 1-80/
       42.                      ROOMS,# OF ROOMS IN DWELLING UNIT AS OF 1-80/
       43.                      HEAT,AVERAGE YEARLY HEATING BILL AS OF 1-80/
       44.                      ELEC,AVERAGE MONTHLY ELECTRIC BILL AS OF 1-80/
       45.                      PHONE,AVERAGE MONTHLY PHONE BILL AS OF 1-80/
       46.                      INCOM,TOTAL FAMILY INCOME IN 1979/
       47.                      PEPLE,NUMBER OF PEOPLE IN HOUSEHOLD AS OF 1-80/
       48.                      CARS,NUMBER OF CARS IN HOUSEHOLD AS OF 1-80/
       49.                      GAS,AVERAGE BIMONTHLY GAS BILL AS OF 1-80/
       50.                      GASCA,AVE. BIMONTHLY GAS PER CAR AS OF 1-80/
```

Exhibit 15–6 (*continued*)

```
51.                    GASTR,AVE. MONTHLY TRIPS FOR GAS AS OF 1-80/
52.                    REPAR,FAVORITE PLACE FOR CAR REPAIRS AS OF 1-80/
53.                    FAVGA,FAVORITE GAS STATION AS OF 1-80/
54.                    HOSP,AVE YRLY HOSP. TRIPS PER HOME AS OF 1-80/
55.                    EAT,AVERAGE WEEKLY FOOD BILL AS OF 1-80/
56.                    EATPL,AVE WKLY FOOD BILL PER PERSON AS OF 1-80/
57.                    FEAT,FAVORITE SUPERMARKET AS OF 1-80/
58.                    LSODA,AVE WKLY DIET SODA PURCHASE AS OF 1-80/
59.                    HSODA,AVE WKLY REG. SODA PURCHASE AS OF 1-80/
60.                    BEER,AVE WKLY BEER PURCHASE AS OF 1-80
61.  COMMENT           LSODA,HSODA, AND BEER PURCHASES ARE IN SIX PACK UNITS.
62.  VALUE LABELS      ZONE (1) ZONE 1 (2) ZONE 2 (3) ZONE 3 (4) ZONE 4/
63.                    DWELL (0)APARTMENT (1)HOUSE/
64.                    REPAR (0)PERFORMS OWN REPAIRS (1)REPAIRS DONE BY STATION
65.                    (2)REPAIRS DONE BY DEALER/
66.                    FAVGA (1)PAUL'S TEXACO STATION (2)HOWIE'S GULF STATION/
67.                    FEAT (1)FOODFAIR (2)GRAND UNION (3)A & P
68.  SAMPLE            100 FROM 1373
69.  LIST CASES        CASES=1373/VARIABLES=ID,CARS,PEPLE
70.  REGRESSION        VARIABLES=CARS,PEPLE/
71.                    REGRESSION=CARS WITH PEPLE(2)
72.  STATISTICS        ALL
```

***** REGRESSION PROBLEM REQUIRES   48 WORDS WORKSPACE, NOT INCLUDING RESIDUALS *****

73. READ INPUT DATA

STAT CITY SPSS DRIVER PROGRAM

FILE   NONAME   (CREATION DATE = 05/18/81)

CASE-N	ID	PEPLE	CARS	CASE-N	ID	PEPLE	CARS
1	14.	7.	2.	51	724.	2.	2.
2	98.	2.	2.	52	727.	3.	2.
3	102.	3.	2.	53	760.	8.	1.
4	108.	4.	2.	54	772.	4.	2.
5	214.	2.	1.	55	781.	5.	1.
6	231.	4.	3.	56	785.	4.	2.
7	233.	3.	2.	57	791.	5.	1.
8	238.	4.	2.	58	829.	1.	1.
9	243.	4.	2.	59	836.	4.	1.
10	250.	5.	3.	60	849.	7.	1.
11	263.	4.	2.	61	878.	3.	2.
12	270.	2.	1.	62	928.	4.	2.
13	289.	3.	2.	63	929.	3.	2.
14	305.	4.	2.	64	934.	3.	1.
15	317.	4.	3.	65	936.	2.	2.
16	326.	5.	3.	66	958.	7.	3.
17	349.	6.	4.	67	967.	6.	2.
18	352.	7.	3.	68	978.	6.	4.
19	355.	4.	2.	69	990.	4.	2.
20	358.	6.	4.	70	1008.	4.	3.
21	364.	3.	2.	71	1035.	4.	2.
22	373.	3.	1.	72	1053.	4.	1.
23	375.	5.	3.	73	1071.	5.	1.
24	377.	4.	3.	74	1075.	6.	1.
25	407.	4.	2.	75	1084.	2.	1.
26	431.	4.	3.	76	1097.	2.	2.
27	446.	8.	2.	77	1130.	6.	2.
28	475.	7.	2.	78	1134.	8.	1.
29	494.	6.	4.	79	1161.	6.	2.
30	517.	3.	2.	80	1164.	2.	2.
31	522.	3.	2.	81	1180.	3.	2.
32	528.	5.	3.	82	1190.	3.	2.
33	536.	3.	2.	83	1219.	4.	1.
34	544.	3.	2.	84	1230.	2.	2.
35	549.	5.	3.	85	1244.	3.	1.
36	583.	4.	3.	86	1261.	5.	2.
37	615.	3.	2.	87	1274.	4.	1.
38	627.	6.	2.	88	1290.	6.	1.
39	630.	5.	2.	89	1293.	2.	2.
40	641.	5.	3.	90	1294.	9.	2.
41	651.	4.	2.	91	1299.	4.	2.
42	657.	4.	4.	92	1309.	4.	2.
43	663.	5.	4.	93	1310.	4.	3.
44	665.	3.	2.	94	1312.	2.	2.
45	667.	7.	1.	95	1314.	6.	2.
46	672.	5.	1.	96	1339.	6.	2.
47	676.	5.	1.	97	1340.	4.	1.
48	687.	5.	1.	98	1348.	4.	1.
49	698.	4.	2.	99	1349.	7.	2.
50	715.	2.	2.	100	1367.	5.	1.

Exhibit 15–6 (*concluded*)

```
                                                      CORRELATION COEFFICIENTS
                                                      A VALUE OF 99.00000 IS PRINTED
STAT CITY SPSS DRIVER PROGRAM                         IF A COEFFICIENT CANNOT BE COMPUTED.
FILE   NONAME   (CREATION DATE = 05/18/81)


VARIABLE            MEAN       STANDARD DEV    CASES                CARS        PEPLE
CARS              2.0100            .8102       100    CARS       1.00000       .12724
PEPLE             4.3300           1.6333       100    PEPLE       .12724      1.00000


* * * * * * * * * * * * * * * * * * * * * MULTIPLE  REGRESSION * * * * * * * * * * * * * *   VARIABLE LIST  1
DEPENDENT VARIABLE..    CARS       NUMBER OF CARS IN HOUSEHOLD AS OF 1-80                     REGRESSION LIST  1
VARIABLE(S) ENTERED ON STEP NUMBER  1..     PEPLE      NUMBER OF PEOPLE IN HOUSEHOLD AS OF 1-80

MULTIPLE R           .12724                       ANALYSIS OF VARIANCE    DF     SUM OF SQUARES      MEAN SQUARE           F
R SQUARE             .01619                       REGRESSION               1.        1.05217          1.05217          1.61270
ADJUSTED R SQUARE    .00615                       RESIDUAL                98.        63.93783          .65243
STANDARD ERROR       .80773

---------------- VARIABLES IN THE EQUATION ------------------       ------------ VARIABLES NOT IN THE EQUATION --------------
VARIABLE        B         BETA      STD ERROR B        F            VARIABLE     BETA IN    PARTIAL   TOLERANCE        F
PEPLE      .6311764-001   .12724      .04970         1.613
(CONSTANT) .1736701+001

MAXIMUM STEP REACHED

STATISTICS WHICH CANNOT BE COMPUTED ARE PRINTED AS ALL NINES.


                                                  SUMMARY TABLE
VARIABLE                                 MULTIPLE R   R SQUARE   RSQ CHANGE   SIMPLE R             B              BETA
PEPLE     NUMBER OF PEOPLE IN HOUSEHOLD AS OF 1-80  .12724   .01619   .01619   .12724      .6311764-001       .12724
(CONSTANT)                                                                                .1736701+001


TRANSPACE REQUIRED..      30 WORDS
       1 TRANSFORMATIONS
       0 RECODE VALUES + LAG VARIABLES
       1 IF/COMPUTE OPERATIONS

CPU TIME REQUIRED..    1.92 SECONDS


       74.  FINISH

    NORMAL END OF JOB.
       74 CONTROL CARDS WERE PROCESSED.
        0 ERRORS WERE DETECTED.
```

5.   ONEWAY. The ONEWAY control card initiates a statistical program that computes and prints a oneway analysis of variance table with supporting statistics.

An example of the format of the ONEWAY card is:

Card column 1 ↓	Card column 16 ↓
SAMPLE	150 FROM 1373
LIST CASES	CASES = 1373/VARIABLES = ID, INCOM, ZONE
ONEWAY	INCOM BY ZONE (1, 4)

The above card sequence will draw and print a simple random sample of 150 Stat City dwelling units. Further, an analysis of variance table will be computed to test if there is a statistically significant difference in mean incomes between the four residential housing zones in Stat City.

An SPSS printout for the above ONEWAY procedure can be seen in Exhibit 15–7.

One final note on data modification and task definition control cards is that multiple statistical procedures can be run in one SPSS program. If an asterisk (*) appears before a data modification card, the data modification is only in effect for the statistical procedure that immediately follows the card. For example, if a researcher wanted to compute the average income

Exhibit 15–7

SPSS BATCH SYSTEM

SPSS FOR SPERRY UNIVAC 1100 EXEC 8, VERSION H, RELEASE 8.0-UW1.0, JUNE 1979

```
SPACE ALLOCATION..          ALLOWS FOR..    20 TRANSFORMATIONS
WORKSPACE      4375 WORDS                    82 RECODE VALUES + LAG VARIABLES
TRANSPACE       625 WORDS                   164 IF/COMPUTE OPERATIONS

    1.  RUN NAME         STAT CITY SPSS DRIVER PROGRAM
    2.  VARIABLE LIST    NAME1,NAME2,NAME3,NAME4,ADDR1,ADDR2,ADDR3,ADDR4,
    3.                   BLOCK,ID,ZONE,DWELL,HCOST,ASSES,ROOMS,HEAT,ELEC,
    4.                   PHONE,INCOM,PEPLE,CARS,GAS,GASCA,GASTR,REPAR,
    5.                   FAVGA,HOSP,EAT,EATPL,FEAT,LSODA,HSODA,BEER
    6.  INPUT FORMAT     FIXED (1X,4A3,1X,4A4,F3.0,2X,F4.0,1X,F1.0,1X,
    7.                   F1.0,1X,F4.0,1X,F6.0,1X,F2.0,1X,F4.0,1X,
    8.                   F4.0,1X,F3.0,1X,F6.0,1X/F2.0,1X,F1.0,1X,F3.0,1X,
    9.                   F3.0,1X,F2.0,1X,F1.0,1X,F1.0,1X,
   10.                   F2.0,1X,F3.0,1X,F2.0,1X,F1.0,1X,F2.0,
   11.                   1X,F2.0,1X,F2.0)
```

ACCORDING TO YOUR INPUT FORMAT, VARIABLES ARE TO BE READ AS FOLLOWS

```
VARIABLE   FORMAT   RECORD     COLUMNS

NAME1       A 3       1         2-    4
NAME2       A 3       1         5-    7
NAME3       A 3       1         8-   10
NAME4       A 3       1        11-   13
ADDR1       A 4       1        15-   18
ADDR2       A 4       1        19-   22
ADDR3       A 4       1        23-   26
ADDR4       A 4       1        27-   30
BLOCK       F 3. 0    1        31-   33
ID          F 4. 0    1        36-   39
ZONE        F 1. 0    1        41-   41
DWELL       F 1. 0    1        43-   43
HCOST       F 4. 0    1        45-   48
ASSES       F 6. 0    1        50-   55
ROOMS       F 2. 0    1        57-   58
HEAT        F 4. 0    1        60-   63
ELEC        F 4. 0    1        65-   68
PHONE       F 3. 0    1        70-   72
INCOM       F 6. 0    1        74-   79
PEPLE       F 2. 0    2         1-    2
CARS        F 1. 0    2         4-    4
GAS         F 3. 0    2         6-    8
GASCA       F 3. 0    2        10-   12
GASTR       F 2. 0    2        14-   15
REPAR       F 1. 0    2        17-   17
FAVGA       F 1. 0    2        19-   19
HOSP        F 2. 0    2        21-   22
EAT         F 3. 0    2        24-   26
EATPL       F 2. 0    2        28-   29
FEAT        F 1. 0    2        31-   31
LSODA       F 2. 0    2        33-   34
HSODA       F 2. 0    2        36-   37
BEER        F 2. 0    2        39-   40
```

THE INPUT FORMAT PROVIDES FOR 33 VARIABLES.   33 WILL BE READ
IT PROVIDES FOR 2 RECORDS ('CARDS') PER CASE. A MAXIMUM OF   80 'COLUMNS' ARE USED ON A RECORD.

```
   12.  N OF CASES       1373
   13.  INPUT MEDIUM     CARD
   14.  COMMENT          THE ENTIRE STAT CITY DATA BASE WAS COLLECTED VIA
   15.  COMMENT          QUESTIONNAIRE SURVEY DURING FEBRUARY AND MARCH OF
   16.  COMMENT          1980.  THE DATA BASE REFLECTS THE CHARACTERISTICS
   17.  COMMENT          OF STAT CITY DWELLING UNITS AS OF JANUARY 1980.
   18.  COMMENT
   19.  COMMENT
   20.  COMMENT
   21.  COMMENT          THE ELEMENTARY UNIT FOR ANALYSIS IN STAT CITY IS THE
   22.  COMMENT          DWELLING UNIT, NOT THE INDIVIDUAL.  ALL PROBLEMS WILL
   23.  COMMENT          CONSIDER THE DWELLING UNIT (FAMILY OR HOME) AS THE
   24.  COMMENT          BASIC UNIT FOR ANALYSIS.
   25.  COMMENT
   26.  COMMENT
   27.  COMMENT
   28.  VAR LABELS       NAME1,FIRST 4 LETTERS OF LAST NAME/
   29.                   NAME2,SECOND 4 LETTERS OF LAST NAME/
   30.                   NAME3,THIRD 4 LETTERS OF LAST NAME/
   31.                   NAME4,FOURTH 4 LETTERS OF LAST NAME/
   32.                   ADDR1,FIRST 4 LETTERS OF ADDRESS/
   33.                   ADDR2,SECOND 4 LETTERS OF ADDRESS/
   34.                   ADDR3,THIRD 4 LETTERS OF ADDRESS/
   35.                   ADDR4,FOURTH 4 LETTERS OF ADDRESS/
   36.                   BLOCK,BLOCK DWELLING UNIT IS LOCATED ON/
   37.                   ID,IDENTIFICATION NUMBER/
   38.                   ZONE,RESIDENTIAL HOUSING ZONE/
   39.                   DWELL,DWELLING TYPE/
   40.                   HCOST,HOUSING COST AS OF 1-80/
   41.                   ASSES,ASSESSED VALUE OF HOUSE AS OF 1-80/
   42.                   ROOMS,# OF ROOMS IN DWELLING UNIT AS OF 1-80/
   43.                   HEAT,AVERAGE YEARLY HEATING BILL AS OF 1-80/
   44.                   ELEC,AVERAGE MONTHLY ELECTRIC BILL AS OF 1-80/
   45.                   PHONE,AVERAGE MONTHLY PHONE BILL AS OF 1-80/
   46.                   INCOM,TOTAL FAMILY INCOME IN 1979/
   47.                   PEPLE,NUMBER OF PEOPLE IN HOUSEHOLD AS OF 1-80/
   48.                   CARS,NUMBER OF CARS IN HOUSEHOLD AS OF 1-80/
   49.                   GAS,AVERAGE BIMONTHLY GAS BILL AS OF 1-80/
   50.                   GASCA,AVE. BIMONTHLY GAS PER CAR AS OF 1-80/
```

Exhibit 15–7 (*continued*)

```
51.                    GASTR,AVE. MONTHLY TRIPS FOR GAS AS OF 1-80/
52.                    REPAR,FAVORITE PLACE FOR CAR REPAIRS AS OF 1-80/
53.                    FAVGA,FAVORITE GAS STATION AS OF 1-80/
54.                    HOSP,AVE YRLY HOSP. TRIPS PER HOME AS OF 1-80/
55.                    EAT,AVERAGE WEEKLY FOOD BILL AS OF 1-80/
56.                    EATPL,AVE WKLY FOOD BILL PER PERSON AS OF 1-80/
57.                    FEAT,FAVORITE SUPERMARKET AS OF 1-80/
58.                    LSODA,AVE WKLY DIET SODA PURCHASE AS OF 1-80/
59.                    HSODA,AVE WKLY REG. SODA PURCHASE AS OF 1-80/
60.                    BEER,AVE WKLY BEER PURCHASE AS OF 1-80
61.    COMMENT         LSODA,HSODA, AND BEER PURCHASES ARE IN SIX PACK UNITS.
62.    VALUE LABELS    ZONE (1) ZONE 1 (2) ZONE 2 (3) ZONE 3 (4) ZONE 4/
63.                    DWELL (0)APARTMENT (1)HOUSE/
64.                    REPAR (0)PERFORMS OWN REPAIRS (1)REPAIRS DONE BY STATION
65.                       (2)REPAIRS DONE BY DEALER/
66.                    FAVGA (1)PAUL'S TEXACO STATION (2)HOWIE'S GULF STATION/
67.                    FEAT  (1)FOODFAIR (2)GRAND UNION (3)A & P
68.    SAMPLE          150 FROM 1373
69.    LIST CASES      CASES=1373/VARIABLES=ID,INCOM,ZONE
70.    ONEWAY          INCOM BY ZONE(1,4)
```

***** ONEWAY PROBLEM REQUIRES   30 WORDS WORKSPACE *****


```
        71.    READ INPUT DATA
```

STAT CITY SPSS DRIVER PROGRAM

FILE   NONAME   (CREATION DATE = 05/18/81)

CASE-N	ID	ZONE	INCOM	CASE-N	ID	ZONE	INCOM
1	19.	1.	49234.	66	690.	4.	17302.
2	47.	1.	51671.	67	709.	4.	9416.
3	56.	1.	15467.	68	719.	4.	17045.
4	65.	1.	29381.	69	722.	4.	12872.
5	69.	1.	33369.	70	726.	4.	16743.
6	74.	1.	43249.	71	735.	4.	12741.
7	90.	1.	57205.	72	743.	4.	16473.
8	107.	1.	41403.	73	763.	4.	8571.
9	110.	1.	15895.	74	775.	4.	3168.
10	113.	1.	26802.	75	779.	4.	7844.
11	117.	1.	32502.	76	781.	4.	8618.
12	118.	1.	51411.	77	797.	4.	13705.
13	125.	1.	48674.	78	799.	4.	19071.
14	129.	1.	63849.	79	800.	4.	22691.
15	136.	2.	32648.	80	812.	4.	24504.
16	149.	2.	89656.	81	824.	4.	18092.
17	158.	2.	42314.	82	838.	4.	10291.
18	203.	2.	76402.	83	841.	4.	10857.
19	210.	2.	146165.	84	846.	4.	16995.
20	220.	2.	60585.	85	859.	4.	18060.
21	222.	2.	126365.	86	863.	4.	16916.
22	236.	2.	74242.	87	876.	4.	8308.
23	237.	2.	61602.	88	885.	4.	10873.
24	238.	2.	72824.	89	886.	4.	13897.
25	239.	2.	33411.	90	918.	4.	6305.
26	241.	2.	47650.	91	935.	4.	2574.
27	248.	2.	83195.	92	950.	4.	6162.
28	255.	2.	129329.	93	959.	4.	24312.
29	264.	2.	43708.	94	994.	4.	34964.
30	272.	2.	66364.	95	995.	4.	23331.
31	281.	2.	109434.	96	996.	4.	18157.
32	293.	3.	15318.	97	997.	4.	17201.
33	338.	3.	33850.	98	1009.	4.	19477.
34	341.	3.	21742.	99	1015.	4.	18300.
35	356.	3.	25999.	100	1017.	4.	10581.
36	375.	3.	31998.	101	1018.	4.	14985.
37	394.	3.	29338.	102	1019.	4.	15404.
38	399.	3.	37679.	103	1021.	4.	14160.
39	403.	3.	31758.	104	1035.	4.	17928.
40	409.	3.	35005.	105	1040.	4.	13837.
41	451.	3.	24255.	106	1041.	4.	10987.
42	454.	3.	37265.	107	1051.	4.	12125.
43	462.	3.	42231.	108	1055.	4.	13371.
44	472.	3.	31856.	109	1060.	4.	7288.
45	480.	3.	12976.	110	1087.	4.	19207.
46	492.	3.	28104.	111	1105.	4.	25132.
47	510.	3.	32901.	112	1107.	4.	31397.
48	518.	3.	32458.	113	1108.	4.	19621.
49	530.	3.	45583.	114	1116.	4.	34432.
50	536.	3.	35394.	115	1125.	4.	17724.
51	554.	3.	44108.	116	1144.	4.	12597.
52	559.	3.	39167.	117	1145.	4.	13043.
53	561.	3.	43705.	118	1152.	4.	11171.
54	562.	3.	43309.	119	1159.	4.	18141.
55	563.	3.	21226.	120	1183.	4.	14584.
56	570.	3.	44546.	121	1184.	4.	6549.
57	603.	3.	24343.	122	1196.	4.	5985.
58	607.	3.	54820.	123	1199.	4.	8189.
59	610.	3.	29398.	124	1207.	4.	12451.
60	635.	4.	20943.	125	1210.	4.	16986.
61	649.	4.	18425.	126	1212.	4.	22910.
62	650.	4.	25052.	127	1219.	4.	19757.
63	651.	4.	27781.	128	1231.	4.	19576.
64	659.	4.	28077.	129	1232.	4.	14331.
65	667.	4.	8495.	130	1254.	4.	13827.

Exhibit 15–7 (concluded)

```
STAT CITY SPSS DRIVER PROGRAM
FILE   NONAME   (CREATION DATE = 05/18/81)
       CASE-N        ID        ZONE       INCOM
           131      1262.        4.       11557.
           132      1266.        4.       20685.
           133      1268.        4.       24872.
           134      1270.        4.        9179.
           135      1273.        4.        6798.
           136      1276.        4.        3342.
           137      1291.        4.       14120.
           138      1294.        4.       14656.
           139      1296.        4.       20703.
           140      1313.        4.        6971.
           141      1318.        4.       14303.
           142      1323.        4.       19772.
           143      1325.        4.       13694.
           144      1328.        4.       18497.
           145      1333.        4.        8807.
           146      1336.        4.       10960.
           147      1354.        4.        8312.
           148      1356.        4.        8544.
           149      1365.        4.        9715.
           150      1372.        4.        7732.
```

- O N E W A Y -

```
VARIABLE  INCOM       TOTAL FAMILY INCOME IN 1979
                             ANALYSIS OF VARIANCE
```

SOURCE	D.F.	SUM OF SQUARES	MEAN SQUARES	F RATIO	F PROB.
BETWEEN GROUPS	3	57408403349.3086	19136134400.0000	98.604	.0000
WITHIN GROUPS	146	28334333632.0000	194070778.0000		
TOTAL	149	85742736384.0000			

```
TRANSPACE REQUIRED..    30 WORDS
        1 TRANSFORMATIONS
        0 RECODE VALUES + LAG VARIABLES
        1 IF/COMPUTE OPERATIONS

CPU TIME REQUIRED..    2.02 SECONDS


        72.  FINISH
    NORMAL END OF JOB.
        72 CONTROL CARDS WERE PROCESSED.
         0 ERRORS WERE DETECTED.
```

in Stat City and in each zone of Stat City, the following task definition card sequence would be used:

```
Card                   Card
column 1               column 16
   ↓                      ↓
CONDESCRIPTIVE         INCOM
STATISTICS             1
READ INPUT DATA

        Stat City Data Base

*SELECT IF             (ZONE EQ 1)
CONDESCRIPTIVE         INCOM
STATISTICS             1
*SELECT IF             (ZONE EQ 2)
CONDESCRIPTIVE         INCOM
STATISTICS             1
*SELECT IF             (ZONE EQ 3)
CONDESCRIPTIVE         INCOM
STATISTICS             1
*SELECT IF             (ZONE EQ 4)
CONDESCRIPTIVE         INCOM
STATISTICS             1
FINISH
```

An SPSS printout for the above problem can be seen in Exhibit 15–8.

Exhibit 15–8

SPSS BATCH SYSTEM

SPSS FOR SPERRY UNIVAC 1100 EXEC 8, VERSION H, RELEASE 8.0-UW1.0, JUNE 1979

```
SPACE ALLOCATION..          ALLOWS FOR..      20 TRANSFORMATIONS
WORKSPACE    4375 WORDS                        82 RECODE VALUES + LAG VARIABLES
TRANSPACE     625 WORDS                       164 IF/COMPUTE OPERATIONS
         1.    RUN NAME        STAT CITY SPSS DRIVER PROGRAM
         2.    VARIABLE LIST   NAME1,NAME2,NAME3,NAME4,ADDR1,ADDR2,ADDR3,ADDR4,
         3.                    BLOCK,ID,ZONE,DWELL,HCOST,ASSES,ROOMS,HEAT,ELEC,
         4.                    PHONE,INCOM,PEPLE,CARS,GAS,GASCA,GASTR,REPAR,
         5.                    FAVGA,HOSP,EAT,EATPL,FEAT,LSODA,HSODA,BEER
         6.    INPUT FORMAT    FIXED (1X,4A3,1X,4A4,F3.0,2X,F4.0,1X,F1.0,1X,
         7.                    F1.0,1X,F4.0,1X,F6.0,1X,F2.0,1X,F4.0,1X,
         8.                    F4.0,1X,F3.0,1X,F6.0,1X/F2.0,1X,F1.0,1X,F3.0,1X,
         9.                    F3.0,1X,F2.0,1X,F1.0,1X,F1.0,1X,
        10.                    F2.0,1X,F3.0,1X,F2.0,1X,F1.0,1X,F2.0,
        11.                    1X,F2.0,1X,F2.0)
```

```
                  ACCORDING TO YOUR INPUT FORMAT, VARIABLES ARE TO BE READ AS FOLLOWS

                  VARIABLE    FORMAT    RECORD      COLUMNS

                  NAME1        A 3        1         2-    4
                  NAME2        A 3        1         5-    7
                  NAME3        A 3        1         8-   10
                  NAME4        A 3        1        11-   13
                  ADDR1        A 4        1        15-   18
                  ADDR2        A 4        1        19-   22
                  ADDR3        A 4        1        23-   26
                  ADDR4        A 4        1        27-   30
                  BLOCK      F 3. 0       1        31-   33
                  ID         F 4. 0       1        36-   39
                  ZONE       F 1. 0       1        41-   41
                  DWELL      F 1. 0       1        43-   43
                  HCOST      F 4. 0       1        45-   48
                  ASSES      F 6. 0       1        50-   55
                  ROOMS      F 2. 0       1        57-   58
                  HEAT       F 4. 0       1        60-   63
                  ELEC       F 4. 0       1        65-   68
                  PHONE      F 3. 0       1        70-   72
                  INCOM      F 6. 0       1        74-   79
                  PEPLE      F 2. 0       2         1-    2
                  CARS       F 1. 0       2         4-    4
                  GAS        F 3. 0       2         6-    8
                  GASCA      F 3. 0       2        10-   12
                  GASTR      F 2. 0       2        14-   15
                  REPAR      F 1. 0       2        17-   17
                  FAVGA      F 1. 0       2        19-   19
                  HOSP       F 2. 0       2        21-   22
                  EAT        F 3. 0       2        24-   26
                  EATPL      F 2. 0       2        28-   29
                  FEAT       F 1. 0       2        31-   31
                  LSODA      F 2. 0       2        33-   34
                  HSODA      F 2. 0       2        36-   37
                  BEER       F 2. 0       2        39-   40
```

THE INPUT FORMAT PROVIDES FOR 33 VARIABLES.   33 WILL BE READ
IT PROVIDES FOR 2 RECORDS ('CARDS') PER CASE. A MAXIMUM OF   80 'COLUMNS' ARE USED ON A RECORD.

```
        12.   N OF CASES       1373
        13.   INPUT MEDIUM     CARD
        14.   COMMENT          THE ENTIRE STAT CITY DATA BASE WAS COLLECTED VIA
        15.   COMMENT          QUESTIONNAIRE SURVEY DURING FEBRUARY AND MARCH OF
        16.   COMMENT          1980.  THE DATA BASE REFLECTS THE CHARACTERISTICS
        17.   COMMENT          OF STAT CITY DWELLING UNITS AS OF JANUARY 1980.
        18.   COMMENT
        19.   COMMENT
        20.   COMMENT
        21.   COMMENT          THE ELEMENTARY UNIT FOR ANALYSIS IN STAT CITY IS THE
        22.   COMMENT          DWELLING UNIT, NOT THE INDIVIDUAL.  ALL PROBLEMS WILL
        23.   COMMENT          CONSIDER THE DWELLING UNIT (FAMILY OR HOME) AS THE
        24.   COMMENT          BASIC UNIT FOR ANALYSIS.
        25.   COMMENT
        26.   COMMENT
        27.   COMMENT
        28.   VAR LABELS       NAME1,FIRST 4 LETTERS OF LAST NAME/
        29.                    NAME2,SECOND 4 LETTERS OF LAST NAME/
        30.                    NAME3,THIRD 4 LETTERS OF LAST NAME/
        31.                    NAME4,FOURTH 4 LETTERS OF LAST NAME/
        32.                    ADDR1,FIRST 4 LETTERS OF ADDRESS/
        33.                    ADDR2,SECOND 4 LETTERS OF ADDRESS/
        34.                    ADDR3,THIRD 4 LETTERS OF ADDRESS/
        35.                    ADDR4,FOURTH 4 LETTERS OF ADDRESS/
        36.                    BLOCK,BLOCK DWELLING UNIT IS LOCATED ON/
        37.                    ID,IDENTIFICATION NUMBER/
        38.                    ZONE,RESIDENTIAL HOUSING ZONE/
        39.                    DWELL,DWELLING TYPE/
        40.                    HCOST,HOUSING COST AS OF 1-80/
        41.                    ASSES,ASSESSED VALUE OF HOUSE AS OF 1-80/
        42.                    ROOMS,# OF ROOMS IN DWELLING UNIT AS OF 1-80/
        43.                    HEAT,AVERAGE YEARLY HEATING BILL AS OF 1-80/
        44.                    ELEC,AVERAGE MONTHLY ELECTRIC BILL AS OF 1-80/
        45.                    PHONE,AVERAGE MONTHLY PHONE BILL AS OF 1-80/
```

Exhibit 15–8 (*continued*)

```
46.                      INCOM,TOTAL FAMILY INCOME IN 1979/
47.                      PEPLE,NUMBER OF PEOPLE IN HOUSEHOLD AS OF 1-80/
48.                      CARS,NUMBER OF CARS IN HOUSEHOLD AS OF 1-80/
49.                      GAS,AVERAGE BIMONTHLY GAS BILL AS OF 1-80/
50.                      GASCA,AVE. BIMONTHLY GAS PER CAR AS OF 1-80/
51.                      GASTR,AVE. MONTHLY TRIPS FOR GAS AS OF 1-80/
52.                      REPAR,FAVORITE PLACE FOR CAR REPAIRS AS OF 1-80/
53.                      FAVGA,FAVORITE GAS STATION AS OF 1-80/
54.                      HOSP,AVE YRLY HOSP. TRIPS PER HOME AS OF 1-80/
55.                      EAT,AVERAGE WEEKLY FOOD BILL AS OF 1-80/
56.                      EATPL,AVE WKLY FOOD BILL PER PERSON AS OF 1-80/
57.                      FEAT,FAVORITE SUPERMARKET AS OF 1-80/
58.                      LSODA,AVE WKLY DIET SODA PURCHASE AS OF 1-80/
59.                      HSODA,AVE WKLY REG. SODA PURCHASE AS OF 1-80/
60.                      BEER,AVE WKLY BEER PURCHASE AS OF 1-80
61.    COMMENT           LSODA,HSODA, AND BEER PURCHASES ARE IN SIX PACK UNITS.
62.    VALUE LABELS      ZONE (1) ZONE 1 (2) ZONE 2 (3) ZONE 3 (4) ZONE 4/
63.                      DWELL (0)APARTMENT (1)HOUSE/
64.                      REPAR (0)PERFORMS OWN REPAIRS (1)REPAIRS DONE BY STATION
65.                           (2)REPAIRS DONE BY DEALER/
66.                      FAVGA (1)PAUL'S TEXACO STATION (2)HOWIE'S GULF STATION/
67.                      FEAT (1)FOODFAIR (2)GRAND UNION (3)A & P
68.    CONDESCRIPTIVE INCOM
```

```
***** GIVEN WORKSPACE ALLOWS FOR  336 VARIABLES FOR CONDESCRIPTIVE PROBLEM *****

        69.   STATISTICS      1
        70.   READ INPUT DATA
```

```
VARIABLE  INCOM     TOTAL FAMILY INCOME IN 1979

MEAN        29394.572

VALID OBSERVATIONS -    1373            MISSING OBSERVATIONS -        0

CPU TIME REQUIRED..    2.10 SECONDS
```

```
        71.   *SELECT IF     (ZONE EQ 1)
        72.   CONDESCRIPTIVE INCOM
```

```
 ***** GIVEN WORKSPACE ALLOWS FOR  336 VARIABLES FOR CONDESCRIPTIVE PROBLEM *****

        73.   STATISTICS      1
```

```
VARIABLE  INCOM     TOTAL FAMILY INCOME IN 1979

MEAN        39492.608

VALID OBSERVATIONS -     130            MISSING OBSERVATIONS -        0

TRANSPACE REQUIRED..      30 WORDS
     1 TRANSFORMATIONS
     0 RECODE VALUES + LAG VARIABLES
     3 IF/COMPUTE OPERATIONS

CPU TIME REQUIRED..     .23 SECONDS
```

```
        74.   *SELECT IF     (ZONE EQ 2)
        75.   CONDESCRIPTIVE INCOM
```

```
 ***** GIVEN WORKSPACE ALLOWS FOR  336 VARIABLES FOR CONDESCRIPTIVE PROBLEM *****

        76.   STATISTICS      1
```

```
VARIABLE  INCOM     TOTAL FAMILY INCOME IN 1979

MEAN        74483.745

VALID OBSERVATIONS -     157            MISSING OBSERVATIONS -        0
```

Exhibit 15–8 (*concluded*)

STAT CITY SPSS DRIVER PROGRAM

TRANSPACE REQUIRED..   30 WORDS
        1 TRANSFORMATIONS
        0 RECODE VALUES + LAG VARIABLES
        3 IF/COMPUTE OPERATIONS

CPU TIME REQUIRED..   .24 SECONDS

        77.   *SELECT IF   (ZONE EQ 3)
        78.   CONDESCRIPTIVE INCOM

***** GIVEN WORKSPACE ALLOWS FOR 336 VARIABLES FOR CONDESCRIPTIVE PROBLEM *****

        79.   STATISTICS   1

FILE   NONAME   (CREATION DATE = 05/18/81)

VARIABLE INCOM   TOTAL FAMILY INCOME IN 1979

MEAN   35438.234

VALID OBSERVATIONS -   338   MISSING OBSERVATIONS -   0

TRANSPACE REQUIRED..   30 WORDS
        1 TRANSFORMATIONS
        0 RECODE VALUES + LAG VARIABLES
        3 IF/COMPUTE OPERATIONS

CPU TIME REQUIRED..   .30 SECONDS

        80.   *SELECT IF   (ZONE EQ 4)
        81.   CONDESCRIPTIVE INCOM

***** GIVEN WORKSPACE ALLOWS FOR 336 VARIABLES FOR CONDESCRIPTIVE PROBLEM *****

        82.   STATISTICS   1

VARIABLE INCOM   TOTAL FAMILY INCOME IN 1979

MEAN   15444.702

VALID OBSERVATIONS -   748   MISSING OBSERVATIONS -   0

TRANSPACE REQUIRED..   30 WORDS
        1 TRANSFORMATIONS
        0 RECODE VALUES + LAG VARIABLES
        3 IF/COMPUTE OPERATIONS

CPU TIME REQUIRED..   .45 SECONDS

        83.   FINISH

        NORMAL END OF JOB.
        83 CONTROL CARDS WERE PROCESSED.
        0 ERRORS WERE DETECTED.

APPENDIX A   REFERENCE EXHIBITS

Exhibit A–1   Information collected on Stat City families

NUMBER	VARIABLE NAME	VARIABLE DESCRIPTION
1	NAME	Last name of the head of the household
2	ADDR	Street address
3	BLOCK	Block location of dwelling unit
4	ID	Dwelling unit's identification code
5	ZONE	Residential housing zone
6	DWELL	Type of dwelling unit 0 = apartment 1 = house
7	HCOST	Housing cost as of January 1980 Rent if apartment Mortgage if house
8	ASST	Assessed value of home ($0.00 if apartment) as of January 1980
9	ROOMS	Numbers of rooms in dwelling unit as of January 1980
10	HEAT	Average total yearly heating bill as of January 1980 (includes all types of heat—electric, gas, etc.)
11	ELEC	Average monthly electric bill as of January 1980
12	PHONE	Average monthly telephone bill as of January 1980
13	INCOM	Total family income for 1979
14	PEPLE	Number of people in household as of Januray 1980
15	CARS	Number of cars in household as of January 1980
16	GAS	Average bimonthly automobile gas bill as of January 1980
17	GASCA	Average bimonthly automobile gas bill per car as of January 1980
18	GASTR	Average number of trips to the gas station per month as of January 1980
19	REPAR	Favorite place to have automotive repairs performed as of January 1980 0 = performs own repairs 1 = repairs done by service station 2 = repairs done by dealer
20	FAVGA	Favorite gas station as of January 1980 1 = Paul's Texaco (11th St. & 7th Ave.) 2 = Howie's Gulf (11th St. & Division St.)
21	HOSP	Average yearly trips to the hospital by all members of a dwelling unit as of January 1980
22	EAT	Average weekly supermarket bill as of January 1980
23	EATPL	Average weekly supermarket bill per person as of January 1980
24	FEAT	Favorite supermarket as of January 1980 1 = Food Fair 2 = Grand Union 3 = A&P
25	LSODA	Average weekly purchase of six-packs of diet soda as of January 1980
26	HSODA	Average weekly purchase of six-packs of regular soda as of January 1980
27	BEER	Average weekly purchase of six-packs of beer as of January 1980

STAT CITY QUESTIONNAIRE
(mailed JANUARY 1980)
(To be filled out by the head of the household!)

1. Name of Head of Household: _____

2. Local Address: _____
   _____

3. Block: _____

4. Identification Code: _____
   (Assigned by Census Bureau)

5. Zone: _____

6. What type of building do you live in?
   apartment building [0]   house [1]

7. If you live in an apartment, what is your monthly rent?
   $ _____

8. If you live in a house:
   a. What is your monthly mortgage? $ _____
   b. What is the assessed value of your house? $ _____

9. How many rooms are in your home (count all rooms including bathrooms)? _____

10. What is your average total yearly heating bill (lump all types of heat used—gas, oil, or electric)? $ _____

11. What is your year-round average monthly electric bill?
    $ _____

12. What is your year-round average monthly telephone bill?
    $ _____

13. What was your gross yearly family income in 1979? $ _____

14. How many people live in your household (include yourself)? _____

15. How many automobiles are registered at your household? _____

16. What is the year-round average bimonthly gas bill for the automobiles registered at your household? _____

17. What is the year-round average monthly number of trips to a gas station the automobiles registered at your household make? _____

18. In general, where are simple repairs and maintenance performed on *your* automobiles?
    Perform own repairs [0]   Repairs done by service station [1]   Repairs done by dealer [2]

19. Which gas station gets more of *your* business?
    a. Pauls' Texaco
       11th St. and 7th Avenue   [station 1]
    b. Howie's Gulf
       11th St. and Division St.   [station 2]

20. What is the yearly average number of trips to the hospital made by all members of your household? _____

21. What is your year-round average weekly supermarket bill? $ _____

22. Which supermarket gets more of *your* business?
    a. Food Fair
       Park St. between 8th and 9th Ave.   [supermarket 1]
    b. Grand Union
       9th St. and 9th Ave.   [supermarket 2]
    c. A&P
       E. Division St. and 5th Ave.   [supermarket 3]

23. What is the year-round average weekly purchase of six-packs of diet soda for your household? _____ six-packs

24. What is the year-round average weekly purchase of six-packs of regular soda for your household? _____ six-packs

25. What is the year-round average weekly purchase of six-packs of beer for your household? _____ six-packs

PLEASE ENCLOSE YOUR COMPLETED QUESTIONNAIRE IN THE STAMPED, ADDRESSED ENVELOPE AND MAIL TO:
STAT CITY CHAMBER OF COMMERCE
CITY HALL
E. DIVISION ST. AND 7TH AVE.
STAT CITY
NO LATER THAN February 25, 1980.
THANK YOU FOR YOUR COOPERATION.

5347	8111	9803	1221	5952	4023	4057	3935	4321	6925
9734	7032	5811	9196	2624	4464	8328	9739	9282	7757
6602	3827	7452	7111	8489	1395	9889	9231	6578	5964
9977	7572	0317	4311	8308	8198	1453	2616	2489	2055
3017	4897	9215	3841	4243	2663	8390	4472	6921	6911
8187	8333	1498	9993	1321	3017	4796	9379	8669	9885
1983	9063	7186	9505	5553	6090	8410	5534	4847	6379
0933	3343	5386	5276	1880	2582	9619	6651	7831	9701
3115	5829	4082	4133	2109	9388	4919	4487	4718	8142
6761	5251	0303	8169	1710	6498	6083	8531	4781	0807
6194	4879	1160	8304	2225	1183	0434	9554	2036	5593
0481	6489	9634	7906	2699	4396	6348	9357	8075	9658
0576	3960	5614	2551	8615	7865	0218	2971	0433	1567
7326	5687	4079	1394	9628	9018	4711	6680	6184	4468
5490	0997	7658	0264	3579	4453	6442	3544	2831	9900
4258	3633	6006	0404	2967	1634	4859	2554	6317	7522
2726	2740	9752	2333	3645	3369	2367	4588	4151	0475
4984	1144	6668	3605	3200	7860	3692	5996	6819	6258
2931	4046	2707	6923	5142	5851	4992	0390	2659	3306
3046	2785	6779	1683	7427	0579	0290	6349	0078	3509
2870	8408	6553	4425	3386	8253	9839	2638	0283	3683
1318	5065	9487	2825	7854	5528	3359	6196	5172	1421
6079	7663	3015	4029	9947	2833	1536	4248	6031	4277
1348	4691	6468	0741	7784	0190	4779	6579	4423	7723
3491	9450	3937	3418	5750	2251	0406	9451	4461	1048
2810	0481	8517	8649	3569	0348	5731	6317	7190	7118
5923	4502	0117	0884	8192	7149	9540	3404	0485	6591
8743	8275	7109	3683	5358	2598	4600	4284	8168	2145
2904	0130	5534	6573	7871	4364	4624	5320	9486	4871
6203	7188	9450	1526	6143	1036	4205	6825	1438	7943
3885	8004	5997	7336	5287	4767	4102	8229	2643	8737
4066	4332	8737	8641	9584	2559	5413	9418	4230	0736
4058	9008	3772	0866	3725	2031	5331	5098	3290	3209
7823	8655	5027	2043	0024	0230	7102	4993	2324	0086
9824	6747	7145	6954	0116	0332	6701	9254	9797	5272
6997	7855	6543	3262	2831	6181	1459	7972	5569	9134
3984	2307	4081	0371	2189	9635	9680	2459	2620	2600
6288	8727	9989	9996	3437	4255	1167	9960	9801	4886
5613	6492	2945	5296	8662	6242	3016	7618	9531	3926
9080	5602	4899	6456	6746	6018	1297	0384	6258	9385
0966	4467	7476	3335	6730	8054	9765	1134	7877	4501
3475	5040	7663	1276	3222	3454	1810	5351	1452	7212
1215	7332	7419	2666	7808	5363	5230	0000	0570	6353
6938	0773	9445	7642	1612	0930	6741	6858	8793	3884
9335	6456	4376	4504	4493	6997	1696	0827	6775	6029
3887	3554	9956	8540	0491	6254	7840	0101	8618	2207
5831	6029	7239	6966	1247	9305	0205	2980	6364	1279
8356	1022	9947	7472	2207	1023	2157	2032	2131	5712
2806	9115	4056	3370	6451	0706	6437	2633	7965	3114
0573	7555	9316	8092	5587	5410	3480	8315	0453	8136

Exhibit A–3 2500 four-digit random numbers

Source: Compiled from Rand Corporation, *A Million Random Digits with 100,000 Normal Deviates* (Glencoe, Ill.: The Free Press, 1955). Used with permission.

Exhibit A–3 (*continued*)

2668	7422	4354	4569	9446	8212	3737	2396	6892	3766
6067	7516	2451	1510	0201	1437	6518	1063	6442	6674
4541	9863	8312	9855	0995	6025	4207	4093	9799	9308
6987	4802	8975	2847	4413	5997	9106	2876	8596	7717
0376	8636	9953	4418	2388	8997	1196	5158	1803	5623
8468	5763	3232	1986	7134	4200	9699	8437	2799	2145
9151	4967	3255	8518	2802	8815	6289	9549	2942	3813
1073	4930	1830	2224	2246	1000	9315	6698	4491	3046
5487	1967	5836	2090	3832	0002	9844	3742	2289	3763
4896	4957	6536	7430	6208	3929	1030	2317	7421	3227
9143	7911	0368	0541	2302	5473	9155	0625	1870	1890
9256	2956	4747	6280	7342	0453	8639	1216	5964	9772
4173	1219	7744	9241	6354	4211	8497	1245	3313	4846
2525	7811	5417	7824	0922	8752	3537	9069	5417	0856
9165	1156	6603	2852	8370	0995	7661	8811	7835	5087
0014	8474	6322	5053	5015	6043	0482	4957	8904	1616
5325	7320	8406	5962	6100	3854	0575	0617	8019	2646
2558	1748	5671	4974	7073	3273	6036	1410	5257	3939
0117	1218	0688	2756	7545	5426	3856	8905	9691	8890
8353	1554	4083	2029	8857	4781	9654	7946	7866	2535
1990	9886	3280	6109	9158	3034	8490	6404	6775	8763
9651	7870	2555	3518	2906	4900	2984	6894	5050	4586
9941	5617	1984	2435	5184	0379	7212	5795	0836	4319
7769	5785	9321	2734	2890	3105	6581	2163	4938	7540
3224	8379	9952	0515	2724	4826	6215	6246	9704	1651
1287	7275	6646	1378	6433	0005	7332	0392	1319	1946
6389	4191	4548	5546	6651	8248	7469	0786	0972	7649
1625	4327	2654	4129	3509	3217	7062	6640	0105	4422
7555	3020	4181	7498	4022	9122	6423	7301	8310	9204
4177	1844	3468	1389	3884	6900	1036	8412	0881	6678
0927	0124	8176	0680	1056	1008	1748	0547	8227	0690
8505	1781	7155	3635	9751	5414	5113	8316	2737	6860
8022	8757	6275	1485	3635	2330	7045	2106	6381	2986
8390	8802	5674	2559	7934	4788	7791	5202	8430	0289
3630	5783	7762	0223	5328	7731	4010	3845	9221	5427
9154	6388	6053	9633	2080	7269	0894	0287	7489	2259
1441	3381	7823	8767	9647	4445	2509	2929	5067	0779
8246	0778	0993	6687	7212	9968	8432	1453	0841	4595
2730	3984	0563	9636	7202	0127	9283	4009	3177	4182
9196	8276	0233	0879	3385	2184	1739	5375	5807	4849
5928	9610	9161	0748	3794	9683	1544	1209	3669	5831
1042	9600	7122	2135	7868	5596	3551	9480	2342	0449
6552	4103	7957	0510	5958	0211	3344	5678	1840	3627
5968	4307	9327	3197	0876	8480	5066	1852	8323	5060
4445	1018	4356	4653	9302	0761	1291	6093	5340	1840
8727	8201	5980	7859	6055	1403	1209	9547	4273	0857
9415	9311	4996	2775	8509	7767	6930	6632	7781	2279
2648	7639	9128	0341	6875	8957	6646	9783	6668	0317
3707	3454	8829	6863	1297	5089	1002	2722	0578	7753
8383	8957	5595	9395	3036	4767	8300	3505	0710	6307

5503	8121	9056	8194	1124	8451	1228	8986	0076	7615
2552	9953	4323	4878	4922	0696	3156	2145	8819	0631
8542	7274	9724	6638	0013	0566	9644	3738	5767	2791
6121	4839	4734	3041	3939	9136	5620	7920	0533	3119
2023	0314	5885	1165	2841	1282	5893	3050	6598	2667
9577	8320	5614	5595	8978	6442	0844	4570	8036	6026
0760	1734	0114	8330	9695	6502	3171	8901	7955	4975
0064	1745	7874	3900	3602	9880	7266	5448	6826	3882
6295	8316	6150	3155	8059	4789	7236	7272	0839	3367
7935	1027	8193	2634	0806	6781	0665	8791	7416	8551
4833	6983	5904	8217	9201	5844	6959	5620	9570	8621
0584	0843	7983	5095	3205	3291	1584	1391	4136	8011
2585	0220	0730	5994	7138	7615	1126	3878	6154	2260
2527	1615	8232	7071	9808	3863	9195	4990	7625	3397
7300	2905	1760	4929	4767	9044	6891	0567	2382	8489
8131	9443	2266	0658	3814	0014	1749	5111	6145	6579
1002	4471	5983	8072	6371	6788	2510	4534	5574	6761
8467	5280	8912	3769	2089	8233	2262	0614	0577	0354
2929	5816	2185	3373	9405	8880	5460	0038	6634	6923
5177	9407	7063	4128	9058	8768	1396	5562	2367	3510
4216	5625	6077	5167	3603	7727	8521	1481	9075	2367
7835	6704	2249	5152	3116	3045	2760	4442	9638	2677
0955	5134	3386	8901	7341	8153	7739	3044	9774	1815
1577	6312	3484	0566	0615	4897	5569	6181	9176	2082
1323	9905	9375	3673	4428	4432	1572	3750	4726	1333
5058	0357	3847	7323	6761	7278	7817	1871	9909	6411
9948	5733	1063	7490	9067	1964	6990	6095	1796	3721
5467	3952	7378	4886	6983	6279	6520	6918	0557	7474
9934	7154	1024	7603	3170	7686	8890	6957	2764	0033
3549	4023	3486	5535	1284	6809	5264	3273	6701	4678
9817	2538	0384	2392	4795	1035	7011	1117	6329	9990
0267	8615	5686	0259	0164	4220	7995	3776	8234	7195
3693	4287	8163	7995	0706	4162	9680	9238	8886	6858
5685	1277	2430	7366	8426	2466	1668	0223	6602	6413
0546	2889	1427	2377	8859	1708	3388	8878	3901	5711
1502	2023	6338	7112	0662	0741	9498	3232	7942	7038
9561	0803	8146	9106	8885	5658	0122	2809	1972	7146
0902	4037	0573	5512	7429	4919	3166	4260	3036	9642
8143	9995	5246	6766	9732	6980	2124	6592	1262	9289
2143	5933	5862	9482	6548	0964	4101	8510	1611	3207
9583	7614	1163	8028	1778	9793	1282	7389	6600	2752
9981	4463	4374	9979	8682	1211	3170	0502	2815	0420
7721	3114	5054	1160	5093	0249	0918	9587	8584	7195
1326	0260	7983	6605	8027	0853	2867	3753	7053	8235
4428	7173	2662	5469	1490	5213	8111	7454	7885	3199
7052	4595	7963	5737	0505	3196	3337	1323	8566	8661
8838	1122	2508	7146	0981	4600	1906	6898	1831	7417
8316	7399	1720	7944	6409	4979	1193	4486	8697	3453
5021	7172	3385	4514	0569	2993	1282	0159	0845	5282
9768	2934	6774	8064	1362	2394	4939	8368	3730	9535

Exhibit A–3 (*continued*)

1236	2389	3150	9072	1871	8914	5859	9942	2284	0826
3889	3023	3423	2257	7442	2273	2693	4060	1078	8012
8078	5541	3977	9331	1827	2114	5208	7809	8563	8114
0239	7758	0885	2356	3354	4579	1097	4472	2478	0969
7372	7018	6911	7188	8014	7287	3898	2340	6395	4475
6138	1722	5523	1896	3900	9350	1827	4981	5280	6967
3916	4428	1497	9749	2597	3360	6014	3003	7767	4929
8090	7448	3988	1988	3731	0420	4967	3959	0105	4399
0905	6567	6366	3403	0657	8783	2812	4888	5048	5573
3342	2422	3204	6008	2041	8504	5357	3255	6409	5232
7265	6947	7364	7153	5545	1957	1555	2057	1212	5003
0414	3209	8358	6182	3548	3273	6340	9149	3719	0276
8522	1419	5221	6074	2441	5785	3188	5126	8229	7355
5488	0357	9167	5950	0861	3379	2901	8519	6226	2868
3325	5151	8203	4523	3935	3322	5946	6554	7680	1698
7597	1595	3240	8208	0221	5714	3352	4719	9452	7325
9063	7531	3538	3445	4924	1146	2510	7148	8988	9970
6506	1549	9334	3356	1942	6682	0304	9736	0815	4748
6442	0742	8223	9781	3957	0776	6584	2998	1553	9011
2717	1738	7696	7511	4558	9990	4716	5536	2566	2540
3221	3009	8727	5689	1562	3259	8066	0808	1942	8071
5420	5804	7235	8982	0270	1681	8998	3738	4403	5936
5928	6696	8484	7154	6755	3386	8301	6621	6937	2390
8387	5816	0122	9555	2219	6590	3878	0135	4748	2817
8331	5708	0336	8001	3960	4069	5643	6405	0249	5088
6454	2950	1335	7864	9262	1935	6047	5733	5213	0711
3926	0007	5548	0152	7656	2257	2032	8462	3018	4390
2976	0567	2819	6551	1195	7859	6390	2134	1921	9028
0631	0299	0146	2773	9028	1769	6451	3955	3469	0321
9754	4760	5765	5910	2185	4444	0797	5429	8467	7875
8296	8571	1161	9772	5351	5378	9894	3840	7093	1131
7687	3472	1252	9064	1692	1366	1742	8448	6830	8524
8739	7888	8723	9208	9563	6684	2290	6498	8695	5470
7404	1273	5961	3369	1259	4489	6798	7297	8979	1058
4789	4141	6643	7004	5079	4592	9656	6795	5636	4472
8777	7169	6414	5436	9211	3403	5906	6205	6204	3352
9697	6314	7221	8004	1199	4769	9562	7299	2904	8589
4382	1328	7781	8169	2993	7075	0202	3237	0055	8668
5720	8396	4009	3923	6595	5991	9141	5557	8842	4557
4906	7217	8093	0601	9032	6368	0793	9958	4901	2645
9425	8427	9579	1347	8013	2633	5516	7341	4076	4517
6814	8138	8238	1867	4045	9282	3004	3741	4342	4513
1220	9780	3361	2886	4164	1673	8886	3263	4198	8461
8831	8970	2611	1241	1943	6566	6098	5976	1141	1825
5672	8035	2961	6305	1525	4468	6468	4235	5102	7768
0713	1232	0107	1930	8704	5892	2845	8106	9397	6665
2118	6455	5561	3608	2433	8439	1602	1220	7755	7566
0215	1225	8873	4391	0365	2109	6080	6324	2684	3581
9095	8523	3277	0730	3618	4742	1968	3318	4138	0324
8010	9130	1285	4129	0032	1501	1957	9113	1272	9260

9263	7824	1926	9545	5349	2389	3770	7986	7647	6641
7944	7873	7154	4484	2610	6731	0070	3498	6675	9972
5965	7196	2738	5000	0535	9403	2928	1854	5242	0608
3152	4958	7661	3978	1353	4808	5948	6068	8467	5301
0634	7693	9037	5139	5588	7101	0920	7915	2444	3024
2870	5170	9445	4839	7378	0643	8664	6923	5766	8018
6810	8926	9473	9576	7502	4846	6554	9658	1891	1639
9993	9070	9362	6633	3339	9526	9534	5176	9161	3323
9154	7319	3444	6351	8383	9941	5882	4045	6926	4856
4210	0278	7392	5629	7267	1224	2527	3667	2131	7576
1713	2758	2529	2838	5135	6166	3789	0536	4414	4267
2829	1428	5452	2161	9532	3817	6057	0808	9499	7846
0933	5671	5133	0628	7534	0881	8271	5739	2525	3033
3129	0420	9371	5128	0575	7939	8739	5177	3307	9706
3614	1556	2759	4208	9928	5964	1522	9607	0996	0537
2955	1843	1363	0552	0279	8101	4902	7903	5091	0939
2350	2264	6308	0819	8942	6780	5513	5470	3294	6452
5788	8584	6796	0783	1131	0154	4853	1714	0855	6745
5533	7126	8847	0433	6391	3639	1119	9247	7054	2977
1008	1007	5598	6468	6823	2046	8938	9380	0079	9594
3410	8127	6609	8887	3781	7214	6714	5078	2138	1670
5336	4494	6043	2283	1413	9659	2329	5620	9267	1592
8297	6615	8473	1943	5579	6922	2866	1367	9931	7687
5482	8467	2289	0809	1432	8703	4289	2112	3071	4848
2546	5909	2743	8942	8075	8992	1909	6773	8036	0879
6760	6021	4147	8495	4013	0254	0957	4568	5016	1560
4492	7092	6129	5113	4759	8673	3556	7664	1821	6344
3317	3097	9813	9582	4978	1330	3608	8076	3398	6862
8468	8544	0620	1765	5133	0287	3501	6757	6157	2074
7188	5645	3656	0939	9695	3550	1755	3521	6910	0167
0047	0222	7472	1472	4021	2135	0859	4562	8398	6374
2599	3888	6836	5956	4127	6974	4070	3799	0343	1887
9288	5317	9919	9380	5698	5308	5052	5590	4302
2513	2681	0709	1567	6068	0441	2450	3789	6718	6282
8463	7188	1299	8302	8248	9033	9195	7457	0353	9012
3400	9232	1279	6145	4812	7427	2836	6656	7522	3590
5377	4574	0573	8616	4276	7017	9731	7389	8860	1999
5931	9788	7280	5496	6085	1193	3526	7160	5557	6771
2047	6655	5070	2699	0985	5259	1406	3021	1989	1929
8618	8493	2545	2604	0222	5201	2182	5059	5167	6541
2145	6800	7271	4026	6128	1317	6381	4897	5173	5411
9806	6837	8008	2413	7235	9542	1180	2974	8164	8661
0178	6442	1443	9457	7515	9457	6139	9619	0322	3225
6246	0484	4327	6870	0127	0543	2295	1894	9905	4169
9432	3108	8415	9293	9998	8950	9158	0280	6947	6827
0579	4398	2157	0990	7022	1979	5157	3643	3349	7988
1039	1428	5218	0972	2578	3856	5479	0489	5901	8925
3517	5698	2554	5973	6471	5263	3110	6238	4948	1140
2563	8961	7588	9825	0212	7209	5718	5588	0932	7346
1646	4828	9425	4577	4515	6886	1138	1178	2269	4198

APPENDIX B REFERENCE TABLES

Table B–1 Binomial distribution

$$P(x) = \frac{n!}{x!\,(n-x)!}\,\pi^x(1-\pi)^{n-x}$$

n	r	.05	.10	.15	.20	.25 (p)	.30	.35	.40	.45	.50
1	0	.9500	.9000	.8500	.8000	.7500	.7000	.6500	.6000	.5500	.5000
	1	.0500	.1000	.1500	.2000	.2500	.3000	.3500	.4000	.4500	.5000
2	0	.9025	.8100	.7225	.6400	.5625	.4900	.4225	.3600	.3025	.2500
	1	.0950	.1800	.2550	.3200	.3750	.4200	.4550	.4800	.4950	.5000
	2	.0025	.0100	.0225	.0400	.0625	.0900	.1225	.1600	.2025	.2500
3	0	.8574	.7290	.6141	.5120	.4219	.3430	.2746	.2160	.1664	.1250
	1	.1354	.2430	.3251	.3840	.4219	.4410	.4436	.4320	.4084	.3750
	2	.0071	.0270	.0574	.0960	.1406	.1890	.2389	.2880	.3341	.3750
	3	.0001	.0010	.0034	.0080	.0156	.0270	.0429	.0640	.0911	.1250
4	0	.8145	.6561	.5220	.4096	.3164	.2401	.1785	.1296	.0915	.0625
	1	.1715	.2916	.3685	.4096	.4219	.4116	.3845	.3456	.2995	.2500
	2	.0135	.0486	.0975	.1536	.2109	.2646	.3105	.3456	.3675	.3750
	3	.0005	.0036	.0115	.0256	.0469	.0756	.1115	.1536	.2005	.2500
	4	.0000	.0001	.0005	.0016	.0039	.0081	.0150	.0256	.0410	.0625
5	0	.7738	.5905	.4437	.3277	.2373	.1681	.1160	.0778	.0503	.0312
	1	.2036	.3280	.3915	.4096	.3955	.3602	.3124	.2592	.2059	.1562
	2	.0214	.0729	.1382	.2048	.2637	.3087	.3364	.3456	.3369	.3125
	3	.0011	.0081	.0244	.0512	.0879	.1323	.1811	.2304	.2757	.3125
	4	.0000	.0004	.0022	.0064	.0146	.0284	.0488	.0768	.1128	.1562
	5	.0000	.0000	.0001	.0003	.0010	.0024	.0053	.0102	.0185	.0312
6	0	.7351	.5314	.3771	.2621	.1780	.1176	.0754	.0467	.0277	.0156
	1	.2321	.3543	.3993	.3932	.3560	.3025	.2437	.1866	.1359	.0938
	2	.0305	.0984	.1762	.2458	.2966	.3241	.3280	.3110	.2780	.2344
	3	.0021	.0146	.0415	.0819	.1318	.1852	.2355	.2765	.3032	.3125
	4	.0001	.0012	.0055	.0154	.0330	.0595	.0951	.1382	.1861	.2344
	5	.0000	.0001	.0004	.0015	.0044	.0102	.0205	.0369	.0609	.0938
	6	.0000	.0000	.0000	.0001	.0002	.0007	.0018	.0041	.0083	.0156
7	0	.6983	.4783	.3206	.2097	.1335	.0824	.0490	.0280	.0152	.0078
	1	.2573	.3720	.3960	.3670	.3115	.2471	.1848	.1306	.0872	.0547
	2	.0406	.1240	.2097	.2753	.3115	.3177	.2985	.2613	.2140	.1641
	3	.0036	.0230	.0617	.1147	.1730	.2269	.2679	.2903	.2918	.2734
	4	.0002	.0026	.0109	.0287	.0577	.0972	.1442	.1935	.2388	.2734
	5	.0000	.0002	.0012	.0043	.0115	.0250	.0466	.0774	.1172	.1641
	6	.0000	.0000	.0001	.0004	.0013	.0036	.0084	.0172	.0320	.0547
	7	.0000	.0000	.0000	.0000	.0001	.0002	.0006	.0016	.0037	.0078
8	0	.6634	.4305	.2725	.1678	.1002	.0576	.0319	.0168	.0084	.0039
	1	.2793	.3826	.3847	.3355	.2670	.1977	.1373	.0896	.0548	.0312
	2	.0515	.1488	.2376	.2936	.3115	.2065	.2587	.2090	.1569	.1094

Source: Extracted from "Tables of the Binomial Probability Distributions." U.S. Department of Commerce, National Bureau of Standards, Applied Mathematics Series 6 (1952).

n	r	.05	.10	.15	.20	.25	P .30	.35	.40	.45	.50
8	3	.0054	.0331	.0839	.1468	.2076	.2541	.2786	.2787	.2568	.2188
	4	.0004	.0046	.0185	.0459	.0865	.1361	.1875	.2322	.2627	.2734
	5	.0000	.0004	.0026	.0092	.0231	.0467	.0808	.1239	.1719	.2188
	6	.0000	.0000	.0002	.0011	.0038	.0100	.0217	.0413	.0403	.1094
	7	.0000	.0000	.0000	.0001	.0004	.0012	.0033	.0079	.0164	.0312
	8	.0000	.0000	.0000	.0000	.0000	.0001	.0002	.0007	.0017	.0039
9	0	.6302	.3874	.2316	.1342	.0751	.0404	.0207	.0101	.0046	.0020
	1	.2985	.3874	.3679	.3020	.2253	.1556	.1004	.0605	.0339	.0176
	2	.0629	.1722	.2597	.3020	.3003	.2668	.2162	.1612	.1110	.0703
	3	.0077	.0446	.1069	.1762	.2336	.2668	.2716	.2508	.2119	.1641
	4	.0006	.0074	.0283	.0661	.1168	.1715	.2194	.2508	.2600	.2461
	5	.0000	.0008	.0050	.0165	.0389	.0735	.1181	.1672	.2128	.2461
	6	.0000	.0001	.0006	.0028	.0087	.0210	.0424	.0743	.1160	.1641
	7	.0000	.0000	.0000	.0003	.0012	.0039	.0098	.0212	.0407	.0703
	8	.0000	.0000	.0000	.0000	.0001	.0004	.0013	.0035	.0083	.0176
	9	.0000	0000	0000	0000	0000	0000	0001	.0003	.0008	.0020
10	0	.5987	.3487	.1969	.1074	.0563	.0282	.0135	.0060	.0025	.0010
	1	.3151	.3874	.3474	.2684	.1877	.1211	.0725	.0403	.0207	.0098
	2	.0746	.1937	.2759	.3020	.2816	.2335	.1757	.1209	.0763	.0439
	3	.0105	.0574	.1298	.2013	.2503	.2668	.2522	.2150	.1665	.1172
	4	.0010	.0112	.0401	.0881	.1460	.2001	.2377	.2508	.2384	.2051
	5	.0001	.0015	.0085	.0264	.0584	.1029	.1536	.2007	.2340	.2461
	6	.0000	.0001	.0012	.0055	.0162	.0368	.0689	.1115	.1596	.2051
	7	.0000	.0000	.0001	.0008	.0031	.0090	.0212	.0425	.0746	.1172
	8	.0000	.0000	.0000	.0001	.0004	.0014	.0043	.0106	.0229	.0439
	9	.0000	.0000	.0000	.0000	.0000	.0001	.0005	.0016	.0042	.0098
	10	.0000	.0000	.0000	.0000	.0000	.0000	.0000	.0001	.0003	.0010
11	0	.5688	.3138	.1673	.0859	.0422	.0198	.0088	.0036	.0014	.0005
	1	.3293	.3835	.3248	.2362	.1549	.0932	.0518	.0266	.0125	.0054
	2	.0867	.2131	.2866	.2953	.2581	.1998	.1395	.0887	.0513	.0269
	3	.0137	.0710	.1517	.2215	.2581	.2568	.2254	.1774	.1259	.0806
	4	.0014	.0158	.0536	.1107	.1721	.2201	.2428	.2365	.2060	.1611
	5	.0001	.0025	.0132	.0388	.0803	.1321	.1830	.2207	.2360	.2256
	6	.0000	.0003	.0023	.0097	.0268	.0566	.0985	.1471	.1931	.2256
	7	.0000	.0000	.0003	.0017	.0064	.0173	.0379	.0701	.1128	.1611
	8	.0000	.0000	.0000	.0002	.0011	.0037	.0102	.0234	.0462	.0806
	9	.0000	.0000	.0000	.0000	.0001	.0005	.0018	.0052	.0126	.0269
	10	.0000	.0000	.0000	.0000	.0000	.0000	.0002	.0007	.0021	.0054
	11	.0000	.0000	.0000	.0000	.0000	.0000	.0000	.0000	.0002	.0005
12	0	.5404	.2824	.1422	.0687	.0317	.0138	.0057	.0022	.0008	.0002
	1	.3413	.3766	.3012	.2062	.1267	.0712	.0368	.0174	.0075	.0029

n	r	.05	.10	.15	.20	.25	p .30	.35	.40	.45	.50
12	2	.0988	.2301	.2924	.2835	.2323	.1678	.1088	.0639	.0339	.0161
	3	.0173	.0852	.1720	.2362	.2581	.2397	.1954	.1419	.0923	.0537
	4	.0021	.0213	.0683	.1329	.1936	.2311	.2367	.2128	.1700	.1208
	5	.0002	.0038	.0193	.0532	.1032	.1585	.2039	.2270	.2225	.1934
	6	.0000	.0005	.0040	.0155	.0401	.0792	.1281	.1766	.2124	.2256
	7	.0000	.0000	.0006	.0033	.0115	.0291	.0591	.1009	.1489	.1934
	8	.0000	.0000	.0001	.0005	.0024	.0078	.0199	.0420	.0762	.1208
	9	.0000	.0000	.0000	.0001	.0004	.0015	.0048	.0125	.0277	.0537
	10	.0000	.0000	.0000	.0000	.0000	.0002	.0008	.0025	.0068	.0161
	11	.0000	.0000	.0000	.0000	.0000	.0000	.0001	.0003	.0010	.0029
	12	.0000	.0000	.0000	.0000	.0000	.0000	.0000	.0000	.0001	.0002
13	0	.5133	.2542	.1209	.0550	.0238	.0097	.0037	.0013	.0004	.0001
	1	.3512	.3672	.2774	.1787	.1029	.0540	.0259	.0113	.0045	.0016
	2	.1109	.2448	.2937	.2680	.2059	.1388	.0836	.0453	.0220	.0095
	3	.0214	.0997	.1900	.2457	.2517	.2181	.1651	.1107	.0660	.0349
	4	.0028	.0277	.0838	.1535	.2097	.2337	.2222	.1845	.1350	.0873
	5	.0003	.0055	.0266	.0691	.1258	.1803	.2154	.2214	.1989	.1571
	6	.0000	.0008	.0063	.0230	.0559	.1030	.1546	.1968	.2169	.2095
	7	.0000	.0001	.0011	.0058	.0186	.0442	.0833	.1312	.1775	.2095
	8	.0000	.0001	.0001	.0011	.0047	.0142	.0336	.0656	.1089	.1571
	9	.0000	.0000	.0000	.0001	.0009	.0034	.0101	.0243	.0495	.0873
	10	.0000	.0000	.0000	.0000	.0001	.0006	.0022	.0065	.0162	.0349
	11	.0000	.0000	.0000	.0000	.0000	.0001	.0003	.0012	.0036	.0095
	12	.0000	.0000	.0000	.0000	.0000	.0000	.0000	.0001	.0005	.0016
	13	.0000	.0000	.0000	.0000	.0000	.0000	.0000	.0000	.0000	.0001
14	0	.4877	.2288	.1028	.0440	.0178	.0068	.0024	.0008	.0002	.0001
	1	.3593	.3559	.2539	.1539	.0832	.0407	.0181	.0073	.0027	.0009
	2	.1229	.2570	.2912	.2501	.1802	.1134	.0634	.0317	.0141	.0056
	3	.0259	.1142	.2056	.2501	.2402	.1943	.1366	.0845	.0462	.0222
	4	.0037	.0349	.0998	.1720	.2202	.2290	.2022	.1549	.1040	.0611
	5	.0004	.0078	.0352	.0860	.1468	.1963	.2178	.2066	.1701	.1222
	6	.0000	.0013	.0093	.0322	.0734	.1262	.1759	.2066	.2088	.1833
	7	.0000	.0002	.0019	.0092	.0280	.0618	.1082	.1574	.1952	.2095
	8	.0000	.0000	.0003	.0020	.0082	.0232	.0510	.0918	.1398	.1833
	9	.0000	.0000	.0000	.0003	.0018	.0066	.0183	.0408	.0762	.1222
	10	.0000	.0000	.0000	.0000	.0003	.0014	.0049	.0136	.0312	.0611
	11	.0000	.0000	.0000	.0000	.0000	.0002	.0010	.0033	.0093	.0222
	12	.0000	.0000	.0000	.0000	.0000	.0000	.0001	.0005	.0019	.0056
	13	.0000	.0000	.0000	.0000	.0000	.0000	.0000	.0001	.0002	.0009
	14	.0000	.0000	.0000	.0000	.0000	.0000	.0000	.0000	.0000	.0001
15	0	.4633	.2059	.0874	.0352	.0134	.0047	.0016	.0005	.0001	.0000
	1	.3658	.3432	.2312	.1319	.0668	.0305	.0126	.0047	.0016	.0005
	2	.1348	.2669	.2856	.2309	.1559	.0916	.0476	.0219	.0090	.0032

$\tilde{n}$	r	.05	.10	.15	.20	.25	p .30	.35	.40	.45	.50
15	3	.0307	.1285	.2184	.2501	.2252	.1700	.1110	.0634	.0318	.0139
	4	.0049	.0428	.1156	.1876	.2252	.2186	.1792	.1268	.0780	.0417
	5	.0006	.0105	.0449	.1032	.1651	.2061	.2123	.1859	.1404	.0916
	6	.0000	.0019	.0132	.0430	.0917	.1472	.1906	.2066	.1914	.1527
	7	.0000	.0003	.0030	.0138	.0393	.0811	.1319	.1771	.2013	.1964
	8	.0000	.0000	.0005	.0035	.0131	.0348	.0710	.1181	.1647	.1964
	9	.0000	.0000	.0001	.0007	.0034	.0116	.0298	.0612	.1048	.1527
	10	.0000	.0000	.0000	.0001	.0007	.0030	.0096	.0245	.0515	.0916
	11	.0000	.0000	.0000	.0000	.0001	.0006	.0024	.0074	.0191	.0417
	12	.0000	.0000	.0000	.0000	.0000	.0001	.0004	.0016	.0052	.0139
	13	.0000	.0000	.0000	.0000	.0000	.0000	.0001	.0003	.0010	.0032
	14	.0000	.0000	.0000	.0000	.0000	.0000	.0000	.0000	.0001	.0005
	15	.0000	.0000	.0000	.0000	.0000	.0000	.0000	.0000	.0000	.0000
16	0	.4401	.1853	.0743	.0281	.0100	.0033	.0010	.0003	.0001	.0000
	1	.3706	.3294	.2097	.1126	.0535	.0228	.0087	.0030	.0009	.0002
	2	.1463	.2745	.2775	.2111	.1336	.0732	.0353	.0150	.0056	.0018
	3	.0359	.1423	.2285	.2463	.2079	.1465	.0888	.0468	.0215	.0085
	4	.0061	.0514	.1311	.2001	.2252	.2040	.1553	.1014	.0572	.0278
	5	.0008	.0137	.0555	.1201	.1802	.2099	.2008	.1623	.1123	.0667
	6	.0001	.0028	.0180	.0550	.1649	.1982	.1983	.1684	.1222	
	7	.0000	.0004	.0045	.0197	.0524	.1010	.1524	.1889	.1969	.1746
	8	.0000	.0001	.0009	.0055	.0197	.0487	.0923	.1417	.1812	.1964
	9	.0000	.0000	.0001	.0012	.0058	.0185	.0442	.0840	.1318	.1746
	10	.0000	.0000	.0000	.0002	.0014	.0056	.0167	.0392	.0755	.1222
	11	.0000	.0000	.0000	.0000	.0002	.0013	.0049	.0142	.0337	.0667
	12	.0000	.0000	.0000	.0000	.0000	.0002	.0011	.0040	.0115	.0278
	13	.0000	.0000	.0000	.0000	.0000	.0000	.0002	.0008	.0029	.0085
	14	.0000	.0000	.0000	.0000	.0000	.0000	.0000	.0001	.0005	.0018
	15	.0000	.0000	.0000	.0000	.0000	.0000	.0000	.0000	.0001	.0002
	16	.0000	.0000	.0000	.0000	.0000	.0000	.0000	.0000	.0000	.0000
17	0	.4181	.1668	.0631	.0225	.0075	.0023	.0007	.0002	.0000	.0000
	1	.3741	.3150	.1893	.0957	.0426	.0169	.0060	.0019	.0005	.0001
	2	.1575	.2800	.2673	.1914	.1136	.0581	.0260	.0102	.0035	.0010
	3	.0415	.1556	.2359	.2393	.1893	.1245	.0701	.0341	.0144	.0052
	4	.0076	.0605	.1457	.2093	.2209	.1868	.1320	.0796	.0411	.0182
	5	.0010	.0175	.0668	.1361	.1914	.2081	.1849	.1379	.0875	.0472
	6	.0001	.0039	.0236	.0680	.1276	.1784	.1991	.1839	.1432	.0944
	7	.0000	.0007	.0065	.0267	.0668	.1201	.1685	.1927	.1841	.1484
	8	.0000	.0001	.0014	.0084	.0279	.0644	.1134	.1606	.1883	.1855
	9	.0000	.0000	.0003	.0021	.0093	.0276	.0611	.1070	.1540	.1855

n	r	.05	.10	.15	.20	.25	p .30	.35	.40	.45	.50
17	10	.0000	.0000	.0000	.0004	.0025	.0095	.0263	.0571	.1008	.1484
	11	.0000	.0000	.0000	.0001	.0005	.0026	.0090	.0242	.0525	.0944
	12	.0000	.0000	.0000	.0000	.0001	.0006	.0024	.0081	.0215	.0472
	13	.0000	.0000	.0000	.0000	.0000	.0001	.0005	.0021	.0066	.0182
	14	.0000	.0000	.0000	.0000	.0000	.0000	.0001	.0004	.0016	.0052
	15	.0000	.0000	.0000	.0000	.0000	.0000	.0000	.0001	.0003	.0010
	16	.0000	.0000	.0000	.0000	.0000	.0000	.0000	.0000	.0000	.0001
	17	.0000	.0000	.0000	.0000	.0000	.0000	.0000	.0000	.0000	.0000
18	0	.3972	.1501	.0536	.0180	.0056	.0016	.0004	.0001	.0000	.0000
	1	.3763	.3002	.1704	.0811	.0338	.0126	.0042	.0012	.0003	.0001
	2	.1683	.2835	.2556	.1723	.0958	.0458	.0190	.0069	.0022	.0006
	3	.0473	.1680	.2406	.2297	.1704	.1046	.0547	.0246	.0095	.0031
	4	.0093	.0700	.1592	.2153	.2130	.1681	.1104	.0614	.0291	.0117
	5	.0014	.0218	.0787	.1507	.1988	.2017	.1664	.1146	.0666	.0327
	6	.0002	.0052	.0301	.0816	.1436	.1873	.1941	.1655	.1181	.0708
	7	.0000	.0010	.0091	.0350	.0820	.1376	.1792	.1892	.1657	.1214
	8	.0000	.0002	.0022	.0120	.0376	.0811	.1327	.1734	.1864	.1669
	9	.0000	.0000	.0004	.0033	.0139	.0386	.0794	.1284	.1694	.1855
	10	.0000	.0000	.0001	.0008	.0042	.0149	.0385	.0771	.1248	.1669
	11	.0000	.0000	.0000	.0001	.0010	.0046	.0151	.0374	.0742	.1214
	12	.0000	.0000	.0000	.0000	.0002	.0012	.0047	.0145	.0354	.0708
	13	.0000	.0000	.0000	.0000	.0000	.0002	.0012	.0045	.0134	.0327
	14	.0000	.0000	.0000	.0000	.0000	.0000	.0002	.0011	.0039	.0117
	15	.0000	.0000	.0000	.0000	.0000	.0000	.0000	.0002	.0009	.0031
	16	.0000	.0000	.0000	.0000	.0000	.0000	.0000	.0000	.0001	.0006
	17	.0000	.0000	.0000	.0000	.0000	.0000	.0000	.0000	.0000	.0001
	18	.0000	.0000	.0000	.0000	.0000	.0000	.0000	.0000	.0000	.0000
19	0	.3774	.1351	.0456	.0144	.0042	.0011	.0003	.0001	.0000	.0000
	1	.3774	.2852	.1529	.0685	.0268	.0093	.0029	.0008	.0002	.0000
	2	.1787	.2852	.2428	.1540	.0803	.0358	.0138	.0046	.0013	.0003
	3	.0533	.1796	.2428	.2182	.1517	.0869	.0422	.0175	.0062	.0018
	4	.0112	.0798	.1714	.2182	.2023	.1491	.0909	.0467	.0203	.0074
	5	.0018	.0266	.0907	.1636	.2023	.1916	.1468	.0933	.0497	.0222
	6	.0002	.0069	.0374	.0955	.1574	.1916	.1844	.1451	.0949	.0518
	7	.0000	.0014	.0122	.0443	.0974	.1525	.1844	.1797	.1443	.0961
	8	.0000	.0002	.0032	.0166	.0487	.0981	.1489	.1797	.1771	.1442
	9	.0000	.0000	.0007	.0051	.0198	.0514	.0980	.1464	.1771	.1762
	10	.0000	.0000	.0001	.0013	.0066	.0220	.0528	.0976	.1449	.1762
	11	.0000	.0000	.0000	.0003	.0018	.0077	.0233	.0532	.0970	.1442
	12	.0000	.0000	.0000	.0000	.0004	.0022	.0083	.0237	.0529	.0961
	13	.0000	.0000	.0000	.0000	.0001	.0005	.0024	.0085	.0233	.0518
	14	.0000	.0000	.0000	.0000	.0000	.0001	.0006	.0024	.0082	.0222

n	r	.05	.10	.15	.20	.25	P .30	.35	.40	.45	.50
19	15	.0000	.0000	.0000	.0000	.0000	.0000	.0001	.0005	.0022	.0074
	16	.0000	.0000	.0000	.0000	.0000	.0000	.0000	.0001	.0005	.0018
	17	.0000	.0000	.0000	.0000	.0000	.0000	.0000	.0000	.0001	.0003
	18	.0000	.0000	.0000	.0000	.0000	.0000	.0000	.0000	.0000	.0000
	19	.0000	.0000	.0000	.0000	.0000	.0000	.0000	.0000	.0000	.0000
20	0	.3585	.1216	.0388	.0115	.0032	.0008	.0002	.0000	.0000	.0000
	1	.3774	.2702	.1368	.0576	.0211	.0068	.0020	.0005	.0001	.0000
	2	.1887	.2852	.2293	.1369	.0669	.0278	.0100	.0031	.0008	.0002
	3	.0596	.1901	.2428	.2054	.1339	.0718	.0323	.0123	.0040	.0011
	4	.0133	.0898	.1821	.2182	.1897	.1304	.0738	.0350	.0139	.0046
	5	.0022	.0319	.1028	.1746	.2023	.1789	.1272	.0746	.0365	.0148
	6	.0003	.0089	.0454	.1091	.1686	.1916	.1712	.1244	.0746	.0370
	7	.0000	.0020	.0160	.0545	.1124	.1643	.1844	.1659	.1221	.0739
	8	.0000	.0004	.0046	.0222	.0609	.1144	.1614	.1797	.1623	.1201
	9	.0000	.0001	.0011	.0074	.0271	.0654	.1158	.1597	.1771	.1602
	10	.0000	.0000	.0002	.0020	.0099	.0308	.0686	.1171	.1593	.1762
	11	.0000	.0000	.0000	.0005	.0030	.0120	.0336	.0710	.1185	.1602
	12	.0000	.0000	.0000	.0001	.0008	.0039	.0136	.0355	.0727	.1201
	13	.0000	.0000	.0000	.0000	.0002	.0010	.0045	.0146	.0366	.0739
	14	.0000	.0000	.0000	.0000	.0000	.0002	.0012	.0049	.0150	.0370
	15	.0000	.0000	.0000	.0000	.0000	.0000	.0003	.0013	.0049	.0148
	16	.0000	.0000	.0000	.0000	.0000	.0000	.0000	.0003	.0013	.0046
	17	.0000	.0000	.0000	.0000	.0000	.0000	.0000	.0000	.0002	.0011
	18	.0000	.0000	.0000	.0000	.0000	.0000	.0000	.0000	.0000	.0002
	19	.0000	.0000	.0000	.0000	.0000	.0000	.0000	.0000	.0000	.0000
	20	.0000	.0000	.0000	.0000	.0000	.0000	.0000	.0000	.0000	.0000

x	$\mu t = 0.1$	$\mu t = 0.2$	$\mu t = 0.3$	$\mu t = 0.4$	$\mu t = 0.5$	$\mu t = 0.6$	$\mu t = 0.7$	$\mu t = 0.8$	$\mu t = 0.9$	$\mu t = 1.0$
0	.9048374	.8187308	.7408182	.6703200	.606531	.548812	.496585	.449329	.406570	.367879
1	.0904837	.1637462	.2222455	.2681280	.303265	.329287	347610	.359463	.365913	.367879
2	.0045242	.0163746	.0333368	.0536256	.075816	.098786	.121663	.143785	.164661	.183940
3	.0001508	.0010916	.0033337	.0071501	.012636	.019757	.028388	.038343	.049398	.061313
4	.0000038	.0000546	.0002500	.0007150	.001580	.002964	004968	.007669	.011115	.015328
5	.0000001	.0000022	.0000150	.0000572	.000158	.000356	000696	.001227	.002001	.003066
6		.0000001	.0000008	.0000038	.000013	.000036	.000081	.000164	.000300	.000511
7				.9000002	.000001	.000003	.000008	.000019	.000039	.000073
8							.000001	.000002	.000004	.000009
9										.000001

x	$\mu t = 2.0$	$\mu t = 3.0$	$\mu t = 4.0$	$\mu t = 5.0$	$\mu t = 6.0$	$\mu t = 7.0$	$\mu t = 8.0$	$\mu t = 9.0$	$\mu t = 10.0$
0	.135335	.049787	.018316	.006738	.002479	.000912	.000335	.000123	.000045
1	.270671	.149361	.073263	.033690	.014873	.006383	.002684	.001111	.000454
2	.270671	.224042	.146525	.084224	.044618	.022341	.010735	.004998	.002270
3	.180447	.224042	.195367	.140374	.089235	.052129	.028626	.014994	.007567
4	.090224	.168031	.195367	.175467	.133853	.091226	.057252	.033737	.018917
5	.036089	.100819	.156293	.175467	.160623	.127717	.091604	.060727	.037833
6	.012030	.050409	.104196	.146223	.160623	.149003	.122138	.091090	.063055
7	.003437	.021604	.059540	.104445	.137677	.149003	.139587	.117116	.090079
8	000859	.008102	.029770	.065278	.103258	.130377	.139587	.131756	.112599
9	.000191	.002701	.013231	.036266	.068838	.101405	.124077	.131756	.125110
10	.000038	.000810	.005292	.018133	.041303	.070983	.099262	.118580	.125110
11	.000007	.000221	.001925	.008242	.022529	.045171	.072190	.097020	.113736
12	.000001	.000055	.000642	.003434	.011264	.026350	.048127	.072765	.094780
13		.000013	.000197	.001321	.005199	.014188	.029616	.050376	.072908
14		.000003	.000056	.000472	.002228	.007094	.016924	.032384	.052077
15		.000001	.000015	.000157	.000891	.003311	.009026	.019431	.034718
16			.000004	.000049	.000334	.001448	.004513	.010930	.021699
17			.000001	.000014	.000118	.000596	.002124	.005786	.012764
18				.000004	.000039	.000232	.000944	.002893	.007091
19				.000001	.000012	.000085	.000397	.001370	.003732
20					.000004	.000030	.000159	.000617	.001866
21					.000001	.000010	.000061	.000264	.000889
22						.000003	.000022	.000108	.000404
23						.000001	.000008	000042	.000176
24							.000003	.000016	.000073
25							.000001	.000006	.000029
26								.000002	.000011
27								.000001	.000004
28									.000001
29									.000001

Source: E. C. Molina, *Poisson's Exponential Binomial Limit* (Princeton, N.J.: D. Van Nostrand Co., Inc., 1942). Reprinted by permission of Brooks/Cole Engineering Division of Wadsworth, Inc., Monterey, California 93940.

Table of Areas
Column (2) Shows

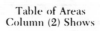

Table of Ordinates
Column (3) Shows

$z = \frac{X-\mu}{\sigma}$	Area Under the Curve between μ and X	Ordinate (Y) of the curve at X	$z = \frac{X-\mu}{\sigma}$	Area Under the Curve between μ and X	Ordinate (Y) of the curve at X
(1)	(2)	(3)	(1)	(2)	(3)
.00	.00000	.39894	.20	.07926	.39104
.01	.00399	.39892	.21	.08317	.39024
.02	.00798	.39886	.22	.08706	.38940
.03	.01197	.39876	.23	.09095	.38853
.04	.01595	.39862	.24	.09483	.38762
.05	.01994	.39844	.25	.09871	.38667
.06	.02392	.39822	.26	.10257	.38568
.07	.02790	.39797	.27	.10642	.38466
.08	.03188	.39767	.28	.11026	.38361
.09	.03586	.39733	.29	.11409	.38251
.10	.03983	.39695	.30	.11791	.38139
.11	.04380	.39654	.31	.12172	.38023
.12	.04776	.39608	.32	.12552	.37903
.13	.05172	.39559	.33	.12930	.37780
.14	.05567	.39505	.34	.13307	.37654
.15	.05962	.39448	.35	.13683	.37524
.16	.06356	.39387	.36	.14058	.37391
.17	.06749	.39322	.37	.14431	.37255
.18	.07142	.39253	.38	.14803	.37115
.19	.07535	.39181	.39	.15173	.36973

Source: J. F. Kenney and E. S. Keeping, *Mathematics of Statistics,* 3d ed. (Princeton, N.J.: D. Van Nostrand Co., Inc., 1954). Reprinted by permission of Brooks/Cole Engineering Division of Wadsworth, Inc., Monterey, California 93940.

$z = \frac{X-\mu}{\sigma}$	Area Under the Curve between μ and X	Ordinate (Y) of the curve at X	$z = \frac{X-\mu}{\sigma}$	Area Under the Curve between μ and X	Ordinate (Y) of the curve at X
(1)	(2)	(3)	(1)	(2)	(3)
.40	.15542	.36827	.90	.31594	.26609
.41	.15910	.36678	.91	.31859	.26369
.42	.16276	.36526	.92	.32121	.26129
.43	.16640	.36371	.93	.32381	.25888
.44	.17003	.36213	.94	.32639	.25647
.45	.17364	.36053	.95	.32894	.25406
.46	.17724	.35889	.96	.33147	.25164
.47	.18082	.35723	.97	.33398	.24923
.48	.18439	.35553	.98	.33646	.24681
.49	.18793	.35381	.99	.33891	.24439
.50	.19146	.35207	1.00	.34134	.24197
.51	.19497	.35029	1.01	.34375	.23955
.52	.19847	.34849	1.02	.34614	.23713
.53	.20194	.34667	1.03	.34850	.23471
.54	.20540	.34482	1.04	.35083	.23230
.55	.20884	.34294	1.05	.35314	.22988
.56	.21226	.34105	1.06	.35543	.22747
.57	.21566	.33912	1.07	.35769	.22506
.58	.21904	.33718	1.08	.35993	.22265
.59	.22240	.33521	1.09	.36214	.22025
.60	.22575	.33322	1.10	.36433	.21785
.61	.22907	.33121	1.11	.36650	.21546
.62	.23237	.32918	1.12	.36864	.21307
.63	.23565	.32713	1.13	.37076	.21069
.64	.23891	.32506	1.14	.37286	.20831
.65	.24215	.32297	1.15	.37493	.20594
.66	.24537	.32086	1.16	.37698	.20357
.67	.24857	.31874	1.17	.37900	.20121
.68	.25175	.31659	1.18	.38100	.19886
.69	.25490	.31443	1.19	.38298	.19652
.70	.25804	.31225	1.20	.38493	.19419
.71	.26115	.31006	1.21	.38686	.19186
.72	.26424	.30785	1.22	.38877	.18954
.73	.26730	.30563	1.23	.39065	.18724
.74	.27035	.30339	1.24	.39251	.18494
.75	.27337	.30114	1.25	.39435	.18265
.76	.27637	.29887	1.26	.39617	.18037
.77	.27935	.29659	1.27	.39796	.17810
.78	.28230	.29431	1.28	.39973	.17585
.79	.28524	.29200	1.29	.40147	.17360
.80	.28814	.28969	1.30	.40320	.17137
.81	.29103	.28737	1.31	.40490	.16915
.82	.29389	.28504	1.32	.40658	.16694
.83	.29673	.28269	1.33	.40824	.16474
.84	.29955	.28034	1.34	.40988	.16256
.85	.30234	.27798	1.35	.41149	.16038
.86	.30511	.27562	1.36	.41309	.15822
.87	.30785	.27324	1.37	.41466	.15608
.88	.31057	.27086	1.38	.41621	.15395
.89	.31327	.26848	1.39	.41774	.15183

$z = \dfrac{X-\mu}{\sigma}$	Area Under the Curve between μ and X	Ordinate (Y) of the curve at X	$z = \dfrac{X-\mu}{\sigma}$	Area Under the Curve between μ and X	Ordinate (Y) of the curve at X
(1)	(2)	(3)	(1)	(2)	(3)
1.40	.41924	.14973	1.90	.47128	.06562
1.41	.42073	.14764	1.91	.47193	.06438
1.42	.42220	.14556	1.92	.47257	.06316
1.43	.42364	.14350	1.93	.47320	.06195
1.44	.42507	.14146	1.94	.47381	.06077
1.45	.42647	.13943	1.95	.47441	.05959
1.46	.42786	.13742	1.96	.47500	.05844
1.47	.42922	.13542	1.97	.47558	.05730
1.48	.43056	.13344	1.98	.47615	.05618
1.49	.43189	.1314ʳ	1.99	.47670	.05508
1.50	.43319	.12952	2.00	.47725	.05399
1.51	.43448	.12758	2.01	.47778	.05292
1.52	.43574	.1256ℇ	2.02	.47831	.05186
1.53	.43699	.12376	2.03	.47882	.05082
1.54	.43822	.12188	2.04	.47932	.04980
1.55	.43943	.12001	2.05	.47982	.04879
1.56	.44062	.11816	2.06	.48030	.04780
1.57	.44179	.11632	2.07	.48077	.04682
1.58	.44295	.11450	2.08	.48124	.04586
1.59	.44408	.11270	2.09	.48169	.04491
1.60	.44520	.11092	2.10	.48214	.04398
1.61	.44630	.10915	2.11	.48257	.04307
1.62	.44738	.10741	2.12	.48300	.04217
1.63	.44845	.10567	2.13	.48341	.04128
1.64	.44950	.10396	2.14	.48382	.04041
1.65	.45053	.10226	2.15	.48422	.03955
1.66	.45154	.10059	2.16	.484ℇ1	.03871
1.67	.45254	.09893	2.17	.48500	.03788
1.68	.45352	.09728	2.18	.48537	.03706
1.69	.45449	.09566	2.19	.48574	.03626
1.70	.45543	.09405	2.20	.48610	.03547
1.71	.45637	.09246	2.21	.48645	.03470
1.72	.45728	.09089	2.22	.48679	.03394
1.73	.45818	.08933	2.23	.48713	.03319
1.74	.45907	.08780	2.24	.48745	.03246
1.75	.45994	.08628	2.25	.48778	.03174
1.76	.46080	.08478	2.26	.48809	.03103
1.77	.46164	.08329	2.27	.48840	.03034
1.78	.46246	.08183	2.28	.48870	.02965
1.79	.46327	.08038	2.29	.48899	.02898
1.80	.46407	.07895	2.30	.48928	.02833
1.81	.46485	.07754	2.31	.48956	.02768
1.82	.46562	.07614	2.32	.48983	.02705
1.83	.46638	.07477	2.33	.49010	.02643
1.84	.46712	.07341	2.34	.49036	.02582
1.85	.46784	.07206	2.35	.49064	.02522
1.86	.46856	.07074	2.36	.49086	.02463
1.87	.46926	.06943	2.37	.49111	.02406
1.88	.46995	.06814	2.38	.49134	.02349
1.89	.47062	.06687	2.39	.49158	.02294

$z = \frac{X-\mu}{\sigma}$	Area Under the Curve between μ and X	Ordinate (Y) of the curve at X	$z = \frac{X-\mu}{\sigma}$	Area Under the Curve between μ and X	Ordinate (Y) of the curve at X
(1)	(2)	(3)	(1)	(2)	(3)
2.40	.49180	.02239	2.90	.49813	.00595
2.41	.49202	.02186	2.91	.49819	.00578
2.42	.49224	.02134	2.92	.49825	.00562
2.43	.49245	.02083	2.93	.49831	.00545
2.44	.49266	.02033	2.94	.49836	.00530
2.45	.49286	.01984	2.95	.49841	.00514
2.46	.49305	.01936	2.96	.49846	.00499
2.47	.49324	.01889	2.97	.49851	.00485
2.48	.49343	.01842	2.98	.49856	.00471
2.49	.49361	.01797	2.99	.49861	.00457
2.50	.49379	.01753	3.00	.49865	.00443
2.51	.49396	.01709	3.01	.49869	.00430
2.52	.49413	.01667	3.02	.49874	.00417
2.53	.49430	.01625	3.03	.49878	.00405
2.54	.49446	.01585	3.04	.49882	.00393
2.55	.49461	.01545	3.05	.49886	.00381
2.56	.49477	.01506	3.06	.49889	.00370
2.57	.49492	.01468	3.07	.49893	.00358
2.58	.49506	.01431	3.08	.49897	.00348
2.59	.49520	.01394	3.09	.49900	.00337
2.60	.49534	.01358	3.10	.49903	.00327
2.61	.49547	.01323	3.11	.49906	.00317
2.62	.49560	.01289	3.12	.49910	.00307
2.63	.49573	.01256	3.13	.49913	.00298
2.64	.49585	.01223	3.14	.49916	.00288
2.65	.49598	.01191	3.15	.49918	.00279
2.66	.49609	.01160	3.16	.49921	.00271
2.67	.49621	.01130	3.17	.49924	.00262
2.68	.49632	.01100	3.18	.49926	.00254
2.69	.49643	.01071	3.19	.49929	.00246
2.70	.49653	.01042	3.20	.49931	.00238
2.71	.49664	.01014	3.21	.49934	.00231
2.72	.49674	.00987	3.22	.49936	.00224
2.73	.49683	.00961	3.23	.49938	.00216
2.74	.49693	.00935	3.24	.49940	.00210
2.75	.49702	.00909	3.25	.49942	.00203
2.76	.49711	.00885	3.26	.49944	.00196
2.77	.49720	.00861	3.27	.49946	.00190
2.78	.49728	.00837	3.28	.49948	.00184
2.79	.49736	.00814	3.29	.49950	.00178
2.80	.49744	.00792	3.30	.49952	.00172
2.81	.49752	.00770	3.31	.49953	.00167
2.82	.49760	.00748	3.32	.49955	.00161
2.83	.49767	.00727	3.33	.49957	.00156
2.84	.49774	.00707	3.34	.49958	.00151
2.85	.49781	.00687	3.35	.43960	.00146
2.86	.49788	.00668	3.36	.49961	.00141
2.87	.49795	.00649	3.37	.49962	.00136
2.88	.49801	.00631	3.38	.49964	.00132
2.89	.49807	.00613	3.39	.49965	.00127

Table B–3 (*continued*)

$z = \dfrac{X-\mu}{\sigma}$	Area Under the Curve between μ and X	Ordinate (Y) of the curve at X	$z = \dfrac{X-\mu}{\sigma}$	Area Under the Curve between μ and X	Ordinate (Y) of the curve at X
(1)	(2)	(3)	(1)	(2)	(3)
3.40	.49966	.00123	3.70	.49989	.00042
3.41	.49968	.00119	3.71	.49990	.00041
3.42	.49969	.00115	3.72	.49990	.00039
3.43	.49970	.00111	3.73	.49990	.00038
3.44	.49971	.00107	3.74	.49991	.00037
3.45	.49972	.00104	3.75	.49991	.00035
3.46	.49973	.00100	3.76	.49992	.00034
3.47	.49974	.00097	3.77	.49992	.00033
3.48	.49975	.00094	3.78	.49992	.00031
3.49	.49976	.00090	3.79	.49992	.00030
3.50	.49977	.00087	3.80	.49993	.00029
3.51	.49978	.00084	3.81	.49993	.00028
3.52	.49978	.00081	3.82	.49993	.00027
3.53	.49979	.00079	3.83	.49994	.00026
3.54	.49980	.00076	3.84	.49994	.00025
3.55	.49981	.00073	3.85	.49994	.00024
3.56	.49981	.00071	3.86	.49994	.00023
3.57	.49982	.00068	3.87	.49995	.00022
3.58	.49983	.00066	3.88	.49995	.00021
3.59	.49983	.00063	3.89	.49995	.00021
3.60	.49984	.00061	3.90	.49995	.00020
3.61	.49985	.00059	3.91	.49995	.00019
3.62	.49985	.00057	3.92	.49996	.00018
3.63	.49986	.00055	3.93	.49996	.00018
3.64	.49986	.00053	3.94	.49996	.00017
3.65	.49987	.00051	3.95	.49996	.00016
3.66	.49987	.00049	3.96	.49996	.00016
3.67	.49988	.00047	3.97	.49996	.00015
3.68	.49988	.00046	3.98	.49997	.00014
3.69	.49989	.00044	3.99	.49997	.00014

df= N−1	Level of significance for one-tailed test					
	.10	.05	.025	.01	.005	.0005
	Level of significance for two-tailed test					
	.20	.10	.05	.02	.01	.001
1	3.078	6.314	12.706	31.821	63.657	636.619
2	1.886	2.920	4.303	6.965	9.925	31.598
3	1.638	2.353	3.182	4.541	5.841	12.941
4	1.533	2.132	2.776	3.747	4.604	8.610
5	1.476	2.015	2.571	3.365	4.032	6.859
6	1.440	1.943	2.447	3.143	3.707	5.959
7	1.415	1.895	2.365	2.998	3.499	5.405
8	1.397	1.860	2.306	2.896	3.355	5.041
9	1.383	1.833	2.262	2.821	3.250	4.781
10	1.372	1.812	2.228	2.764	3.169	4.587
11	1.363	1.796	2.201	2.718	3.106	4.437
12	1.356	1.782	2.179	2.681	3.055	4.318
13	1.350	1.771	2.160	2.650	3.012	4.221
14	1.345	1.761	2.145	2.624	2.977	4.140
15	1.341	1.753	2.131	2.602	2.947	4.073
16	1.337	1.746	2.120	2.583	2.921	4.015
17	1.333	1.740	2.110	2.567	2.898	3.965
18	1.330	1.734	2.101	2.552	2.878	3.922
19	1.328	1.729	2.093	2.539	2.861	3.883
20	1.325	1.725	2.086	2.528	2.845	3.850
21	1.323	1.721	2.080	2.518	2.831	3.819
22	1.321	1.717	2.074	2.508	2.819	3.792
23	1.319	1.714	2.069	2.500	2.807	3.767
24	1.318	1.711	2.064	2.492	2.797	3.745
25	1.316	1.708	2.060	2.485	2.787	3.725
26	1.315	1.706	2.056	2.479	2.779	3.707
27	1.314	1.703	2.052	2.473	2.771	3.690
28	1.313	1.701	2.048	2.467	2.763	3.674
29	1.311	1.699	2.045	2.462	2.756	3.659
30	1.310	1.697	2.042	2.457	2.750	3.646
40	1.303	1.684	2.021	2.423	2.704	3.551
60	1.296	1.671	2.000	2.390	2.660	3.460
120	1.289	1.658	1.980	2.358	2.617	3.373
∞	1.282	1.645	1.960	2.326	2.576	3.291

Source: Abridged by permission of the authors and publishers from Table III of Fisher & Yates, *Statistical Tables for Biological, Agricultural and Medical Research,* published by Longman Group Ltd. London (1974), 6th edition (previously published by Oliver & Boyd Ltd. Edinburgh).

Table B-5 Critical values of the F-distribution at a 5 percent level of significance ($\alpha = 0.05$)

Degrees of freedom for numerator

Degrees of freedom for denominator	1	2	3	4	5	6	7	8	9	10	12	15	20	24	30	40	60	120	∞
1	161	200	216	225	230	234	237	239	241	242	244	246	248	249	250	251	252	253	254
2	18.5	19.0	19.2	19.2	19.3	19.3	19.4	19.4	19.4	19.4	19.4	19.4	19.4	19.5	19.5	19.5	19.5	19.5	19.5
3	10.1	9.55	9.28	9.12	9.01	8.94	8.89	8.85	8.81	8.79	8.74	8.70	8.66	8.64	8.62	8.59	8.57	8.55	8.53
4	7.71	6.94	6.59	6.39	6.26	6.16	6.09	6.04	6.00	5.96	5.91	5.86	5.80	5.77	5.75	5.72	5.69	5.66	5.63
5	6.61	5.79	5.41	5.19	5.05	4.95	4.88	4.82	4.77	4.74	4.68	4.62	4.56	4.53	4.50	4.46	4.43	4.40	4.37
6	5.99	5.14	4.76	4.53	4.39	4.28	4.21	4.15	4.10	4.06	4.00	3.94	3.87	3.84	3.81	3.77	3.74	3.70	3.67
7	5.59	4.74	4.35	4.12	3.97	3.87	3.79	3.73	3.68	3.64	3.57	3.51	3.44	3.41	3.38	3.34	3.30	3.27	3.23
8	5.32	4.46	4.07	3.84	3.69	3.58	3.50	3.44	3.39	3.35	3.28	3.22	3.15	3.12	3.08	3.04	3.01	2.97	2.93
9	5.12	4.26	3.86	3.63	3.48	3.37	3.29	3.23	3.18	3.14	3.07	3.01	2.94	2.90	2.86	2.83	2.79	2.75	2.71
10	4.96	4.10	3.71	3.48	3.33	3.22	3.14	3.07	3.02	2.98	2.91	2.85	2.77	2.74	2.70	2.66	2.62	2.58	2.54
11	4.84	3.98	3.59	3.36	3.20	3.09	3.01	2.95	2.90	2.85	2.79	2.72	2.65	2.61	2.57	2.53	2.49	2.45	2.40
12	4.75	3.89	3.49	3.26	3.11	3.00	2.91	2.85	2.80	2.75	2.69	2.62	2.54	2.51	2.47	2.43	2.38	2.34	2.30
13	4.67	3.81	3.41	3.18	3.03	2.92	2.83	2.77	2.71	2.67	2.60	2.53	2.46	2.42	2.38	2.34	2.30	2.25	2.21
14	4.60	3.74	3.34	3.11	2.96	2.85	2.76	2.70	2.65	2.60	2.53	2.46	2.39	2.35	2.31	2.27	2.22	2.18	2.13
15	4.54	3.68	3.29	3.06	2.90	2.79	2.71	2.64	2.59	2.54	2.48	2.40	2.33	2.29	2.25	2.20	2.16	2.11	2.07
16	4.49	3.63	3.24	3.01	2.85	2.74	2.66	2.59	2.54	2.49	2.42	2.35	2.28	2.24	2.19	2.15	2.11	2.06	2.01
17	4.45	3.59	3.20	2.96	2.81	2.70	2.61	2.55	2.49	2.45	2.38	2.31	2.23	2.19	2.15	2.10	2.06	2.01	1.96
18	4.41	3.55	3.16	2.93	2.77	2.66	2.58	2.51	2.46	2.41	2.34	2.27	2.19	2.15	2.11	2.06	2.02	1.97	1.92
19	4.38	3.52	3.13	2.90	2.74	2.63	2.54	2.48	2.42	2.38	2.31	2.23	2.16	2.11	2.07	2.03	1.98	1.93	1.88
20	4.35	3.49	3.10	2.87	2.71	2.60	2.51	2.45	2.39	2.35	2.28	2.20	2.12	2.08	2.04	1.99	1.95	1.90	1.84
21	4.32	3.47	3.07	2.84	2.68	2.57	2.49	2.42	2.37	2.32	2.25	2.18	2.10	2.05	2.01	1.96	1.92	1.87	1.81
22	4.30	3.44	3.05	2.82	2.66	2.55	2.46	2.40	2.34	2.30	2.23	2.15	2.07	2.03	1.98	1.94	1.89	1.84	1.78
23	4.28	3.42	3.03	2.80	2.64	2.53	2.44	2.37	2.32	2.27	2.20	2.13	2.05	2.01	1.96	1.91	1.86	1.81	1.76
24	4.26	3.40	3.01	2.78	2.62	2.51	2.42	2.36	2.30	2.25	2.18	2.11	2.03	1.98	1.94	1.89	1.84	1.79	1.73
25	4.24	3.39	2.99	2.76	2.60	2.49	2.40	2.34	2.28	2.24	2.16	2.09	2.01	1.96	1.92	1.87	1.82	1.77	1.71
30	4.17	3.32	2.92	2.69	2.53	2.42	2.33	2.27	2.21	2.16	2.09	2.01	1.93	1.89	1.84	1.79	1.74	1.68	1.62
40	4.08	3.23	2.84	2.61	2.45	2.34	2.25	2.18	2.12	2.08	2.00	1.92	1.84	1.79	1.74	1.69	1.64	1.58	1.51
60	4.00	3.15	2.76	2.53	2.37	2.25	2.17	2.10	2.04	1.99	1.92	1.84	1.75	1.70	1.65	1.59	1.53	1.47	1.39
120	3.92	3.07	2.68	2.45	2.29	2.18	2.09	2.02	1.96	1.91	1.83	1.75	1.66	1.61	1.55	1.50	1.43	1.35	1.25
∞	3.84	3.00	2.60	2.37	2.21	2.10	2.01	1.94	1.88	1.83	1.75	1.67	1.57	1.52	1.46	1.39	1.32	1.22	1.00

Source: M. Merrington and C. M. Thompson, "Tables of Percentage Points of the Inverted Beta (F) Distribution," *Biometrika,* vol. 33 (1943), by permission of the Biometrika trustees.

Degrees of freedom for numerator

	1	2	3	4	5	6	7	8	9	10	12	15	20	24	30	40	60	120	∞
1	4,052	5,000	5,403	5,625	5,764	5,859	5,928	5,982	6,023	6,056	6,106	6,157	6,209	6,235	6,261	6,287	6,313	6,339	6,366
2	98.5	99.0	99.2	99.2	99.3	99.3	99.4	99.4	99.4	99.4	99.4	99.4	99.4	99.5	99.5	99.5	99.5	99.5	99.5
3	34.1	30.8	29.5	28.7	28.2	27.9	27.7	27.5	27.3	27.2	27.1	26.9	26.7	26.6	26.5	26.4	26.3	26.2	26.1
4	21.2	18.0	16.7	16.0	15.5	15.2	15.0	14.8	14.7	14.5	14.4	14.2	14.0	13.9	13.8	13.7	13.7	13.6	13.5
5	16.3	13.3	12.1	11.4	11.0	10.7	10.5	10.3	10.2	10.1	9.89	9.72	9.55	9.47	9.38	9.29	9.20	9.11	9.02
6	13.7	10.9	9.78	9.15	8.75	8.47	8.26	8.10	7.98	7.87	7.72	7.56	7.40	7.31	7.23	7.14	7.06	6.97	6.88
7	12.2	9.55	8.45	7.85	7.46	7.19	6.99	6.84	6.72	6.62	6.47	6.31	6.16	5.99	5.99	5.91	5.82	5.74	5.65
8	11.3	8.65	7.59	7.01	6.63	6.37	6.18	6.03	5.91	5.81	5.67	5.52	5.36	5.28	5.20	5.12	5.03	4.95	4.86
9	10.6	8.02	6.99	6.42	6.06	5.80	5.61	5.47	5.35	5.26	5.11	4.96	4.81	4.73	4.65	4.57	4.48	4.40	4.31
10	10.0	7.56	6.55	5.99	5.64	5.39	5.20	5.06	4.94	4.85	4.71	4.56	4.41	4.33	4.25	4.17	4.08	4.00	3.91
11	9.65	7.21	6.22	5.67	5.32	5.07	4.89	4.74	4.63	4.54	4.40	4.25	4.10	4.02	3.94	3.86	3.78	3.69	3.60
12	9.33	6.93	5.95	5.41	5.06	4.82	4.64	4.50	4.39	4.30	4.16	4.01	3.86	3.78	3.70	3.62	3.54	3.45	3.36
13	9.07	6.70	5.74	5.21	4.86	4.62	4.44	4.30	4.19	4.10	3.96	3.82	3.66	3.59	3.51	3.43	3.34	3.25	3.17
14	8.86	6.51	5.56	5.04	4.70	4.46	4.28	4.14	4.03	3.94	3.80	3.66	3.51	3.43	3.35	3.27	3.18	3.09	3.00
15	8.68	6.36	5.42	4.89	4.56	4.32	4.14	4.00	3.89	3.80	3.67	3.52	3.37	3.29	3.21	3.13	3.05	2.96	2.87
16	8.53	6.23	5.29	4.77	4.44	4.20	4.03	3.89	3.78	3.69	3.55	3.41	3.26	3.18	3.10	3.02	2.93	2.84	2.75
17	8.40	6.11	5.19	4.67	4.34	4.10	3.93	3.79	3.68	3.59	3.46	3.31	3.16	3.08	3.00	2.92	2.83	2.75	2.65
18	8.29	6.01	5.09	4.58	4.25	4.01	3.84	3.71	3.60	3.51	3.37	3.23	3.08	3.00	2.92	2.84	2.75	2.66	2.57
19	8.19	5.93	5.01	4.50	4.17	3.94	3.77	3.63	3.52	3.43	3.30	3.15	3.00	2.92	2.84	2.76	2.67	2.58	2.49
20	8.10	5.85	4.94	4.43	4.10	3.87	3.70	3.56	3.46	3.37	3.23	3.09	2.94	2.86	2.78	2.69	2.61	2.52	2.42
21	8.02	5.78	4.87	4.37	4.04	3.81	3.64	3.51	3.40	3.31	3.17	3.03	2.88	2.80	2.72	2.64	2.55	2.46	2.36
22	7.95	5.72	4.82	4.31	3.99	3.76	3.59	3.45	3.35	3.26	3.12	2.98	2.83	2.75	2.67	2.58	2.50	2.40	2.31
23	7.88	5.66	4.76	4.26	3.94	3.71	3.54	3.41	3.30	3.21	3.07	2.93	2.78	2.70	2.62	2.54	2.45	2.35	2.26
24	7.82	5.61	4.72	4.22	3.90	3.67	3.50	3.36	3.26	3.17	3.03	2.89	2.74	2.66	2.58	2.49	2.40	2.31	2.21
25	7.77	5.57	4.68	4.18	3.86	3.63	3.46	3.32	3.22	3.13	2.99	2.85	2.70	2.62	2.53	2.45	2.36	2.27	2.17
30	7.56	5.39	4.51	4.02	3.70	3.47	3.30	3.17	3.07	2.98	2.84	2.70	2.55	2.47	2.39	2.30	2.21	2.11	2.01
40	7.31	5.18	4.31	3.83	3.51	3.29	3.12	2.99	2.89	2.80	2.66	2.52	2.37	2.29	2.20	2.11	2.02	1.92	1.80
60	7.08	4.98	4.13	3.65	3.34	3.12	2.95	2.82	2.72	2.63	2.50	2.35	2.20	2.12	2.03	1.94	1.84	1.73	1.60
120	6.85	4.79	3.95	3.48	3.17	2.96	2.79	2.66	2.56	2.47	2.34	2.19	2.03	1.95	1.86	1.76	1.66	1.53	1.38
∞	6.63	4.61	3.78	3.32	3.02	2.80	2.64	2.51	2.41	2.32	2.18	2.04	1.88	1.79	1.70	1.59	1.47	1.32	1.00

Degrees of freedom for denominator

Table B–6 Critical values of chi-square

Percentages represent areas in right-hand end of distribution. Example:

For 9 degrees of freedom:
$P[x^2 > 16.92] = 0.05$

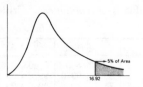

Degrees of freedom	Probability that chi-square value will be exceeded									
	0.995	0.990	0.975	0.950	0.900	0.100	0.050	0.025	0.010	0.005
1	0.0⁴393	0.0³157	0.0³982	0.0²393	0.0158	2.71	3.84	5.02	6.63	7.88
2	0.0100	0.0201	0.0506	0.103	0.211	4.61	5.99	7.38	9.21	10.60
3	0.072	0.115	0.216	0.352	0.584	6.25	7.81	9.35	11.34	12.84
4	0.207	0.297	0.484	0.711	1.064	7.78	9.49	11.14	13.28	14.86
5	0.412	0.554	0.831	1.145	1.61	9.24	11.07	12.83	15.09	16.75
6	0.676	0.872	1.24	1.64	2.20	10.64	12.59	14.45	16.81	18.55
7	0.989	1.24	1.69	2.17	2.83	12.02	14.07	16.01	18.48	20.28
8	1.34	1.65	2.18	2.73	3.49	13.36	15.51	17.53	20.09	21.96
9	1.73	2.09	2.70	3.33	4.17	14.68	16.92	19.02	21.67	23.59
10	2.16	2.56	3.25	3.94	4.87	15.99	18.31	20.48	23.21	25.19
11	2.60	3.05	3.82	4.57	5.58	17.28	19.68	21.92	24.72	26.76
12	3.07	3.57	4.40	5.23	6.30	18.55	21.03	23.34	26.22	28.30
13	3.57	4.11	5.01	5.89	7.04	19.81	22.36	24.74	27.69	29.82
14	4.07	4.66	5.63	6.57	7.79	21.06	23.68	26.12	29.14	31.32
15	4.60	5.23	6.26	7.26	8.55	22.31	25.00	27.49	30.58	32.80
16	5.14	5.81	6.91	7.96	9.31	23.54	26.30	28.85	32.00	34.27
17	5.70	6.41	7.56	8.67	10.09	24.77	27.59	30.19	33.41	35.72
18	6.26	7.01	8.23	9.39	10.86	25.99	28.87	31.53	34.81	37.16
19	6.84	7.63	8.91	10.12	11.65	27.20	30.14	32.85	36.19	38.58
20	7.43	8.26	9.59	10.85	12.44	28.41	31.41	34.17	37.57	40.00
21	8.03	8.90	10.28	11.59	13.24	29.62	32.67	35.48	38.93	41.40
22	8.64	9.54	10.98	12.34	14.04	30.81	33.92	36.78	40.29	42.80
23	9.26	10.20	11.69	13.09	14.85	32.01	35.17	38.08	41.64	44.18
24	9.89	10.86	12.40	13.85	15.66	33.20	36.42	39.36	42.98	45.56
25	10.52	11.52	13.12	14.61	16.47	34.38	37.65	40.65	44.31	46.93
26	11.16	12.20	13.84	15.38	17.29	35.56	38.89	41.92	45.64	48.29
27	11.81	12.88	14.57	16.15	18.11	36.74	40.11	43.19	46.96	49.64
28	12.46	13.56	15.31	16.93	18.94	37.92	41.34	44.46	48.28	50.99
29	13.12	14.26	16.05	17.71	19.77	39.09	42.56	45.72	49.59	52.34
30	13.79	14.95	16.79	18.49	20.60	40.26	43.77	46.98	50.89	53.67
40	20.71	22.16	24.43	26.51	29.05	51.80	55.76	59.34	63.69	66.77
50	27.99	29.71	32.36	34.76	37.69	63.17	67.50	71.42	76.15	79.49
60	35.53	37.48	40.48	43.19	46.46	74.40	79.08	83.30	88.38	91.95
70	43.28	45.44	48.76	51.74	55.33	85.53	90.53	95.02	100.4	104.22
80	51.17	53.54	57.15	60.39	64.28	96.58	101.9	106.6	112.3	116.32
90	59.20	61.75	65.65	69.13	73.29	107.6	113.1	118.1	124.1	128.3
100	67.33	70.06	74.22	77.93	82.36	118.5	124.3	129.6	135.8	140.2
z_α	−2.58	−2.33	−1.96	−1.64	−1.28	+1.28	+1.64	+1.96	+2.33	+2.58

Source: Adapted from E. S. Pearson and H. O. Hartley, eds. *Biometrika Tables for Statisticians,* vol. I, Table 18, pp. 160–63. Published for the Biometrika trustees by Cambridge University, 1954, by permission of the trustees of Biometrika.

One-tailed:	$\alpha = .05$	$\alpha = .025$	$\alpha = .01$	$\alpha = .005$
Two-tailed:	$\alpha = .10$	$\alpha = .05$	$\alpha = .02$	$\alpha = .01$
m	(lower, upper)			
5	0,15	—,—	—,—	—,—
6	2,19	0,21	—,—	—,—
7	3,25	2,26	0,28	—,—
8	5,31	3,33	1,35	0,36
9	8,37	5,40	3,42	1,44
10	10,45	8,47	5,50	3,52
11	13,53	10,56	7,59	5,61
12	17,61	13,65	10,68	7,71
13	21,70	17,74	12,79	10,81
14	25,80	21,84	16,89	13,92
15	30,90	25,95	19,101	16,104
16	35,101	29,107	23,113	19,117
17	41,112	34,119	27,126	23,130
18	47,124	40,131	32,139	27,144
19	53,137	46,144	37,153	32,158
20	60,150	52,158	43,167	37,173

Table B–7 Lower and upper critical values W of Wilcoxon one-sample signed ranks test

Source: Adapted from Table 2 of F. Wilcoxon and R. A. Wilcox, *Some Rapid Approximate Statistical Procedures* (Pearl River, N.Y.: Lederle Laboratories, 1964) with permission of the American Cyanamid Company. (m used instead of n).

n_2	α One-tailed	α Two-tailed	n_1 4	5	6	7	8	9	10
4	.05	.10	11,25						
	.025	.05	10,26						
	.01	.02	—,—						
	.005	.01	—,—						
5	.05	.10	12,28	19,36					
	.025	.05	11,29	17,38					
	.01	.02	10,30	16,39					
	.005	.01	—,—	15,40					
6	.05	.10	13,31	20,40	28,50				
	.025	.05	12,32	18,42	26,52				
	.01	.02	11,33	17,43	24,54				
	.005	.01	10,34	16,44	23,55				
7	.05	.10	14,34	21,44	29,55	39,66			
	.025	.05	13,35	20,45	27,57	36,69			
	.01	.02	11,37	18,47	25,59	34,71			
	.005	.01	10,38	16,49	24,60	32,73			
8	.05	.10	15,37	23,47	31,59	41,71	51,85		
	.025	.05	14,38	21,49	29,61	38,74	49,87		
	.01	.02	12,40	19,51	27,63	35,77	45,91		
	.005	.01	11,41	17,53	25,65	34,78	43,93		
9	.05	.10	16,40	24,51	33,63	43,76	54,90	66,105	
	.025	.05	14,42	22,53	31,65	40,79	51,93	62,109	
	.01	.02	13,43	20,55	28,68	37,82	47,97	59,112	
	.005	.01	11,45	18,57	26,70	35,84	45,99	56,115	
10	.05	.10	17,43	26,54	35,67	45,81	56,96	69,111	82,128
	.025	.05	15,45	23,57	32,70	42,84	53,99	65,115	78,132
	.01	.02	13,47	21,59	29,73	39,87	49,103	61,119	74,136
	.005	.01	12,48	19,61	27,75	37,89	47,105	58,122	71,139

Table B–8 Lower and upper critical values T_{n_1} of Wilcoxon rank sum test

Source: Adapted from Table 1 of F. Wilcoxon and R. A. Wilcox, *Some Rapid Approximate Statistical Procedures* (Pearl River, N.Y.: Lederle Laboratories, 1964) with permission of the American Cyanamid Company.

APPENDIX C   STAT CITY DATA BASE

```
                                                              VARIABLES
  0        0             0   0 0 0  0      0    0  0    1    1 1    1    1 1 1   1    1 1 2   2    2    2   2  2  2
  1        2             3   4 5 6  7      8    9  0    1    2 3    4    5 6 7   8    9 0 1   2    3    4   5  6  7
*************************************************************************************************************************
KILFOYLE   406 6TH ST       43   1 1 1   502  88264  10 1199  88  90  56419  6 3 165  55   6 1 2   6 128 21  3  2  0  5
GROGER     405 7TH ST       43   2 1 1   819  89452   9 1060  82  12  23280  5 3 191  64   6 1 2   6  81 16  2  2  0  0
SMAY       406 7TH ST       44   3 1 1   644  76737   8  930  66  62  33350  4 2 159  80   7 2 1   3  95 24  3  1  0  0
POTTER     405 8TH ST       44   4 1 1   716  88671   9 1051  78  15  45826  5 3 187  62   4 2 1   3 104 21  2  0  6  0
STOUT      406 8TH ST       45   5 1 1   442  55659   6  674  54 107  40603  2 2 133  67   5 0 1   2  52 26  3  2  0  1

DIERCKS    405 9TH ST       45   6 1 1   340  48452   5  563  42  24  45454  2 2 133  76   8 1 1   4  50 25  3  0  0  1
READ       406 9TH ST       46   7 1 1   372  38403   4  520  30 103  19469  2 1  65  65   8 1 1   4 107 27  3  0  4  1
NELSON     404 7TH ST       43   8 1 1   631  93538   9 1088  80  86  35118  5 3 199  66   8 2 2   5 106 21  3  0  0  1
HILL       403 7TH ST       43   9 1 1   796  92083   9 1101  85  13  38822  4 3 177  59   6 1 1   3  95 24  3  0  0  1
GOLDEN     404 7TH ST       44  10 1 1   690  71316   8  909  80  78  21030  3 2 152  76   4 1 2   1  60 20  3  0  4  3

ABLES      403 8TH ST       44  11 1 1   446  65276   7  839  62  60  54135  3 2 151  76   4 2 1   3  67 22  3  0  0  6
CURRY      404 8TH ST       45  12 1 1  1022  94549  11 1320  95  20  57872  2 5 234  47   4 2 1   3 122 20  3  0  3  6
RIVAS      403 9TH ST       45  13 1 1   520  54886   6  694  52  14  31098  2 2 127  64   4 2 2   1  54 27  3  0  0  0
KNIGHT     404 9TH ST       46  14 1 1   917 109640  11 1394 101  15  28180  2 2 120  60   6 1 2   7 155 22  3  0  0  1
HALTER     403 DIVISION ST  46  15 1 1   696  97060   9 1070  78  70  23328  4 3 183  61   6 1 2   4  90 23  3  0  1  0

MATTHEWS   402 6TH ST       43  16 1 1   549  65679   7  817  63  79  23094  4 3 177  59   7 2 2   8  92 23  3  0  1 10
RODRIGUEZ  401 7TH ST       43  17 1 1   683  78597   8  907  72  47  32756  4 2 149  75   5 2 2   4  77 19  3  1  0  0
MATUS      402 7TH ST       44  18 1 1   568  63206   7  821  64  17  55162  4 3 183  61   5 1 2   6 101 25  3  0  5  3
BROWN      401 8TH ST       44  19 1 1   634  86482   9 1016  84  13  49234  5 3 185  62   9 0 2   1 101 20  3  0  0  0
SOLLA      402 8TH ST       45  20 1 1   542  70656   7  898  61  35  41890  2 2 137  69   6 2 2   1  49 25  3  0  4  0

KOATIK     401 9TH ST       45  21 1 1   681  83328   8  983  69  19  52729  4 2 146  73   6 2 2   6  72 18  3  0  0  0
UNDERWOOD  402 9TH ST       46  22 1 1   446  65397   7  797  59  49  56263  3 2 123  62   9 2 1   3  74 25  3  0  0  3
DUNN       401 10TH ST      46  23 1 1   599  89070   9  981  86  52  36304  5 3 186  62   5 1 1   5 115 23  3  0  0  2
KELLY      402 DIVISION ST  47  24 1 1   727  84227   9 1135  81  18  26977  3 2 174  58  10 0 1   5  91 23  3  0  0  0
MCGIBNEY   306 6TH ST       32  25 1 1   652  79381   9 1019  87  60  63040  5 4 201  50   6 2 2   4  90 18  3  0  5  1

SMULIN     305 7TH ST       32  26 1 1   647  99762  10 1226  86  23  33693  5 4 206  52   3 1 1   5 121 24  3  1  0  9
BRENNER    306 7TH ST       33  27 1 1   511  63601   7  846  61  47  52135  2 1 113 113   3 1 2   1  45 23  3  1  0  0
KLIEMAN    305 8TH ST       33  28 1 1   449  73180   7  834  62  13  38567  3 2 159  80   9 1 2   1  64 21  3  2  0  0
JONES      306 8TH ST       34  29 1 1   598  64743   7  778  68  19  54086  3 2 133  67   6 2 1   2  75 25  3  0  0  0
BAUER      305 9TH ST       34  30 1 1   767  86624   9 1136  83  14  29053  4 2 144  72   7 1 1   5  79 20  2  1  0  2

CLARK      306 9TH ST       35  31 1 1   766  80786   9  995  77  25  51304  5 3 197  66   8 2 1   4  99 20  3  0  0  0
REID       305 10TH ST      35  32 1 1   657  82577   8  984  81  17  31608  5 3 211  53   6 2 1   4  95 19  3  0  0  3
FRENCH     306 10TH ST      36  33 1 1   503  88014   9 1004  84  70  49645  4 3 183  61   4 2 1   1  83 21  3  0  0  0
LILLEY     304 6TH ST       32  34 1 1   882  97466  11 1303 103  11  44956  7 2 120  60   9 1 2   7 176 25  3  0  3  1
EHREN      303 7TH ST       32  35 1 1   882  97642  10 1143  89  55  29053  6 3 165  55   4 0 1   3 115 19  3  0  0  0

KILPATRICK 304 7TH ST       33  36 1 1   611  79019   8  894  74  41  65173  3 3 175  58   5 1 2   4  74 25  3  2  0  3
SMITH      303 8TH ST       33  37 1 1   820  94974   9 1069  83  91  36483  3 3 178  59   5 1 2   4  83 21  2  1  0  0
AYCOCK     304 8TH ST       34  38 1 1   655  91327  10 1204  99  46  49754  4 2 217  54   9 1 2   9 108 18  3  0  2  0
HARRIS     303 9TH ST       34  39 1 1   514  79306   8 1035  74  14  45814  3 3 180  61   5 2 1   1  68 23  2  0  1  3
OEKLER     304 9TH ST       35  40 1 1   683  90484  10 1105  93  17  30000  6 4 220  55   7 2 1   8 116 19  3  0  4  1

PAGE       303 10TH ST      35  41 1 1   263  46025   4  259  27  82  19505  1 2 120  60   6 1 1   2  23 23  2  0  0  0
NEWCOMB    304 10TH ST      36  42 1 1   616  39420   8  639  65  25  21756  3 2 144  72   4 2 1   0  72 24  2  2  0  0
COLLINS    304 DIVISION ST  36  43 1 1   545  68259   7  876  64  17  38757  4 3 183  61   5 0 2   5 111 28  3  1  0  5
HUGHES     302 10TH ST      32  44 1 1   687  76909   9 1147  88  16  50588  4 2 147  74  11 1 2   1  76 19  3  1  0  0
VALENTI    301 7TH ST       32  45 1 1   626  77460   9 1063  91  17  30237  4 2 144  72   6 2 1   2  88 22  3  0  1  8

MARKLEY    302 7TH ST       33  46 1 1   795  83364   9 1115  85  68  35457  4 3 172  57   8 1 2   6  75 19  3  2  0  4
CRIBBS     301 8TH ST       33  47 1 1   610  83133   9 1107  84  20  51671  2 4 162  81   6 2 2   3  80 20  3  0  0  6
BOGUS      302 8TH ST       34  48 1 1   645  79711   8 1029  73  21  29375  3 2 165  83   6 1 2   4  58 19  3  0  0  0
CLEMA      301 9TH ST       34  49 1 1   327  45077   4  513  35  12  19626  4 2 120  60   7 0 1   3  82 21  3  0  1 11
HAYES      302 9TH ST       35  50 1 1   666  75622   8  925  66  19  45756  4 2 211  53   8 2 2   2  96 24  3  1  0  0

BARR       301 10TH ST      35  51 1 1   332  42456   5  556  35  99  41062  1 1  65  65   5 2 2   1  22 22  2  0  0  0
SUMMERS    302 10TH ST      36  52 1 1   775  91355   9 1084  85  13  24718  4 3 120  60   5 2 2   8 110 18  2  0  6  0
PHILLIPS   302 DIVISION ST  36  53 1 1   433  60875   6  751  51  25  43451  1 1  65  65   6 1 2   1  27 27  3  2  0  0
FLYTHE     206 6TH ST       20  54 1 1   566  66700   7  802  64  55  42770  3 2 124  62   6 1 2   1  69 23  3  0  4  0
KING       205 7TH ST       20  55 1 1   678  86592   9 1065  83  38  33910  5 3 187  62   5 2 2   0  91 18  3  1  0  8

BRYANT     206 7TH ST       21  56 1 1   321  42926   5  432  37  41  15467  4 1  65  65   8 1 1   3  74 19  3  1  0  3
MCLEAN     205 8TH ST       21  57 1 1   445  57500   5  852  59  15  23580  4 3 166  65   4 2 2   3  78 20  3  0  0  0
CLOSSEY    206 8TH ST       22  58 1 1   597  99761  11 1276  91  63  30830  6 4 217  54   5 1 2   8 125 21  3  1  0  3
BARTLEY    205 9TH ST       22  59 1 1   567  78597   8  901  76  23  33536  4 2 154  74   4 1 1   3  88 22  3  0  5  0
COX        206 9TH ST       23  60 1 1   670 109025  11 1351 109  15  33200  7 3 165  55   5 1 1   2 157 22  3  2  0  0

NIXON      205 10TH ST      23  61 1 1   485  63638   7  875  58  81  47583  3 2 163  82   6 1 1   2  70 23  3  0  0  4
JOHNSTONE  206 10TH ST      24  62 1 1   573  73912   9  915  76  69  29501  3 2 184  61  10 2 2   5 108 22  3  0  0  0
SIBLEY     205 11TH ST      24  63 1 1   653  67194   7  857  65  49  35468  3 2 134  67   5 2 2   2  67 22  3  0  0  0
LUNN       206 11TH ST      25  64 1 1   362  60679   7  821  61  34  30624  3 2 167  56   6 1 1   2  86 22  3  1  0  0
BOYER      204 6TH ST       20  65 1 1   623  70850   7  833  57  53  29381  3 2 123  62   4 1 2   2  69 23  1  1  0  0

EVERHAM    203 7TH ST       20  66 1 1   661  85978   9 1071  78  18  72453  5 3 188  63   7 2 2   5  93 19  3  1  0  0
LANDER     204 7TH ST       21  67 1 1   667  90105   9 1060  81  24  55787  5 4 202  61   7 2 1   5 111 22  3  1  0  0
CHALMERS   203 8TH ST       21  68 1 1   305  49752   5  568  39  21  41836  1 1  65  65   8 1 1   1  27 22  3  1  0  2
BROWNING   204 8TH ST       22  69 1 1   815  92235   9 1034  82  59  33369  5 4 210  53   8 1 2   1 108 22  3  0  0  0
STRONG     203 9TH ST       22  70 1 1   480  60991   6  709  53  44  54578  3 2 149  75   8 2 2   2  58 19  3  0  1  0

PRICE      204 9TH ST       23  71 1 1   696  76904   8  883  67  27  22348  4 3 181  60   4 2 1   2  97 24  3  0  0  0
DESOSA     203 10TH ST      23  72 1 1   618  69089   7  825  70  39  29910  3 2 131  66   4 2 1   3  79 26  3  0  4  0
DILLON     204 10TH ST      24  73 1 1   685  93546   9 1152  82  69  39119  6 4 214  54   6 2 2   1 113 19  3  0  0  0
DELVECCHIO 203 11TH ST      24  74 1 1   732  84303   8  853  77  70  43249  3 2 129  65   2 1 1   1  69 23  3  0  0  0
STALEY     204 11TH ST      25  75 1 1   358  45419   5  548  44  70  28812  1 1  85  85   2 1 1   1  36 36  3  0  0  4

LEONARD    204 DIVISION ST  25  76 1 1   698 107868  11 1253 100  11  46185  6 4 219  55   6 2 2   6 108 18  3  2  0  0
HASTINGS   202 6TH ST       20  77 1 1   781  86713   9 1021  90  53  59055  5 3 204  51   6 2 2   3  91 18  3  0  0  0
MICHAEL    201 7TH ST       20  78 1 1   456  56732   6  732  54  16  60633  2 1 117 117   4 2 2   1  41 21  3  0  1  0
IRELAN     202 7TH ST       21  79 1 1   615  75608   8 1004  76  59  41305  4 2 160  80   4 2 2   2  72 18  3  1  0  7
OELZE      201 8TH ST       21  80 1 1   509  89782   9 1048  81  21  39197  4 2 150  75   6 1 2   6  74 19  3  2  0  1

COOK       202 8TH ST       22  81 1 1   840  86669   9 1086  87  14  33260  5 3 193  64   8 1 2   4  97 19  2  0  1  2
POKORNEY   201 9TH ST       22  82 1 1   263  48297   4  338  28  57  17046  1 1  65  65   9 0 1   1 103 21  2  1  0  4
HAMATY     202 9TH ST       23  83 1 1   802  82304   9 1056  86  20  23483  5 3 197  66   8 1 2   0  96 19  3  0  1  4
WARREN     201 10TH ST      23  84 1 1   461  67962   7  769  64  25  53950  3 2 163  82   5 0 1   1  73 24  3  0  0  0
SAUNDERS   202 10TH ST      24  85 1 1   566  73351   7  864  65  19  39171  3 2 156  78   3 0 1   1  51 17  3  0  0  0

CAGLE      201 11TH ST      24  86 1 1   880  97313  10 1194 102  17  42818  6 4 219  55   7 0 1   1 145 24  3  0  3  0
MANLEY     202 11TH ST      25  87 1 1   506  71024   8  910  73  16  37523  4 3 211  53   7 0 1   1 107 21  3  0  3  0
SLOOP      201 12TH ST      25  88 1 1   359  50844   5  646  42  19  40283  2 1  99  99   2 1 2   2  41 21  3  1  3  5
RADLEY     202 DIVISION ST  25  89 1 1   837  96634  10 1146  92  19  34680  6 3 120  60   7 2 1  10 146 24  3  1  0  0
ALLISON    106 6TH ST        7  90 1 1   402  51569   5  641  40  15  57205  1 1  79  79   3 2 2   1  24 24  1  0  0  0

TOURSH     105 7TH ST        7  91 1 1   510  58524   8 1074  70  18  44914  4 3 177  59   4 1 2   3  99 25  3  0  0  0
IMPERI     106 7TH ST        8  92 1 1   629  80079   9 1081  84  19  52548  5 3 190  63   4 1 2   3  99 21  3  0  0  0
SOMMER     105 8TH ST        8  93 1 1   662  82695   9 1019  82  22  56518  5 3 231  46   4 1 2   3 106 19  3  0  0  0
GEBHARDT   106 8TH ST        9  94 1 1   530  66490   7  954  69  25  36155  3 2 158  79   6 1 2   1  57 19  3  0  2  5
ANGUEIRA   105 9TH ST        9  95 1 1   469  52686   5  585  42  23  27039  1 1  75  75   2 1 1   1  24 24  3  0  2  5
```

```
                                                    VARIABLES
  0         0          0    0 0 0   0      0      0  0  1     1  1   1   1 1 1  1  1 1 2  2    2   2 2 2 2
  1         2          3    4 5 6   7      8      9  0  1     2  3   4   5 6 7   8  9 0 1  2    3   4 5 6 7
**************************************************************************************************

RABIN     106 9TH ST       10   96 1 1  537  65852  6   778  55  28  52221  2 2 140  70  9 1 1 1  58 29 3 0 1 2
JOHNSON   105 10TH ST      10   97 1 1  411  59277  6   805  53  71  57553  2 1  68  68  4 2 2 1  28 28 3 0 5 7
LONGUEIRA 106 10TH ST      11   98 1 1  547  66167  6   752  53  18  57092  2 2 142  71  7 2 2 2  63 32 3 2 0 7
DEGEORGE  105 11TH ST      11   99 1 1  551  78553  8   912  74  16  22771  2 2 164  82 10 2 2 2  84 21 3 0 4 8
BARNES    106 11TH ST      12  100 1 1  819  91319  9  1061  85  84  28863  4 3 170  57  6 2 2 1  99 25 3 0 0 3

LUND      105 12TH ST      12  101 1 1  580  75224  8   957  71  21  30853  3 2 164  82  6 0 2 2  66 22 3 0 0 0
QUINTANA  106 12TH ST      13  102 1 1  504  67603  7   767  66  73  35399  4 2 165  83  7 2 1 2  64 21 3 0 4 5
SWICK     104 6TH ST        7  103 1 1  696  69735  8   948  73  15  57061  4 3 166  55  7 2 1 2  65 16 3 0 0 2
GRIMANY   103 7TH ST        7  104 1 1  487  63368  6   706  53  80  41494  3 2 136  68  9 2 1 2  51 26 3 0 0 3
GRIFFITH  104 7TH ST        8  105 1 1  563  73015  7   802  64  21  49068  3 2 147  74  6 1 2 2  50 17 3 0 0 3

CAPPER    103 8TH ST        8  106 1 1  560  78595  9  1138  85  20  32150  5 4 200  50  8 0 2 3 114 23 3 0 5 1
BECTON    104 8TH ST        9  107 1 1  559  74627  8  1072  73  19  41403  4 2 150  75  6 0 2 2  77 19 3 0 4 6
KAUFMAN   103 9TH ST        9  108 1 1  483  85407  9  1017  84  34  40120  4 2 149  75  6 2 1 2  82 13 3 0 4 6
LINTON    104 9TH ST       10  109 1 1  611  82737  8  1017  75  13  56580  4 2 161  81  9 2 2 1  91 23 3 0 0 4
NEUMAN    103 10TH ST      10  110 1 1  445  57964  6   809  52  23  15895  6 1  65  65  4 1 1 6 126 21 1 0 4 1

ALONSO    104 10TH ST      11  111 1 1  490  76738  8   973  75  14  51931  5 3 193  64  5 2 2 0  99 20 3 0 0 11
KEENBURG  103 11TH ST      11  112 1 1  462  74980  8   940  70  17  32375  3 2 158  79  3 2 2 0  56 19 3 0 0 2
GETTIS    104 11TH ST      12  113 1 1  484  83877  8   971  67  17  26802  5 5 230  46  5 1 2 1 111 22 3 2 0 0
ROTHMAN   103 12TH ST      12  114 1 1  658  79453  8   925  67  20  33910  4 3 177  59  5 1 2 1 109 17 3 2 0 3
UPTON     104 12TH ST      13  115 1 1  897  94751  9  1157  83  33  23375  4 4 214  54  5 2 1 1 106 21 3 1 0 3

STUART    104 DIVISION ST  13  116 1 1  527  76193  9  1077  83  13  31244  6 4 221  55  6 2 2 4 105 18 3 0 3 0
WEEKES    102 6TH ST        7  117 1 1  598  74658  7   809  62  37  32502  3 2 145  73  5 1 2 2  56 19 3 0 4 2
PHILLIPS  101 7TH ST        7  118 1 1  549  77118  8  1027  71  26  51411  4 3 183  61  3 2 2 2  87 22 3 0 4 3
ROTH      102 7TH ST        8  119 1 1  389  57601  6   635  57  17  15300  6 1  65  65  4 2 1 2 126 21 3 0 4 1
DEROSE    101 8TH ST        8  120 1 1  509  68227  7   908  54  87  51526  2 2 139  70  8 2 1 1  45 23 1 2 0 0

RITZIE    102 8TH ST        9  121 1 1  757 101541 10  1130  96  16  24839  5 3 194  65  4 1 1 5  96 19 1 1 0 0
MILLER    101 9TH ST        9  122 1 1  537  83132  9  1036  86  49  62205  5 3 192  65  5 2 2 0 105 21 3 0 0 0
FLETCHER  102 9TH ST       10  123 1 1  298  37598  4   483  32  15  56479  1 1  65  65  5 2 2 0  25 25 3 0 0 0
GITLOW    101 10TH ST      10  124 1 1  448  52737  6   739  53 101  60229  1 1  86  86  1 0 1 1  28 28 2 0 0 10
SUGRUE    102 10TH ST      11  125 1 1  590  73690  8   972  74  15  48674  4 2 163  80  2 2 1 2  82 21 3 0 2 0

NELSON    101 11TH ST      11  126 1 1  456  47385  5   628  40  21  26316  2 2 139  70  6 0 1 2  52 26 3 0 5 3
FLEMING   102 11TH ST      12  127 1 1  411  46942  5   633  41  21  38611  2 1 115 115  3 2 2 1  47 34 3 0 1 0
BOYD      101 12TH ST      12  128 1 1  803  87319  9  1097  84  15  24917  4 2 150  75  4 2 2 0  78 30 3 0 1 0
SCHERER   102 12TH ST      13  129 1 1  606  70174  7   874  62  21  63849  3 2 122  61  5 4 2 1  75 25 3 0 1 0
ZANCA     102 DIVISION ST  13  130 1 1  590  86491  9  1067  86  11  31286  5 3 193  64  3 1 2 4 113 23 2 2 0 4

WOODSON   505 1ST ST       48  131 2 1  607  75135  6   781  57  20  51589  2 2 134  67  6 2 1 1  53 27 3 1 0 0
RIVAS     506 1ST ST       49  132 2 1  486  62746  7   541  66  52  20234  4 3 167  56  7 1 1 4  69 17 3 0 1 0
HERN      505 2ND ST       49  133 2 1  744 101344  8  1042  67 101  50001  4 3 170  57  6 2 2 0  79 20 3 2 0 0
KATZ      506 2ND ST       50  134 2 1  498  78077  7  1005  72  25  63330  3 2 125  63  5 1 1 0  65 22 3 2 0 0
MORALES   505 3RD ST       50  135 2 1  554  78850  7   987  58  84  70831  4 3 180  60  9 0 1 1  86 22 3 1 0 0

VEYHL     506 3RD ST       51  136 2 1  884 107094  9  1228  84  13  32648  5 3 190  63  5 0 2 2  89 18 3 2 0 9
SANDT     505 4TH ST       51  137 2 1  880  94070  8  1154  74  18  83345  3 2 135  68  6 2 1 2  70 23 3 0 6 0
WARREN    506 4TH ST       52  138 2 1  754  98666  8  1184  65  61  94249  3 2 139  70  7 1 1 2  78 26 3 0 0 0
GOLDBERG  503 1ST ST       48  139 2 1  742 108917  9  1253  84  22  90435  4 2 150  75  3 0 0 0  67 17 3 2 0 0
BOLDING   504 1ST ST       49  140 2 1  401  72312  6   475  50  51  23576  6 2 120  60  4 2 1 4 129 22 3 0 6 4

BRYAN     503 2ND ST       49  141 2 1  708  84778  7  1002  62 102  54903  3 2 133  67  7 2 1 2  68 23 3 0 1 0
GREENE    504 2ND ST       50  142 2 1  702  92637  8  1034  75  12  61106  4 3 162  81  3 2 1 0  75 19 3 1 0 0
ALVAREZ   503 3RD ST       50  143 2 1  832 103104  9  1265  91  85  76450  4 2 162  81  3 2 1 0  82 21 1 0 0 0
MCADAMS   504 3RD ST       51  144 2 1  582  84689  7   914  68  20  58321  3 2 149  75  5 2 1 1  58 19 1 0 0 0
ALEXANDER 503 4TH ST       51  145 2 1 1031 131033 11  1436 101 104  83669  5 4 203  51  8 2 1 3 109 22 3 0 1 0

WILKES    504 4TH ST       52  146 2 1  870 115109  9  1279  84  21  62771  5 4 216  54  4 2 2 1 100 20 3 2 0 0
O'CONNELL 502 1ST ST       48  147 2 1  842 112528  9  1261  81  12 103709  4 3 188  63  4 2 1 5  78 20 3 0 1 5
LOWE      501 1ST ST       49  148 2 1  956 113924  9  1268  84  10 120030  4 2 156  78  6 2 1 5  85 21 2 0 0 1
LEE       501 2ND ST       49  149 2 1  733  85027  8  1070  72  18  89656  3 2 162  81  4 2 1 1  72 24 3 2 0 10
RIVET     502 2ND ST       50  150 2 1  593  79421  7  1010  65  86  69898  2 1 118 118  7 1 2 1  43 22 3 2 0 11

FOGELSON  501 3RD ST       50  151 2 1  763  90176  8  1122  72  91  60310  5 3 194  65  2 1 2 1 104 21 3 0 0 1
JONES     502 3RD ST       51  152 2 1  880  99279  8  1174  74  15  83484  5 3 172  57  6 2 1 1  75 25 3 1 0 1
RIVERS    501 4TH ST       51  153 2 1  771  88803  7   944  64  82  80217  2 2 140  70  7 1 2 1  53 27 1 2 0 3
FERNANDEZ 502 4TH ST       52  154 2 1  840  92755  7  1018  62  21  70927  2 2 161  81 11 2 2 0  84 21 1 2 0 6
ALDIN     405 1ST ST       37  155 2 1  730  94470  8  1083  75  20  56172  4 2 144  72  5 1 2 5  84 21 1 2 0 2

CAIN      406 1ST ST       38  156 2 1 1102 129056 10  1510  84  40  82770  4 3 180  60  8 2 2 4  98 25 3 1 0 0
KNIGHT    405 2ND ST       38  157 2 1  806  93774  9  1006  79  47  52841  4 2 147  74  9 0 2 2  75 19 3 0 0 0
PHELPS    406 2ND ST       39  158 2 1 1021 126065 10  1443  99  26  42314  7 2 120  60  6 1 1 9 182 26 3 0 0 0
ROMERO    405 3RD ST       39  159 2 1  970  99652  8  1089  82  15  68809  5 4 203  51  4 1 1 6  99 20 3 0 0 6
BOIX      406 3RD ST       40  160 2 1  734  89888  7   891  65  40  78086  2 2 125  63  6 2 2 1  45 23 1 0 0 0

JEFFREY   405 4TH ST       40  161 2 1  672  93422  8  1053  75  18  67110  4 3 173  58  6 1 2 3  92 23 1 1 0 11
BOBSON    406 4TH ST       41  162 2 1  785  95599  8  1189  68  23  58966  4 2 157  79  4 2 1 1  79 20 3 1 0 0
MIRANDA   406 5TH ST       41  163 2 1  692  89142  7   954  66  18  69817  4 2 151  76  7 2 1 1  77 19 3 0 1 0
INFANTE   406 5TH ST       42  164 2 1  827 103777  8  1194  70  25  68773  3 2 175  58  4 1 2 1  71 24 1 0 0 0
RASSEL    405 6TH ST       42  165 2 1  683  98453  9  1288  91  26  31189  6 4 214  54  6 2 2 3 143 24 1 0 0 0

KARD      403 1ST ST       37  166 2 1  740  84686  8  1176  72  22 108774  3 2 156  78  7 2 1 1  54 18 3 0 0 0
SHELTON   403 1ST ST       38  167 2 1  975 121338 11  1492 115  47  91910  5 4 211  53  7 2 1 1 112 22 3 1 0 0
VERHOVAN  403 2ND ST       38  168 2 1  693  83080  7  1081  61  86  52722  3 2 150  75  4 2 1 1  68 23 3 1 0 0
BOREK     404 2ND ST       39  169 2 1  781 102110  9  1237  91  44  47989  3 3 190  63  5 0 2 0 112 22 3 0 1 0
GAINES    403 3RD ST       39  170 2 1  632  86855  7   984  60  18  88572  1 1  65  65  4 1 1 2  25 25 3 0 1 0

KARSEN    404 3RD ST       40  171 2 1 1012 117967 10  1308  96  21  81374  4 3 175  77  8 1 2 4  79 20 1 0 5 1
CABILI    404 4TH ST       40  172 2 1  654  90436  8  1146  72  24  91161  4 3 175  58  4 0 2 3  95 24 3 0 0 0
KUSNICK   404 4TH ST       41  173 2 1  767 102531  8  1152  74  26  69621  4 3 199  60  6 2 1 2 112 28 3 0 4 0
ALDA      403 5TH ST       41  174 2 1  888  91700  9  1309  80  20  35840  4 2 211  53  4 2 1 3 111 22 3 0 2 0
PEEPLES   404 5TH ST       42  175 2 1  852 112325  9  1272  79  21  76434  2 2 156  78  3 2 2 1  95 24 1 0 0 0

QUEENE    403 6TH ST       42  176 2 1  664  91731  8  1097  72  41  60262  5 4 208  52  8 2 2 0 116 23 1 0 3 4
LYSINGER  401 1ST ST       37  177 2 1  639  97080  9  1121  72  48  60080  3 3 167  56  8 2 2 1  93 23 1 0 2 0
JUNCO     402 1ST ST       38  178 2 1 1101 114558  9  1247  82  17  91058  3 2 142  71  3 2 2 0  98 21 1 0 0 0
HAMBLEY   401 2ND ST       38  179 2 1  619  75661  6   845  47  43  56944  2 1 105 105  3 2 1 0  43 22 3 0 0 0
SCHWARTZ  402 2ND ST       39  180 2 1  541  70408  6   867  48  12  65255  1 1  65  65  7 2 2 0  24 24 1 0 1 2

TINDALL   401 3RD ST       39  181 2 1  945 119960 10  1476  86  32  64578  5 3 199  66  7 2 1 5 103 21 3 0 0 7
GREENBERG 402 3RD ST       40  182 2 1  818 101114  9  1123  74  15  74106  3 2 183  61  6 2 1 2  89 22 3 0 0 0
JAKUBEK   401 4TH ST       40  183 2 1  645  88804  9   935  64  18  77238  3 3 144  72  6 2 1 2  88 22 3 0 1 0
BUTLLR    402 4TH ST       41  184 2 1  868 103436  9  1284  84  58  35859  3 3 190  63  4 0 2 0  93 19 3 0 0 2
DIAZ      401 5TH ST       41  185 2 1  438  56743  5   618  40 107  45174  2 1 118 118  4 2 2 1  41 21 3 0 2 0

MIRSKY    402 5TH ST       42  186 2 1  766 106234  9  1286  85  25  73136  4 3 170  57  7 2 2 5  86 22 3 0 3 3
EDMONDS   401 6TH ST       42  187 2 1  805 111800 10  1399  85  23  63817  4 3 150  75  6 2 1 2 139 23 1 0 0 0
HEACOCK   305 1ST ST       26  188 2 1  606  88689  8  1187  65  18 101290  3 3 136  68  6 2 1 0  56 28 1 1 1 0
SANDSTROM 306 1ST ST       27  189 2 1  556  76680  7   960  67  22  56692  2 2 130  70  5 2 1 1  56 28 1 1 0 5
NOVITSKY  305 2ND ST       27  190 2 1  501  69822  6   910  50  64  61570  2 2 125  63  5 2 1 0  52 26 3 1 0 0
```

C Stat City data base

VARIABLES

01	02	03	04	05	06	07	08	09	10	11	12	13	14	15	16	17	18	19	20	21	22	23	24	25	26	27
HEARIN	306 2ND ST	28	191	2	1	522	59099	5	627	41	64	54855	1	1	75	75	6	1	1	1	31	31	1	0	0	0
ROWE	305 3RD ST	28	192	2	1	887	93933	8	1137	68	24	89496	3	3	179	60	6	2	1	1	80	27	3	0	4	2
ROBINSON	306 3RD ST	29	193	2	1	857	113574	9	1357	85	20	87855	4	3	175	58	6	1	2	2	93	23	3	0	0	0
HOFFMAN	305 4TH ST	29	194	2	1	651	83975	8	1084	75	19	62459	4	2	144	72	7	1	2	4	73	18	3	0	0	3
BRADSTREET	306 4TH ST	30	195	2	1	728	92815	8	1207	72	16	60714	4	2	160	80	8	1	1	2	79	20	3	1	0	0
DERR	305 5TH ST	30	196	2	1	771	99445	8	1129	78	14	39201	4	2	145	73	12	1	1	4	80	20	3	0	1	3
GLADSTONE	306 5TH ST	31	197	2	1	600	82927	7	1029	64	43	38527	4	3	170	57	7	0	1	3	90	23	1	0	0	0
WAYNE	305 6TH ST	31	198	2	1	734	73395	6	764	51	66	49668	2	1	112	112	11	2	1	1	43	22	3	1	0	0
GRAY	303 1ST ST	26	199	2	1	734	107481	9	1382	74	46	95761	4	2	163	82	7	2	2	4	76	19	3	0	0	0
KAHN	304 1ST ST	27	200	2	1	777	100028	8	1153	76	24	62806	4	3	177	59	4	2	1	3	89	22	3	0	0	0
HOTCHNER	303 2ND ST	27	201	2	1	714	96955	8	1177	75	21	96448	3	2	142	71	7	2	2	1	75	25	3	1	0	0
RODALE	304 2ND ST	28	202	2	1	1089	138976	11	1560	97	19	70538	7	3	165	55	7	1	1	5	156	22	1	1	0	0
PEARL	303 3RD ST	28	203	2	1	546	62786	5	656	40	13	76402	1	1	74	74	5	2	1	1	28	28	3	2	0	2
HEMINGWAY	304 3RD ST	29	204	2	1	739	100182	8	1125	74	34	61103	3	2	132	66	8	2	2	1	65	22	3	0	4	3
CLARK	303 4TH ST	29	205	2	1	675	94743	8	1124	77	21	57598	3	2	149	75	3	2	2	4	54	18	3	1	0	1
DEAN	304 4TH ST	30	206	2	1	1018	111431	9	1225	81	14	70492	4	3	168	56	7	1	2	2	98	25	3	0	6	0
PORTER	303 5TH ST	30	207	2	1	631	69956	6	813	51	12	34286	2	1	116	116	3	2	1	1	44	22	3	0	3	0
MONTAGU	304 5TH ST	31	208	2	1	767	99957	8	1177	75	20	94297	3	2	122	61	6	1	2	1	73	24	1	0	0	0
HAMLYN	303 6TH ST	31	209	2	1	914	116695	9	1289	81	35	80920	4	2	148	74	2	2	2	2	83	21	1	2	0	4
SPINK	301 1ST ST	26	210	2	1	730	88228	8	1125	73	17	146165	3	2	121	61	7	1	2	1	63	21	2	0	0	5
MILLER	302 1ST ST	27	211	2	1	675	97688	8	1059	71	59	96618	3	2	148	74	7	2	1	2	58	19	3	0	5	5
RASMUSSON	301 2ND ST	27	212	2	1	1099	112194	9	1305	85	24	81605	3	2	142	71	7	2	1	3	72	24	3	2	0	1
ABRAMS	302 2ND ST	28	213	2	1	711	89235	7	953	66	16	117671	3	2	148	74	11	0	1	3	54	18	3	0	1	0
FIELDS	301 3RD ST	28	214	2	1	714	81858	7	1059	65	22	67797	2	1	104	104	4	2	1	1	43	22	3	0	0	0
WICKERSHAM	302 3RD ST	29	215	2	1	611	81634	7	1053	58	79	78511	4	2	163	82	8	2	2	5	95	24	1	0	2	0
STEINWAY	301 4TH ST	29	216	2	1	928	125871	10	1440	92	12	75292	5	3	195	65	5	2	2	4	94	19	3	0	6	0
FROST	302 4TH ST	30	217	2	1	513	60894	5	709	43	24	72563	2	2	129	65	7	1	1	1	50	25	1	0	0	0
BOARDMAN	301 5TH ST	30	218	2	1	785	87717	7	989	59	67	60222	2	1	112	112	6	2	2	3	34	17	3	0	0	1
HAUGEN	302 5TH ST	31	219	2	1	662	92710	8	1185	69	54	35701	4	3	178	59	3	1	2	3	96	24	3	0	0	0
BLOOMFIELD	301 6TH ST	31	220	2	1	731	115390	10	1431	100	19	60585	6	4	214	54	9	2	1	6	115	19	3	0	5	4
WEISS	205 1ST ST	14	221	2	1	733	90652	8	1121	69	25	132830	3	2	163	82	6	1	2	1	61	20	2	1	0	4
KRESSEL	208 1ST ST	15	222	2	1	754	85080	8	1112	71	15	126365	2	2	127	64	7	2	2	2	45	23	1	2	0	2
REISMAN	205 2ND ST	15	223	2	1	710	107242	9	1249	83	22	91255	4	2	150	75	8	2	1	2	60	15	3	0	0	0
BOAZ	206 2ND ST	16	224	2	1	699	93133	7	923	62	88	91208	3	2	130	65	6	1	2	1	68	23	2	0	0	4
MEAD	205 3RD ST	16	225	2	1	501	53910	8	747	69	13	30043	4	3	178	59	9	1	2	2	95	24	2	0	6	0
SPELLER	206 3RD ST	17	226	2	1	699	81235	7	1038	61	16	72187	3	2	154	77	6	1	1	1	68	23	1	2	0	6
HOFFMAN	205 4TH ST	17	227	2	1	674	85248	7	980	67	69	85373	2	1	110	110	5	2	2	1	45	23	3	0	0	0
BAXTER	206 4TH ST	18	228	2	1	639	86562	7	1064	63	117	80710	2	1	105	105	4	1	1	1	42	21	3	1	0	1
ROSS	205 5TH ST	18	229	2	1	624	94756	8	1126	76	76	49047	5	3	188	63	6	1	2	1	105	21	3	0	1	1
KING	206 5TH ST	19	230	2	1	818	104945	8	1031	67	25	66701	3	2	130	65	6	1	2	3	73	24	3	0	0	0
VIGIL	205 6TH ST	19	231	2	1	915	97712	8	1003	69	27	56476	4	3	173	58	8	2	1	3	80	20	3	0	0	9
DAVIS	203 1ST ST	14	232	2	1	763	105395	9	1356	89	38	80995	4	4	206	52	5	2	1	6	97	24	3	0	5	0
ROYCE	206 1ST ST	15	233	2	1	670	93724	8	1139	64	37	70074	3	2	138	69	4	1	2	1	65	22	3	2	0	0
JOHANSON	204 1ST ST	15	234	2	1	528	73921	6	646	52	93	166518	2	1	111	111	7	1	2	1	39	20	3	0	6	1
LANDRY	203 2ND ST	15	235	2	1	520	57906	5	632	44	25	78359	2	2	134	67	5	1	2	2	44	22	3	2	0	0
DEITZ	204 2ND ST	16	236	2	1	1054	115666	9	1254	79	59	74242	4	2	153	77	6	2	1	3	74	19	1	0	0	0
CALHOUN	203 3RD ST	16	237	2	1	950	105530	8	1047	78	66	61602	4	3	181	60	4	2	2	4	93	23	3	2	0	5
FLANAGHAN	204 3RD ST	17	238	2	1	909	100493	8	1201	80	34	72824	4	2	159	80	5	1	2	1	91	23	1	0	1	1
BROWN	203 4TH ST	17	239	2	1	371	65392	6	493	45	78	33411	6	2	120	60	6	2	2	4	126	21	2	0	3	3
COHN	204 4TH ST	18	240	2	1	506	74311	6	753	52	19	62170	3	2	133	67	5	2	2	4	62	21	3	0	4	1
KANDELL	203 5TH ST	18	241	2	1	768	91563	7	1052	62	19	47650	3	2	137	69	8	2	2	4	76	25	1	2	0	0
PFEIFFER	204 5TH ST	19	242	2	1	699	100303	8	1092	69	18	62162	3	3	181	60	4	1	2	1	72	24	1	0	4	0
MENDOZA	203 6TH ST	19	243	2	1	779	99421	8	1033	69	22	46337	4	2	156	78	5	2	1	4	78	20	1	0	0	5
RODRIGUEZ	201 1ST ST	14	244	2	1	867	100532	9	1270	88	16	107165	4	3	190	63	8	2	1	4	86	22	3	0	0	0
DEHAVEN	202 1ST ST	15	245	2	1	545	70147	6	907	52	42	140379	1	1	89	89	5	1	1	1	33	33	3	0	0	11
MARIUS	201 2ND ST	15	246	2	1	769	95962	9	1181	80	24	111260	3	2	121	61	6	2	2	1	63	21	1	0	2	0
VALDEZ	202 2ND ST	16	247	2	1	731	86440	7	977	63	35	141031	2	1	109	109	4	2	1	3	44	22	3	2	0	2
INNELLO	201 3RD ST	16	248	2	1	736	103402	9	1170	87	16	83195	3	2	155	78	9	1	2	1	59	20	1	2	0	1
PAULIN	202 3RD ST	17	249	2	1	635	77700	7	1018	63	34	51166	2	2	136	68	8	2	2	2	50	25	3	2	0	2
ELINOFF	201 4TH ST	17	250	2	1	797	106539	9	1311	82	16	76231	5	3	197	66	8	1	1	4	105	21	3	2	0	2
CHAVEZ	202 4TH ST	18	251	2	1	618	89219	7	1064	60	15	82865	2	2	134	67	4	2	2	1	37	19	1	0	0	0
MCINTYRE	201 5TH ST	18	252	2	1	1009	112314	9	1341	79	19	67588	4	3	171	57	8	0	1	4	94	24	3	0	4	4
SOUTHERN	202 5TH ST	19	253	2	1	875	100919	8	1179	67	37	80265	4	3	185	62	9	2	1	3	85	21	1	0	0	0
VICIOSO	201 6TH ST	19	254	2	1	846	99056	9	1185	89	38	40217	5	4	209	52	10	0	1	4	125	25	3	0	1	1
FLETCHER	105 1ST ST	1	255	2	1	577	73681	7	968	63	69	129329	2	2	124	62	5	1	2	1	54	27	3	1	0	1
CHEANEY	106 1ST ST	2	256	2	1	627	73920	6	863	53	76	121005	2	2	134	67	6	2	2	1	50	25	3	0	0	0
ANDRES	105 2ND ST	2	257	2	1	662	80570	7	964	59	27	64797	3	2	149	75	8	2	2	0	58	19	1	0	0	0
HOLMES	106 2ND ST	3	258	2	1	748	92419	8	1136	73	24	116203	3	2	127	64	3	2	2	4	63	21	1	0	0	10
QUINTANA	105 3RD ST	3	259	2	1	696	86765	8	1138	64	19	63368	4	3	179	60	2	2	2	3	113	28	3	1	0	3
VALLEJO	106 3RD ST	4	260	2	1	914	102334	9	1300	85	18	78070	5	3	195	65	6	2	1	3	105	21	3	0	1	0
PHILLIPS	105 4TH ST	4	261	2	1	518	70201	6	780	53	81	61120	3	2	125	63	7	0	1	4	68	23	3	0	5	7
HOLMAN	106 4TH ST	5	262	2	1	418	61077	5	667	40	15	52098	1	1	91	91	7	0	2	1	33	33	1	0	6	0
PHIPPS	105 5TH ST	5	263	2	1	766	99977	9	1256	80	10	70732	4	2	162	81	5	2	2	3	86	22	1	1	0	11
WALKER	106 5TH ST	6	264	2	1	1052	117710	10	1274	90	24	43708	6	4	220	55	5	1	2	6	135	23	3	2	0	7
RAMOS	105 6TH ST	6	265	2	1	724	85366	7	957	64	72	71569	3	2	122	61	3	2	2	1	50	17	3	0	0	0
PICKER	103 1ST ST	1	266	2	1	703	83619	7	1016	66	20	120939	2	2	143	72	7	2	2	1	55	28	3	0	0	1
HOWARD	104 1ST ST	2	267	2	1	733	88656	7	990	54	20	113297	2	1	105	105	8	2	1	1	34	17	3	0	0	0
MILLER	103 2ND ST	2	268	2	1	641	74903	7	1092	64	15	86760	3	2	129	65	5	1	1	1	69	23	1	0	5	6
DION	104 2ND ST	3	269	2	1	676	86714	7	979	64	26	122969	1	1	65	65	3	2	2	1	26	26	3	0	0	0
ERNST	103 3RD ST	3	270	2	1	378	61443	5	682	43	18	54379	2	1	94	94	5	2	1	1	32	16	3	0	4	0
KASTER	104 3RD ST	4	271	2	1	738	82477	7	969	64	14	116346	3	2	136	68	3	2	2	3	72	24	3	2	0	0
BOILEN	103 4TH ST	4	272	2	1	891	110652	9	1365	86	20	66364	6	3	165	55	6	2	2	4	137	23	1	0	0	10
HYKIN	104 4TH ST	5	273	2	1	689	70435	6	874	57	17	68609	1	1	65	65	5	2	2	1	22	22	3	0	0	0
SHEELEY	103 5TH ST	5	274	2	1	725	87374	7	959	66	14	57519	3	2	144	72	9	2	2	1	68	23	1	2	0	0
WILLIAMS	104 5TH ST	6	275	2	1	845	109625	9	1194	83	49	81038	3	2	141	71	2	2	2	1	71	24	3	0	5	0
NORDBERG	103 6TH ST	6	276	2	1	863	109886	10	1480	95	38	27835	6	5	237	47	5	0	2	6	124	21	3	0	4	7
HAMRICK	101 1ST ST	1	277	2	1	752	103468	9	1265	88	59	131063	4	2	156	78	6	2	2	4	70	18	1	0	6	4
POE	102 1ST ST	2	278	2	1	1180	124599	10	1407	92	66	105357	4	4	209	52	4	2	2	2	90	23	3	1	0	6
OBREGON	101 2ND ST	2	279	2	1	837	100652	9	1193	87	78	136427	3	2	150	75	5	1	2	1	60	20	1	2	0	0
RIVERO	102 2ND ST	3	280	2	1	871	99899	8	1043	71	39	134443	3	2	141	71	3	2	1	3	78	26	3	0	0	1
MILES	101 3RD ST	3	281	2	1	599	105640	9	1385	78	65	109434	3	2	128	64	6	2	2	1	66	22	3	0	0	7
GURSON	102 3RD ST	4	282	2	1	849	109223	9	1313	84	19	58220	5	4	201	50	6	2	1	6	107	21	3	0	5	1
CRABTREE	101 4TH ST	4	283	2	1	651	98405	8	1162	82	62	57095	4	2	155	78	7	1	2	4	71	18	1	0	4	0
LOWELL	102 4TH ST	5	284	2	1	713	96645	8	1089	74	14	68773	4	3	170	57	5	2	2	1	72	18	3	1	0	1
QUINN	101 5TH ST	5	285	2	1	749	82646	7	916	57	21	50449	3	3	177	59	5	2	1	1	82	27	3	0	0	0
LOPEZ	102 5TH ST	6	286	2	1	621	76856	6	809	53	71	50602	2	1	104	104	9	1	2	1	35	18	1	0	1	11
GRABON	101 6TH ST	6	287	2	1	802	107009	9	1293	81	17	43982	5	3	197	66	7	2	2	0	95	19	3	2	0	9

VARIABLES

01	02	03	04	05	06	07	08	09	10	11	12	13	14	15	16	17	18	19	20	21	22	23	24	25	26	27
POTTS	1205 1ST ST	83	288	3	1	152	19254	4	439	32	56	15340	3	2	120	60	5	1	2	2	56	19	2	0	0	0
ZACCARIA	1206 1ST ST	84	289	3	1	134	13917	3	363	21	30	17624	3	2	120	60	5	0	1	1	76	25	1	0	0	0
RIVERS	1205 2ND ST	84	290	3	1	250	34683	6	564	49	26	30236	2	2	121	61	5	0	2	1	48	24	2	1	0	4
KELLY	1205 3RD ST	85	291	3	1	505	50816	8	727	65	68	21086	4	3	176	59	8	2	1	5	80	20	2	2	0	0
RESNICK	1206 3RD ST	86	292	3	1	181	51823	8	769	73	81	48054	4	4	203	51	7	1	2	1	73	18	2	0	1	0
SMOTRICK	1205 4TH ST	86	293	3	1	176	34510	6	670	50	23	15318	6	1	65	65	8	0	1	6	121	20	1	0	0	0
BORMANN	1206 4TH ST	87	294	3	1	314	37339	7	735	64	22	31873	3	2	138	69	3	1	1	2	66	22	2	0	4	0
FOGELMAN	1205 PARK AVE	87	295	3	1	117	23773	5	538	42	65	25328	2	1	99	99	6	0	1	2	32	16	3	0	3	8
MARCEAU	1206 PARK AVE	88	296	3	1	196	45207	7	654	62	14	29773	3	2	128	64	6	2	2	4	68	23	1	0	5	0
CEBOLLERO	1205 5TH ST	88	297	3	1	275	51953	10	1034	99	23	33745	6	2	120	60	7	2	2	1	132	22	1	0	6	0
JONES	1206 5TH ST	89	298	3	1	668	82355	12	1195	127	88	34378	8	2	120	60	5	2	1	10	178	22	1	0	6	11
MILES	1205 6TH ST	89	299	3	1	335	60394	12	1206	118	56	25780	7	2	120	60	4	0	2	9	102	15	1	0	3	0
VALDES	1206 6TH ST	90	300	3	1	355	38718	8	786	73	15	33796	4	3	177	59	11	2	1	2	86	22	1	0	0	0
WEEDEN	1205 7TH ST	90	301	3	1	412	58924	10	967	96	14	41160	6	2	120	60	3	1	2	4	139	23	1	2	0	3
OLIVER	1206 7TH ST	91	302	3	1	275	39109	7	649	65	55	36110	2	1	104	104	3	2	2	1	37	19	2	2	0	7
HARDY	1205 8TH ST	91	303	3	1	365	35446	7	712	62	39	24357	3	2	139	70	6	2	2	1	75	25	2	0	0	0
LORING	1206 8TH ST	92	304	3	1	389	50679	9	875	83	59	30090	5	3	195	65	7	2	1	1	88	18	1	0	2	1
FLETCHER	1205 9TH ST	92	305	3	1	241	22518	4	436	32	90	17617	4	2	120	60	4	1	1	1	87	22	3	0	4	0
MCIVER	1203 1ST ST	83	306	3	1	297	37465	6	562	53	17	31869	2	1	114	114	6	1	1	1	47	24	1	0	4	1
ROBINS	1204 1ST ST	84	307	3	1	542	59316	10	992	88	63	27342	5	5	228	46	7	2	1	5	100	20	2	0	6	1
EDMOND	1203 2ND ST	84	308	3	1	261	39773	7	705	58	68	31791	4	3	180	60	6	2	2	3	94	24	1	2	0	0
BORGES	1203 3RD ST	85	309	3	1	383	41668	9	907	91	22	25638	6	2	120	60	8	1	2	1	144	24	1	2	0	0
GARCIA	1204 3RD ST	86	310	3	1	314	40706	7	750	56	46	33156	3	2	141	71	9	2	2	4	71	24	1	0	0	0
HALTZMAN	1203 4TH ST	86	311	3	1	204	33659	7	658	71	21	30922	4	2	155	78	9	1	2	2	84	21	1	0	0	0
RAY	1204 4TH ST	87	312	3	1	434	47312	9	970	83	82	27274	4	2	158	79	8	1	2	1	88	22	1	0	0	0
FLYNN	1203 PARK AVE	87	313	3	1	570	61440	11	1079	92	59	38188	6	2	120	60	6	2	1	6	143	24	2	0	3	0
BABBITT	1204 PARK AVE	88	314	3	1	504	55476	10	969	95	103	36939	7	3	165	55	3	2	1	7	153	22	1	0	0	0
ABRONSKI	1203 5TH ST	88	315	3	1	515	52979	9	904	78	50	58703	5	3	197	66	5	2	2	1	116	23	2	0	0	0
CLARK	1204 5TH ST	89	316	3	1	505	54997	10	1009	102	12	28784	6	2	120	60	6	1	1	6	122	20	1	0	1	0
HALL	1203 6TH ST	89	317	3	1	360	46118	7	722	65	30	42701	4	3	175	58	9	1	2	3	96	24	2	0	0	0
OAKLEY	1204 6TH ST	90	318	3	1	512	66265	10	982	104	22	27684	6	2	120	60	5	0	1	4	135	23	1	2	0	0
ACCETTA	1203 7TH ST	90	319	3	1	387	43258	7	689	57	25	34235	3	3	169	56	8	1	1	1	57	19	1	2	0	3
CIVANTOS	1204 7TH ST	91	320	3	1	451	60677	9	910	81	36	31665	5	3	200	67	7	0	2	1	103	21	1	0	3	4
ERIKSON	1203 8TH ST	91	321	3	1	265	34154	6	647	54	21	142251	1	1	86	86	2	2	2	1	34	34	1	2	0	0
LITMAN	1204 8TH ST	92	322	3	1	225	38124	6	573	49	19	13452	6	2	120	60	6	0	2	9	126	21	2	0	0	0
ANGSTADT	1203 9TH ST	92	323	3	1	237	24941	4	388	31	55	15853	4	2	120	60	5	0	2	2	74	19	1	0	4	2
RAYMAN	1201 1ST ST	83	324	3	1	260	32015	7	687	53	91	46033	4	3	168	56	8	2	1	3	97	24	1	0	6	0
FLEMING	1202 1ST ST	84	325	3	1	267	39854	7	692	63	22	34853	4	3	167	56	8	1	2	5	89	22	1	0	0	0
MEYER	1201 2ND ST	84	326	3	1	210	43572	8	776	65	20	25287	5	3	192	64	7	2	1	3	98	20	1	0	0	0
KAFKA	1201 3RD ST	85	327	3	1	386	60478	11	1099	112	20	41507	6	4	220	55	4	2	2	3	110	18	2	0	1	0
ROLNICK	1202 3RD ST	86	328	3	1	275	41424	8	871	73	39	22232	4	4	205	51	9	1	2	3	90	23	1	0	1	4
STEVENSON	1201 4TH ST	86	329	3	1	397	46515	9	880	82	18	25900	6	4	221	55	2	1	2	1	121	20	1	0	0	0
PHELAN	1202 4TH ST	87	330	3	1	317	33731	7	694	60	65	21138	4	3	175	58	4	2	2	2	100	25	1	0	5	1
LARSON	1201 PARK AVE	87	331	3	1	296	48179	10	1062	98	21	33285	6	5	234	47	5	2	1	3	126	21	1	0	4	5
HARANG	1202 PARK AVE	88	332	3	1	209	29694	7	647	62	25	43656	4	2	151	76	8	1	2	0	79	20	2	0	1	0
MICHENER	1201 5TH ST	88	333	3	1	278	43591	8	770	72	13	42494	3	2	135	68	5	2	2	3	70	23	1	2	0	5
RUSSELL	1202 5TH ST	89	334	3	1	294	57954	12	1230	129	20	37439	3	3	165	55	4	1	1	7	148	21	2	0	1	0
MATEO	1201 6TH ST	89	335	3	1	537	58998	10	953	108	16	43767	5	3	199	66	5	2	1	3	102	20	2	1	0	0
HARMON	1202 6TH ST	90	336	3	1	238	36302	9	899	85	53	35425	5	4	211	53	8	2	2	4	102	20	2	0	0	0
BROWN	1201 7TH ST	90	337	3	1	806	74803	12	1133	109	118	29141	7	2	120	60	6	0	1	3	144	21	1	0	0	0
DORFMAN	1202 7TH ST	91	338	3	1	491	38901	9	852	93	39	33850	4	2	155	78	6	2	2	1	91	23	1	0	6	3
KRIMSKY	1201 8TH ST	91	339	3	1	598	33376	9	817	72	21	45259	3	2	135	68	5	1	1	2	77	26	1	0	0	0
FLEISCHER	1202 8TH ST	92	340	3	1	193	25400	4	422	33	15	18485	3	2	120	60	9	0	2	4	86	22	1	2	0	3
NORMAN	1201 9TH ST	92	341	3	1	334	52662	10	1020	92	68	21742	6	4	225	56	5	2	2	8	117	20	2	0	0	0
CRAMER	1105 1ST ST	73	342	3	1	511	64779	11	1179	115	63	43665	6	2	120	60	6	0	2	4	137	23	1	0	0	0
KONWISER	1106 1ST ST	74	343	3	1	483	70465	10	1063	96	94	28525	6	4	225	56	6	1	1	6	111	19	1	0	0	0
TRESCOTT	1105 2ND ST	74	344	3	1	286	43460	8	807	69	20	28264	4	2	144	72	5	2	1	1	78	20	1	2	0	0
DARWIN	1105 3RD ST	75	345	3	1	153	25884	6	620	49	14	33872	2	2	122	61	3	1	2	1	47	24	1	1	0	0
LANIER	1106 3RD ST	76	346	3	1	139	43320	9	862	82	39	37524	4	2	155	78	7	1	2	3	79	20	1	0	1	3
ANGONES	1105 4TH ST	76	347	3	1	194	31384	5	501	42	14	42778	1	1	65	65	7	2	2	1	30	30	2	0	1	0
CRUZ	1106 4TH ST	77	348	3	1	468	46224	9	869	85	77	54212	5	4	207	52	2	2	1	10	113	23	2	0	0	0
MITTLER	1105 PARK AVE	77	349	3	1	438	49312	10	1032	90	17	39124	6	4	224	56	8	1	1	8	153	26	3	0	4	0
SIEGEL	1106 PARK AVE	78	350	3	1	355	58520	10	951	99	89	23889	6	4	216	54	7	1	2	9	104	17	3	0	0	0
PEREZ	1105 5TH ST	78	351	3	1	296	43243	7	758	72	16	29330	3	2	148	74	5	2	1	4	55	18	1	0	1	0
COYLE	1106 5TH ST	79	352	3	1	414	63591	12	1164	120	91	45473	7	3	165	55	6	1	1	3	160	23	2	2	0	10
HOLMES	1105 6TH ST	79	353	3	1	640	55411	9	904	85	21	44163	5	4	209	52	6	1	1	0	111	22	1	0	5	0
DOLANSKY	1106 6TH ST	80	354	3	1	165	25936	5	544	46	20	16117	5	1	65	65	3	1	2	4	120	24	1	0	0	7
MANNING	1105 7TH ST	80	355	3	1	443	51830	8	784	69	27	23584	4	2	161	81	11	1	2	7	89	22	1	2	0	10
SIKES	1106 7TH ST	81	356	3	1	201	46074	9	888	78	18	25999	6	2	120	60	7	1	2	3	121	20	1	0	0	0
TRENZADO	1105 8TH ST	81	357	3	1	586	65599	10	1090	95	20	33578	6	4	214	54	12	2	1	1	134	22	1	2	0	0
GARCIA	1106 8TH ST	82	358	3	1	570	57602	9	875	80	98	21129	6	4	223	56	4	1	1	4	106	18	1	0	0	0
GRIMALDO	1105 9TH ST	82	359	3	1	249	31471	5	524	40	73	18890	4	1	65	65	2	0	2	1	78	20	1	2	0	0
ALDERSON	1103 1ST ST	73	360	3	1	470	46375	7	681	67	18	28224	3	2	140	70	4	2	2	2	82	27	2	0	0	0
GETTLEMAN	1104 1ST ST	74	361	3	1	593	52062	10	998	98	12	37448	6	2	120	60	4	2	1	3	138	23	1	2	0	0
OLIN	1103 2ND ST	74	362	3	1	328	38569	7	666	55	24	42293	3	2	162	81	7	1	1	4	59	20	2	0	0	0
WILLIAMS	1103 3RD ST	75	363	3	1	426	58335	10	1011	99	54	34926	5	3	186	62	6	1	1	4	105	21	1	0	0	0
STEWARD	1104 3RD ST	76	364	3	1	190	29513	6	647	54	59	25067	3	2	143	72	6	2	2	4	69	23	2	0	4	11
MANNION	1103 4TH ST	76	365	3	1	325	55102	10	949	96	13	36839	5	3	198	66	5	0	2	5	90	18	1	0	1	0
PEAKE	1104 4TH ST	77	366	3	1	55	19918	2	179	13	24	16945	2	2	120	60	6	0	2	0	60	30	3	0	0	0
BOLANOS	1103 PARK AVE	77	367	3	1	663	60424	9	906	82	57	32111	5	3	196	65	7	2	1	1	108	22	3	0	3	1
LOWE	1104 PARK AVE	78	368	3	1	297	51382	9	864	78	87	37566	4	2	159	80	6	1	2	2	67	17	2	0	1	0
SHELTON	1103 5TH ST	78	369	3	1	395	42780	8	816	78	23	42447	4	3	177	59	7	1	2	3	78	20	2	0	0	6
CARABALLO	1104 5TH ST	79	370	3	1	177	33211	7	676	61	23	35385	3	2	126	63	3	2	2	2	78	26	1	0	6	0
PARKE	1103 6TH ST	79	371	3	1	295	40735	7	739	56	23	33837	2	2	137	69	6	0	2	2	59	30	2	1	0	5
CLOUTIER	1104 6TH ST	80	372	3	1	241	30965	6	585	59	83	41366	2	2	139	70	5	1	2	1	53	27	2	0	1	3
HOLMAN	1103 7TH ST	80	373	3	1	126	14852	3	293	22	97	16222	3	1	65	65	7	0	1	4	66	22	1	0	4	0
PHINNEY	1104 7TH ST	81	374	3	1	641	59610	9	906	87	31	25648	5	3	198	66	4	1	2	3	109	22	2	0	2	0
SUAREZ	1103 8TH ST	81	375	3	1	419	47682	10	1020	100	78	31998	5	3	184	61	5	0	1	4	87	17	1	0	0	0
GOLDEN	1104 8TH ST	82	376	3	1	561	49679	8	779	75	34	26955	4	2	164	82	7	0	1	0	74	19	2	0	0	2
CLEMENTE	1103 9TH ST	82	377	3	1	386	48626	8	752	67	13	21821	4	3	178	59	7	2	2	1	86	22	2	0	5	0
KUSNICK	1101 1ST ST	73	378	3	1	183	43436	8	848	69	17	34245	3	2	164	82	3	1	1	1	63	21	1	0	0	7
RASHKIND	1102 1ST ST	74	379	3	1	475	46474	7	698	58	21	31622	3	2	159	80	5	2	2	1	57	19	1	0	0	0
TINDELL	1101 2ND ST	74	380	3	1	335	36486	6	600	54	33	36714	3	2	163	82	4	2	2	1	64	21	2	2	0	3
DASEN	1101 3RD ST	75	381	3	1	251	42338	7	719	60	16	46485	4	2	149	75	7	1	1	1	81	20	2	2	0	0
MCINTOSH	1102 3RD ST	76	382	3	1	366	38743	7	730	65	97	33414	4	3	174	58	7	2	2	7	89	22	2	0	0	3
JACOBS	1101 4TH ST	76	383	3	1	445	52551	9	895	85	46	28119	5	4	207	52	9	1	1	1	119	24	2	0	4	5
CARROLL	1102 4TH ST	77	384	3	1	524	80075	12	1198	111	14	53952	7	3	165	55	4	2	2	7	144	21	2	0	1	0
NARDIN	1101 PARK AVE	77	385	3	1	374	49217	9	887	86	47	33217	5	4	216	54	9	1	1	1	112	22	2	0	0	0
SENK	1102 PARK AVE	78	386	3	1	276	39558	8	792	72	23	32667	4	2	160	80	6	0	1	3	88	22	1	2	0	1
HAMBURGER	1101 5TH ST	78	387	3	1	368	43921	8	733	66	13	32136	5	3	187	62	9	1	2	4	103	21	2	0	0	0

VARIABLES

01	02	03	04	05	06	07	08	09	10	11	12	13	14	15	16	17	18	19	20	21	22	23	24	25	26	27
SHAFFER	1102 5TH ST	79	388	3	1	594	61733	9	827	76	20	28381	6	5	235	47	9	2	2	9	136	23	1	2	0	5
KATZMAN	1101 6TH ST	79	389	3	1	340	37227	7	636	54	21	31298	4	3	169	56	5	2	2	3	88	22	2	1	0	0
FAULKNER	1102 6TH ST	80	390	3	1	472	58108	11	1076	111	20	28284	6	4	220	55	5	2	2	8	125	21	1	0	0	0
CAPOTRIO	1101 7TH ST	80	391	3	1	323	37374	8	774	73	23	26049	4	3	174	58	8	2	2	4	91	23	1	0	0	0
DALBERT	1102 7TH ST	81	392	3	1	233	45444	9	796	86	85	32354	4	3	172	57	9	1	2	2	88	22	1	2	0	6
KARRASCH	1101 8TH ST	81	393	3	1	154	16009	4	436	32	11	15598	4	1	65	65	7	0	2	1	84	21	1	0	1	10
GREEN	1102 8TH ST	82	394	3	1	144	27898	5	489	44	45	29338	2	1	105	105	7	2	2	1	43	22	1	0	0	5
OLIVER	1101 9TH ST	82	395	3	1	632	70121	10	1035	97	15	27953	7	2	120	60	5	2	2	5	130	19	1	1	0	2
DONAHUE	1004 1ST ST	70	396	3	0	298	0	5	414	40	24	52318	1	1	65	65	4	2	2	1	25	25	3	0	0	0
PARKER	1004 1ST ST	70	397	3	0	681	0	10	869	83	17	49763	4	2	150	75	6	1	2	2	84	21	1	2	0	0
MARTIN	1004 1ST ST	70	398	3	0	467	0	6	623	52	64	31103	6	4	217	54	4	2	1	4	146	24	2	2	0	0
BOULIS	1004 1ST ST	70	399	3	0	443	0	6	570	48	16	37679	5	4	209	52	6	1	1	4	86	17	1	0	0	0
MATUSEK	1004 1ST ST	70	400	3	0	474	0	6	553	46	11	27444	6	4	215	54	8	0	1	3	118	20	1	0	0	1
ROWLEY	1004 1ST ST	70	401	3	0	637	0	9	859	78	31	37959	4	3	169	56	3	1	2	4	88	22	2	2	0	10
WEEKES	1004 1ST ST	70	402	3	0	434	0	6	593	52	18	32598	5	4	202	51	7	2	1	3	104	21	1	2	0	0
GOODING	1004 1ST ST	70	403	3	0	469	0	6	462	44	66	31758	2	2	130	65	7	1	2	1	56	28	1	0	1	0
AUSTIN	1004 1ST ST	70	404	3	0	582	0	6	571	41	57	33909	6	4	219	55	4	0	2	1	126	21	2	0	0	2
NOWACK	1004 1ST ST	70	405	3	0	591	0	8	678	63	88	38353	3	2	132	66	6	1	1	4	58	19	1	2	0	3
FLORA	1004 1ST ST	70	406	3	0	380	0	5	490	38	27	36635	2	1	116	116	6	2	1	1	44	22	1	0	0	0
VETTER	1003 2ND ST	70	407	3	0	356	0	5	497	41	14	30951	4	2	161	81	8	2	2	2	84	21	3	0	5	9
LANGSAM	1003 2ND ST	70	408	3	0	394	0	6	588	48	14	33300	7	3	165	55	4	2	2	9	152	22	2	2	0	0
DAVIS	1003 2ND ST	70	409	3	0	334	0	5	434	31	11	35005	4	2	149	75	7	0	2	3	86	22	1	0	0	7
ADAMS	1003 2ND ST	70	410	3	0	391	0	6	546	51	24	27779	4	3	182	61	7	1	1	4	92	23	1	0	0	0
GREENBURG	1003 2ND ST	70	411	3	0	525	0	6	555	46	59	27132	4	2	158	79	4	1	1	6	84	21	2	0	5	0
MASKIN	1003 2ND ST	70	412	3	0	546	0	7	624	54	74	27804	5	3	189	63	5	0	2	4	108	22	1	0	0	0
DOBBS	1003 2ND ST	70	413	3	0	406	0	5	431	38	26	34889	4	2	153	77	8	1	1	2	76	19	2	2	0	3
OVERTON	1003 2ND ST	70	414	3	0	668	0	9	867	75	48	30457	2	2	143	72	3	2	1	3	61	31	2	0	5	3
VEINGRAD	1003 2ND ST	70	415	3	0	284	0	5	413	37	22	26972	1	1	65	65	6	1	2	1	24	24	3	0	0	3
ROSENBUSH	1003 2ND ST	70	416	3	0	253	0	5	457	37	17	41549	4	3	174	58	7	2	2	4	85	21	3	2	0	0
SALICHS	1004 2ND ST	71	417	3	0	299	0	6	484	42	75	21512	6	2	120	60	6	1	2	3	163	27	1	0	0	1
MENENDEZ	1004 2ND ST	71	418	3	0	659	0	10	856	84	41	40912	2	2	136	68	3	1	1	2	50	25	2	0	3	2
STRICKLER	1004 2ND ST	71	419	3	0	296	0	5	455	37	18	38626	3	3	168	56	3	0	1	3	58	19	2	0	1	3
HALL	1004 2ND ST	71	420	3	0	408	0	5	491	37	16	31364	4	3	172	57	4	2	2	3	108	27	1	0	0	0
DUBOIS	1004 2ND ST	71	421	3	0	483	0	6	576	46	54	38785	5	3	193	64	5	0	2	4	101	20	2	0	0	0
GOLDFARB	1004 2ND ST	71	422	3	0	276	0	7	680	58	19	28757	3	2	147	74	10	1	2	3	67	22	1	1	0	6
DEITCH	1004 2ND ST	71	423	3	0	356	0	6	575	48	95	36459	3	2	142	71	7	2	2	1	74	25	2	0	1	1
KLEIN	1004 2ND ST	71	424	3	0	672	0	8	671	68	34	21362	4	3	199	66	7	2	1	4	79	20	1	0	0	0
MATEO	1004 2ND ST	71	425	3	0	659	0	9	811	68	16	28988	3	2	148	74	6	0	2	2	58	19	1	2	0	1
SQUILLACE	1004 2ND ST	71	426	3	0	352	0	6	494	46	87	26316	6	4	216	54	4	1	2	4	104	17	1	1	0	0
WEISSER	1004 2ND ST	71	427	3	0	625	0	8	689	66	16	30535	5	3	188	63	8	2	2	5	111	22	2	0	0	0
TOPPING	1003 3RD ST	71	428	3	0	344	0	5	465	37	100	44682	4	3	170	57	5	1	2	3	94	24	1	0	0	2
KAVANAUGH	1003 3RD ST	71	429	3	0	436	0	6	554	48	18	38844	4	3	184	61	7	1	2	0	69	17	2	2	0	0
MAINIERO	1003 3RD ST	71	430	3	0	345	0	5	313	34	28	31781	2	2	129	65	6	1	1	2	46	23	2	2	0	1
HARARI	1003 3RD ST	71	431	3	0	440	0	6	542	42	14	32195	4	3	192	64	6	2	1	4	81	20	2	2	0	1
KOHOUT	1003 3RD ST	71	432	3	0	571	0	8	718	69	93	32862	4	2	157	79	3	1	2	2	86	22	2	0	0	0
COWAN	1003 3RD ST	71	433	3	0	747	0	10	835	85	67	33033	3	2	124	62	5	2	2	3	66	22	1	0	5	2
ARONSON	1003 3RD ST	71	434	3	0	399	0	5	405	33	30	33632	5	3	194	65	7	1	2	5	99	20	3	0	0	6
GEISEL	1003 3RD ST	71	435	3	0	608	0	9	841	72	22	35841	4	2	155	78	4	0	1	3	84	21	2	0	6	3
ESPOLITA	1003 3RD ST	71	436	3	0	270	0	5	477	32	20	31597	1	1	65	65	3	2	1	1	25	25	1	0	4	0
SAMUELS	1003 3RD ST	71	437	3	0	499	0	7	566	53	46	15515	3	2	120	60	3	0	1	3	58	19	1	0	0	0
JOHNSEN	1004 3RD ST	72	438	3	0	451	0	6	584	48	25	21423	7	2	120	60	2	1	2	3	154	22	3	2	0	2
FRIEDMAN	1004 3RD ST	72	439	3	0	615	0	8	721	62	15	25971	4	2	145	73	5	2	2	2	72	18	3	2	0	5
JOHNSEN	1004 3RD ST	72	440	3	0	603	0	8	736	58	13	21729	2	2	141	71	4	1	2	1	63	32	1	0	4	0
HALLISSEY	1004 3RD ST	72	441	3	0	472	0	7	659	62	14	22881	2	2	138	69	3	0	2	3	57	29	1	1	0	1
PHILBRICK	1004 3RD ST	72	442	3	0	574	0	6	524	51	68	18941	7	1	65	65	4	0	2	5	151	22	1	0	0	0
MUTKA	1004 3RD ST	72	443	3	0	479	0	6	582	50	24	20404	5	4	203	51	5	1	2	4	113	23	2	0	0	6
VARGA	1004 3RD ST	72	444	3	0	506	0	6	514	45	23	29151	5	4	211	53	5	2	2	3	103	21	2	0	6	2
SLACK	1004 3RD ST	72	445	3	0	588	0	9	703	71	13	30474	4	2	148	74	4	1	2	2	68	17	1	0	0	2
TAYLOR	1004 3RD ST	72	446	3	0	452	0	6	562	47	23	30436	8	2	120	60	5	2	1	4	166	21	2	2	0	0
MIDONECK	1004 3RD ST	72	447	3	0	273	0	5	453	39	57	44565	3	2	148	74	8	1	1	4	57	19	1	2	0	4
GARDNER	1003 4TH ST	72	448	3	0	396	0	6	514	47	14	31574	5	4	209	52	5	2	2	1	117	23	1	1	0	0
PETERSON	1003 4TH ST	72	449	3	0	562	0	6	545	50	16	29863	6	2	120	60	5	1	2	6	130	22	1	0	0	0
IBANEZ	1003 4TH ST	72	450	3	0	406	0	8	756	66	30	32865	3	2	150	75	6	1	2	3	48	16	1	0	4	3
KASALTA	1003 4TH ST	72	451	3	0	487	0	6	528	48	12	24255	3	3	172	57	6	1	2	2	62	21	1	2	0	1
DODGE	1003 4TH ST	72	452	3	0	572	0	8	681	63	13	23786	4	2	156	78	4	0	2	3	86	22	1	0	1	0
MILNER	1003 4TH ST	72	453	3	0	534	0	9	785	75	23	26582	3	2	131	66	7	2	2	1	73	24	1	2	0	0
HARWELL	1003 4TH ST	72	454	3	0	389	0	6	587	48	47	37265	7	3	165	55	5	0	1	12	172	25	1	2	0	0
SACHER	1003 4TH ST	72	455	3	0	354	0	6	573	52	65	42136	6	2	120	60	6	2	2	1	120	20	1	2	0	4
LAMBERT	1003 4TH ST	72	456	3	0	333	0	5	426	41	22	34392	3	2	139	70	7	1	2	1	72	24	1	0	5	3
WRANGLER	1003 4TH ST	72	457	3	0	437	0	7	608	56	17	26552	5	3	189	63	4	1	1	3	109	22	2	0	1	0
FERRO	1003 4TH ST	72	458	3	0	448	0	6	534	47	43	34061	5	3	196	65	8	2	1	4	100	20	1	0	0	0
CREIG	1002 1ST ST	70	459	3	0	424	0	6	474	53	90	42806	6	2	120	60	6	0	2	3	124	21	2	0	0	2
BERG	1002 1ST ST	70	460	3	0	401	0	6	534	51	11	31137	5	4	203	51	2	1	1	1	108	22	2	2	0	1
KAHL	1002 1ST ST	70	461	3	0	387	0	6	482	48	23	49718	5	3	197	66	9	2	1	3	117	23	1	0	0	0
MANNE	1002 1ST ST	70	462	3	0	612	0	9	854	76	26	42231	2	2	143	72	7	2	2	2	56	28	1	0	0	0
WARREN	1002 1ST ST	70	463	3	0	479	0	7	619	58	12	56220	7	3	165	55	8	1	1	10	95	14	1	0	0	0
ZELIGMAN	1002 1ST ST	70	464	3	0	591	0	9	838	78	15	39873	3	2	140	70	7	1	1	1	85	28	1	2	0	0
KOX	1002 1ST ST	70	465	3	0	348	0	6	609	49	41	44327	4	2	151	76	3	2	2	1	77	19	2	2	0	0
ALLOWAY	1002 1ST ST	70	466	3	0	340	0	5	393	40	94	49898	3	3	179	60	5	0	1	1	64	21	1	1	0	6
PELTIER	1002 1ST ST	70	467	3	0	563	0	8	783	63	20	32548	3	3	167	56	10	2	1	4	69	23	2	2	0	4
SCHERER	1002 1ST ST	70	468	3	0	441	0	6	576	45	21	47583	2	2	134	67	6	1	2	1	53	27	2	0	6	3
VICIANA	1002 1ST ST	70	469	3	0	370	0	6	556	46	16	36907	5	3	197	66	6	1	2	4	99	20	1	0	0	0
TARREN	1001 2ND ST	70	470	3	0	807	0	10	900	87	36	45300	4	3	182	61	6	1	1	3	96	24	2	0	0	0
RICCI	1001 2ND ST	70	471	3	0	314	0	6	553	43	27	33427	6	5	234	47	5	2	2	4	117	20	2	2	0	5
OLIVER	1001 2ND ST	70	472	3	0	288	0	5	418	40	21	31856	1	1	86	86	5	1	2	0	31	31	2	0	1	0
LUNDY	1001 2ND ST	70	473	3	0	574	0	9	825	72	54	31398	3	2	131	66	5	0	2	0	74	25	2	2	0	2
BAY	1001 2ND ST	70	474	3	0	343	0	6	521	51	76	30138	7	3	165	55	4	1	1	3	144	21	3	0	0	0
ARGAIN	1001 2ND ST	70	475	3	0	559	0	6	524	51	22	42768	7	2	120	60	10	2	2	3	148	21	2	2	0	4
WHITE	1001 2ND ST	70	476	3	0	363	0	5	479	35	16	31367	2	1	119	119	2	0	2	3	47	24	1	0	1	2
GOELD	1001 2ND ST	70	477	3	0	379	0	6	566	48	20	29803	6	4	222	56	11	1	2	3	111	19	1	0	0	0
IGLESIAS	1001 2ND ST	70	478	3	0	537	0	6	581	47	30	27430	6	4	219	55	3	2	2	6	96	16	2	0	6	0
LANDMAN	1001 2ND ST	70	479	3	0	540	0	7	597	58	17	21848	5	3	193	64	6	1	1	5	106	21	3	0	1	2
ARMAS	1002 2ND ST	71	480	3	0	404	0	5	303	35	17	12976	5	1	65	65	3	1	2	5	101	20	2	2	0	5
COMAS	1002 2ND ST	71	481	3	0	312	0	7	698	66	39	28259	3	2	127	64	6	2	1	2	74	25	2	0	1	1
FOGELBERG	1002 2ND ST	71	482	3	0	654	0	9	822	73	25	25538	5	3	200	67	5	2	1	8	87	17	2	0	1	2
HOLMER	1002 2ND ST	71	483	3	0	406	0	8	671	64	16	25687	3	2	139	70	8	2	1	2	77	26	1	0	3	1
MCPHEE	1002 2ND ST	71	484	3	0	338	0	5	427	41	82	29990	3	2	132	66	4	2	2	1	61	20	3	0	1	0
ALLEN	1002 2ND ST	71	485	3	0	504	0	8	799	64	13	30802	4	3	179	60	4	1	1	2	103	26	1	0	0	0
ROTHKOPF	1002 2ND ST	71	486	3	0	683	0	8	693	63	23	36237	4	4	203	51	3	2	1	4	102	26	1	0	0	0
WALKER	1002 2ND ST	71	487	3	0	447	0	7	671	58	39	23527	4	2	162	81	7	2	2	4	81	20	2	2	0	9

Appendixes
324

VARIABLES

01	02	03	04	05	06	07	08	09	10	11	12	13	14	15	16	17	18	19	20	21	22	23	24	25	26	27
KARY	1002 2ND ST	71	488	3	0	427	0	6	489	50	14	31762	5	4	207	52	6	1	2	3	120	24	1	0	3	0
ARTEAGA	1002 2ND ST	71	489	3	0	503	0	8	693	76	23	25193	3	2	142	71	6	1	2	1	63	21	3	0	6	8
ALVAREZ	1002 2ND ST	71	490	3	0	433	0	8	757	66	16	27809	4	2	163	82	6	1	2	4	84	21	1	0	0	0
HALUSKA	1001 3RD ST	71	491	3	0	338	0	3	310	22	19	26430	2	2	122	61	5	1	1	3	51	26	1	0	1	0
OWENS	1001 3RD ST	71	492	3	0	647	0	8	811	76	38	28104	2	2	130	65	7	2	1	2	45	23	1	0	4	3
KEUSCO	1001 3RD ST	71	493	3	0	746	0	10	861	87	17	45631	2	2	141	71	7	1	2	1	58	29	1	0	0	0
MITES	1001 3RD ST	71	494	3	0	463	0	6	502	47	17	33673	6	4	224	56	3	1	1	3	140	23	2	0	1	8
HALPERN	1001 3RD ST	71	495	3	0	450	0	6	572	47	26	34054	5	4	201	50	7	2	2	4	118	24	1	0	0	0
SKIDELL	1001 3RD ST	71	496	3	0	429	0	7	679	59	18	38874	3	2	160	80	6	2	2	3	50	17	1	2	0	0
VIAROS	1001 3RD ST	71	497	3	0	448	0	6	538	42	15	34964	4	2	153	77	3	0	2	1	82	21	1	2	0	3
LAWSON	1001 3RD ST	71	498	3	0	424	0	6	546	44	19	41587	5	5	231	46	4	0	1	3	117	23	1	0	0	2
PURNELL	1001 3RD ST	71	499	3	0	407	0	6	474	46	97	31801	6	2	120	60	8	0	1	1	143	24	1	0	0	0
HECK	1001 3RD ST	71	500	3	0	428	0	6	546	51	20	40799	6	4	216	54	7	1	1	6	127	21	2	2	0	5
KOYNER	1002 3RD ST	72	501	3	0	566	0	6	546	51	88	26669	6	4	216	54	9	2	2	4	125	21	2	0	3	0
BRODY	1002 3RD ST	72	502	3	0	610	0	8	689	63	20	23004	4	3	197	66	7	1	2	2	84	21	1	0	1	0
PABLO	1002 3RD ST	72	503	3	0	350	0	6	591	46	17	24262	5	4	208	52	6	1	1	1	103	21	2	0	1	1
WRIGHT	1002 3RD ST	72	504	3	0	475	0	8	729	62	17	22306	3	2	138	69	3	1	1	1	70	23	1	0	1	0
ANDREWS	1002 3RD ST	72	505	3	0	315	0	5	385	38	21	25025	2	1	105	105	3	0	1	1	37	19	2	0	1	0
LYNN	1002 3RD ST	72	506	3	0	241	0	5	416	40	28	26628	2	1	112	112	5	2	2	1	49	25	1	0	6	0
HOBACK	1002 3RD ST	72	507	3	0	400	0	6	495	48	68	19607	5	2	120	60	7	0	2	6	106	21	2	0	0	0
SIMON	1002 3RD ST	72	508	3	0	516	0	6	501	43	21	43502	5	4	209	52	5	2	2	3	109	22	2	0	0	0
ADAMS	1002 3RD ST	72	509	3	0	346	0	5	476	38	75	27390	4	2	147	74	7	2	1	4	66	17	1	1	0	0
EPHRAIM	1002 3RD ST	72	510	3	0	459	0	6	552	42	21	32901	5	4	205	51	6	2	2	5	104	21	1	0	0	0
MANDERSON	1001 4TH ST	72	511	3	0	473	0	6	582	47	16	34899	4	2	157	79	8	1	1	4	90	23	2	0	1	5
HAMILTON	1001 4TH ST	72	512	3	0	614	0	10	892	80	22	28627	3	2	139	70	4	2	1	2	83	28	1	0	1	0
HUMPHREYS	1001 4TH ST	72	513	3	0	435	0	6	574	45	22	41266	3	2	160	80	7	2	2	3	64	21	1	0	1	0
DAUGHERTY	1001 4TH ST	72	514	3	0	236	0	4	370	26	16	44547	2	1	98	98	3	2	1	1	39	20	1	0	0	3
NELSON	1001 4TH ST	72	515	3	0	578	0	6	537	44	103	33009	2	2	132	66	6	0	2	1	61	31	1	0	0	0
TOLEDO	1001 4TH ST	72	516	3	0	404	0	7	617	59	13	25410	4	2	155	78	5	2	2	4	86	22	2	1	0	1
SCANLAN	1001 4TH ST	72	517	3	0	343	0	5	465	36	84	37596	3	2	148	74	6	1	1	3	54	18	2	0	6	0
GORDON	1001 4TH ST	72	518	3	0	172	0	4	363	27	19	32458	2	1	113	113	6	2	2	1	49	25	1	2	0	10
WHITFIELD	1001 4TH ST	72	519	3	0	327	0	5	464	41	14	44058	2	1	116	116	5	1	1	1	47	24	2	2	0	1
CAMERON	1001 4TH ST	72	520	3	0	572	0	8	748	67	19	39430	3	2	135	68	3	2	2	1	66	22	1	2	0	5
RULE	1001 4TH ST	72	521	3	0	609	0	9	822	80	21	27082	4	2	152	76	4	2	2	6	96	24	1	0	0	0
MAINSTER	907 1ST ST	65	522	3	1	353	49317	6	638	38	16	29644	3	2	145	73	7	0	2	0	90	30	3	0	0	0
HASTING	908 1ST ST	66	523	3	1	359	42834	7	729	60	25	51217	4	3	186	62	7	2	2	2	72	18	2	0	5	7
SCATTI	907 2ND ST	66	524	3	1	360	29113	9	885	82	15	28469	5	3	191	64	8	0	2	3	103	21	1	0	1	3
MIRAGLIA	908 2ND ST	67	525	3	1	226	26286	5	543	45	14	23632	1	1	65	65	4	2	2	1	28	28	1	0	0	0
ESPLIN	908 3RD ST	68	526	3	1	677	71141	11	1105	109	21	24351	7	2	120	60	9	2	2	9	153	22	1	2	0	0
KOI	907 4TH ST	68	527	3	1	222	33937	7	726	66	76	44147	3	2	132	66	7	2	2	2	63	21	2	0	4	3
HOWARD	908 4TH ST	69	528	3	1	315	48792	8	779	71	34	41557	5	3	199	66	9	1	2	5	102	20	2	2	0	2
GIROGOSIAN	907 PARK AVE	69	529	3	1	450	45206	8	726	71	25	20543	4	2	164	82	10	1	2	1	100	25	1	0	1	5
HOWELL	905 1ST ST	65	530	3	1	452	48152	7	720	62	99	45583	3	2	147	74	4	2	1	4	54	18	1	0	0	1
BROWN	906 1ST ST	66	531	3	1	172	37227	7	655	61	18	49726	4	3	176	59	3	1	2	2	97	24	2	0	0	0
WEAVER	905 2ND ST	66	532	3	1	332	35538	8	814	76	20	44212	5	3	198	66	8	1	2	8	127	25	1	0	5	1
SCOTT	906 2ND ST	67	533	3	1	405	57459	10	974	88	14	49325	5	4	204	51	8	0	2	4	109	22	2	0	0	0
MONROE	906 3RD ST	68	534	3	1	328	36898	7	675	56	106	25329	3	2	153	77	6	1	2	1	63	21	1	0	0	0
FLASHMAN	905 4TH ST	68	535	3	1	346	48219	8	833	73	12	37221	4	3	184	61	7	2	1	3	82	21	1	0	1	3
CAGIGAS	906 4TH ST	69	536	3	1	254	41578	7	676	59	55	35394	3	2	120	60	6	1	2	4	70	23	2	1	0	9
MCGILL	905 PARK AVE	69	537	3	1	222	44747	8	850	75	18	34667	4	3	175	58	8	1	1	5	95	24	1	0	0	0
WILLIAMS	903 1ST ST	65	538	3	1	589	71624	12	1160	125	15	42033	7	3	165	55	5	2	2	5	170	24	2	2	0	0
RODRIGUEZ	904 1ST ST	66	539	3	1	279	38791	7	737	55	17	49245	3	2	142	71	6	0	2	1	77	26	2	2	0	5
MORRIS	903 2ND ST	66	540	3	1	242	45198	8	825	72	12	36056	4	3	179	60	2	1	2	4	93	23	1	0	3	5
GENET	904 2ND ST	67	541	3	1	438	47259	8	701	80	16	40145	4	3	166	55	3	1	1	4	93	23	1	2	0	1
LOFF	904 3RD ST	68	542	3	1	390	49015	9	814	79	25	40026	5	5	230	46	8	2	2	8	132	26	1	0	0	0
ARONSON	903 4TH ST	68	543	3	1	428	54621	10	996	98	19	36756	6	4	224	56	4	1	2	4	118	20	2	2	0	1
VERA	904 4TH ST	69	544	3	1	242	34326	7	659	58	11	40988	3	2	137	69	6	1	2	4	64	21	3	0	0	3
WOODROW	903 PARK AVE	69	545	3	1	486	55412	8	854	67	37	30647	5	3	198	66	6	2	2	6	109	22	2	0	6	11
KIMBLER	901 1ST ST	65	546	3	1	453	52577	9	889	82	34	43090	5	4	210	53	6	1	2	3	116	23	2	0	0	0
ENGEL	902 1ST ST	66	547	3	1	421	56719	6	515	47	17	37621	6	4	218	55	6	2	2	1	134	22	2	0	4	2
LANE	901 2ND ST	66	548	3	1	458	66894	10	901	83	15	39624	5	3	193	64	4	0	2	1	92	18	1	2	0	6
BARNES	902 2ND ST	67	549	3	1	480	44243	10	1001	97	21	44364	5	3	196	65	6	1	2	5	104	21	1	1	0	4
RINKER	902 3RD ST	68	550	3	1	327	47819	8	851	75	16	45409	4	3	196	65	6	2	2	2	74	19	1	0	5	3
PETERSON	901 4TH ST	68	551	3	1	439	44707	7	709	61	23	44988	2	2	135	68	3	2	2	1	50	25	1	0	4	0
KERR	902 4TH ST	69	552	3	1	557	49224	8	793	70	21	46878	5	4	217	54	12	2	1	6	89	18	1	0	1	1
EAST	901 PARK AVE	69	553	3	1	151	20833	4	463	34	64	14275	5	2	120	60	5	0	1	1	103	21	1	0	3	0
LEVENBERG	807 1ST ST	61	554	3	1	461	45211	8	812	72	22	44108	4	2	152	76	5	2	2	3	61	15	1	0	5	0
PITMAN	808 1ST ST	62	555	3	1	347	43867	8	809	74	97	37546	3	2	146	73	6	2	2	1	60	20	1	2	0	0
MILLER	807 2ND ST	62	556	3	1	357	44160	8	768	71	37	43475	4	3	173	58	6	1	1	2	83	21	1	0	0	1
HOUGH	808 2ND ST	63	557	3	1	170	40685	9	852	81	18	43040	6	4	216	54	4	2	2	9	105	18	1	0	0	0
IMBER	808 3RD ST	64	558	3	1	396	38860	8	792	81	18	57597	3	3	176	59	5	1	1	3	61	20	1	0	1	0
GARCIA	807 4TH ST	64	559	3	1	375	39142	7	642	66	13	39167	3	2	150	75	7	1	1	4	57	19	1	0	0	0
DEJAMES	805 1ST ST	61	560	3	1	233	42074	8	796	80	23	46247	5	3	190	63	5	2	1	4	100	20	2	2	0	5
LAMPERT	806 1ST ST	62	561	3	1	514	51378	9	891	90	50	43705	4	3	169	56	5	2	2	2	65	16	2	0	0	3
SALIM	805 2ND ST	62	562	3	1	564	50267	10	986	95	16	43309	5	4	219	55	5	2	1	1	93	19	1	0	0	0
PURDY	806 2ND ST	63	563	3	1	507	59354	11	1137	103	138	21226	7	2	120	60	6	1	2	5	144	21	2	0	5	0
LAMODA	806 3RD ST	64	564	3	1	270	40776	9	952	85	11	50821	6	3	165	55	12	2	1	8	135	23	1	0	5	0
LOSNER	805 4TH ST	64	565	3	1	397	46147	7	708	60	66	43199	3	2	130	65	6	0	1	2	55	18	2	0	6	0
FRANTZ	803 1ST ST	61	566	3	1	344	39208	6	628	53	14	42614	2	2	142	71	12	2	1	4	59	30	1	2	0	2
COUDRIET	804 1ST ST	62	567	3	1	450	51210	6	548	49	85	43078	6	4	221	55	7	2	2	4	120	20	2	0	0	4
DOBBS	803 2ND ST	62	568	3	1	162	24862	5	552	42	28	51522	2	1	118	118	10	2	2	1	40	20	1	0	3	3
OCONNOR	804 2ND ST	63	569	3	1	331	51051	8	829	77	71	39785	5	4	205	51	2	1	1	3	103	21	2	0	5	0
NATT	804 3RD ST	64	570	3	1	504	46973	10	965	100	95	44546	6	2	120	60	4	2	1	3	125	21	2	0	0	2
LAZARUS	803 4TH ST	64	571	3	1	208	33247	6	633	51	22	30268	2	2	128	64	5	1	2	3	51	26	2	0	5	0
SALSBURY	801 1ST ST	61	572	3	1	384	47297	8	780	71	19	47162	4	2	164	82	4	0	2	3	99	25	1	0	6	0
REEVES	802 1ST ST	62	573	3	1	350	35536	7	729	66	15	49237	3	2	145	73	10	2	2	0	60	20	1	2	0	5
MENENDEZ	801 2ND ST	62	574	3	1	214	39234	7	625	62	22	54133	3	3	166	55	6	2	2	1	74	25	2	0	3	0
JACKSON	802 2ND ST	63	575	3	1	493	49758	10	1037	86	20	48159	5	4	217	54	7	2	2	4	107	21	1	0	4	2
POMER	802 3RD ST	64	576	3	1	696	65376	10	934	101	17	31458	6	2	120	60	5	1	1	4	141	24	2	0	0	0
LONGACRE	801 4TH ST	64	577	3	1	418	40771	7	707	62	15	48544	3	2	143	72	6	1	1	2	74	25	1	0	1	5
HAIMES	707 1ST ST	58	578	3	1	489	49073	10	1058	92	64	73548	5	3	187	62	5	2	1	5	87	17	1	2	0	11
PHILLIPS	708 1ST ST	59	579	3	1	330	36459	8	823	71	16	45016	4	3	182	61	3	2	2	1	112	28	2	0	0	0
JACOBS	707 2ND ST	59	580	3	1	562	68364	10	997	101	25	34903	6	4	222	56	6	2	2	6	112	19	1	1	0	1
KAY	708 2ND ST	60	581	3	1	270	50188	9	820	80	40	41974	5	3	189	63	6	1	2	1	92	18	1	2	0	0
CRUZ	705 1ST ST	58	582	3	1	452	64120	6	505	50	18	33408	4	3	168	56	5	1	2	2	65	16	2	0	2	11
EPSTEIN	706 1ST ST	59	583	3	1	474	48286	8	825	69	16	41420	4	3	172	57	4	1	1	3	92	23	2	1	0	7
MALLEY	705 2ND ST	59	584	3	1	387	46743	8	797	72	23	60473	4	3	183	61	6	2	2	2	103	26	1	0	0	0
PRICE	706 2ND ST	60	585	3	1	304	48560	8	825	75	13	50514	3	2	136	68	5	2	1	2	76	25	1	1	0	0
GARCIA	703 1ST ST	58	586	3	1	197	32592	8	819	82	91	23088	5	4	203	51	5	2	2	4	110	22	1	2	0	0
SHUMP	704 1ST ST	59	587	3	1	162	47148	9	814	84	69	45456	5	3	198	66	7	2	1	3	113	23	1	0	0	0

VARIABLES

01	02	03	04	05	06	07	08	09	10	11	12	13	14	15	16	17	18	19	20	21	22	23	24	25	26	27
KIMBLER	703 2ND ST	59	588	3	1	396	46928	6	494	46	21	44824	4	2	160	80	7	1	2	1	88	22	2	0	1	1
RUDOLPH	704 2ND ST	60	589	3	1	520	71731	12	1257	123	25	58090	7	3	165	55	4	0	1	7	135	19	1	1	0	0
LIPINSKY	701 1ST ST	58	590	3	1	193	29101	5	500	44	72	44993	1	1	65	65	4	2	2	1	25	25	2	0	0	0
WARD	702 1ST ST	59	591	3	1	221	42184	7	734	62	17	51039	3	2	123	62	5	2	2	2	72	24	3	1	0	3
CHAVEZ	701 2ND ST	59	592	3	1	497	41224	10	961	95	71	54468	5	4	214	54	7	2	2	4	104	21	2	0	0	0
MOSKOVITZ	702 2ND ST	60	593	3	1	177	26392	6	598	55	10	49830	2	1	105	105	10	0	2	3	39	20	2	0	0	0
SAMPSON	607 1ST ST	53	594	3	1	470	49256	7	761	67	63	36794	4	2	154	77	9	1	2	1	76	19	2	2	0	0
LOGAN	608 1ST ST	54	595	3	1	434	50269	8	804	75	20	107273	3	2	156	78	6	1	2	1	70	23	2	2	0	0
THOMAS	607 2ND ST	54	596	3	1	230	39216	7	769	61	38	40191	3	2	139	70	3	1	2	1	72	24	1	0	4	0
SPECTOR	608 2ND ST	55	597	3	1	346	47310	9	915	87	67	48075	4	3	183	61	2	1	2	2	84	21	1	0	4	5
HEATH	607 3RD ST	55	598	3	1	331	46275	9	952	81	15	50684	5	4	205	51	7	1	2	3	105	21	2	0	0	1
WYATT	608 3RD ST	56	599	3	1	366	41596	9	943	91	19	34509	4	3	173	58	6	2	2	0	90	23	1	2	0	0
HEYMAN	607 4TH ST	56	600	3	1	538	19500	8	668	65	22	19730	4	2	120	60	7	1	1	0	100	25	3	0	0	0
PERRY	608 4TH ST	57	601	3	1	388	50954	9	976	83	49	27868	5	4	208	52	5	1	2	5	121	24	1	0	0	9
SUSSMAN	605 1ST ST	53	602	3	1	89	18385	2	172	14	12	12805	1	2	120	60	7	1	1	0	19	19	3	0	0	0
MALLEY	606 1ST ST	54	603	3	1	276	33995	7	734	62	49	24343	3	2	137	69	3	1	1	2	76	25	2	2	0	0
UNGER	605 2ND ST	54	604	3	1	323	44674	7	730	65	92	37915	4	2	159	80	7	2	2	2	85	21	1	0	0	0
HOOK	606 2ND ST	55	605	3	1	337	40716	7	650	61	24	50032	4	3	173	58	6	1	1	7	97	24	1	1	0	0
COWAN	605 3RD ST	55	606	3	1	465	41429	8	748	77	41	44766	4	3	173	58	5	0	1	3	85	21	2	0	6	5
TRIBBLE	606 3RD ST	56	607	3	1	357	52540	10	972	93	22	54820	5	4	209	52	8	1	1	5	71	14	1	2	0	9
DEJESUS	605 4TH ST	56	608	3	1	290	39256	6	643	54	17	28593	2	2	136	68	9	2	2	1	48	24	2	1	0	11
WASHINGTON	606 4TH ST	57	609	3	1	596	48997	9	964	82	59	28050	5	3	194	65	6	1	2	1	104	21	2	0	0	0
BARQUIN	603 1ST ST	53	610	3	1	362	42466	7	754	62	14	29398	3	2	160	80	6	0	2	1	60	20	1	2	0	7
GARNET	604 1ST ST	54	611	3	1	287	35145	6	610	48	14	18591	6	2	120	60	7	1	2	4	120	20	1	1	0	0
AVELLO	603 2ND ST	54	612	3	1	231	42844	8	730	73	90	41959	4	3	190	63	5	2	2	2	85	21	1	2	0	2
DEITZ	604 2ND ST	55	613	3	1	358	47436	8	803	72	63	63844	4	2	145	73	5	0	2	1	83	21	1	0	0	9
ZEIGLER	603 3RD ST	55	614	3	1	397	50925	9	943	89	13	46125	4	3	181	60	5	2	2	3	89	22	2	0	4	11
BARNETT	604 3RD ST	56	615	3	1	457	42012	7	651	64	65	57483	3	2	149	75	5	2	2	2	62	21	2	0	0	0
HACH	603 4TH ST	56	616	3	1	484	50323	9	906	84	48	27071	6	4	217	54	5	2	2	4	138	23	1	0	2	4
YOUNG	604 4TH ST	57	617	3	1	611	51380	9	936	75	37	68981	4	3	184	61	5	2	1	6	75	19	1	0	0	3
COOPER	601 1ST ST	53	618	3	1	410	40858	9	976	81	15	28485	5	3	194	65	7	1	1	6	93	19	2	0	4	2
LESHAW	602 1ST ST	54	619	3	1	180	20029	4	422	33	66	31290	1	1	91	91	7	2	2	1	35	35	2	0	0	0
FLANAGHAN	601 2ND ST	54	620	3	1	337	39117	4	395	28	21	24156	1	1	65	65	4	2	2	1	29	29	2	0	0	0
KAPLAN	602 2ND ST	55	621	3	1	318	41702	8	788	73	22	45689	4	3	174	58	3	2	2	3	83	21	2	0	0	0
JACKSON	601 3RD ST	55	622	3	1	300	40127	7	697	65	84	28887	4	2	148	74	2	2	1	4	83	21	1	2	0	2
NOBLE	602 3RD ST	56	623	3	1	334	50065	8	850	75	12	76129	4	4	203	51	5	2	2	2	88	22	1	0	0	2
PERIGO	601 4TH ST	56	624	3	1	178	36624	8	792	69	88	42217	4	4	208	52	5	2	2	2	87	22	2	1	0	2
COPELAND	602 4TH ST	57	625	3	1	427	67105	10	988	100	21	36356	6	4	214	54	8	2	1	8	103	17	2	1	0	0
GARFIELD	1202 9TH ST	93	626	4	0	439	0	6	511	51	83	14081	7	2	120	60	6	0	2	3	136	19	2	0	1	3
AVERSA	1202 9TH ST	93	627	4	0	401	0	5	418	35	20	13735	6	2	120	60	4	0	2	3	122	20	2	0	6	1
LEOPOLD	1202 9TH ST	93	628	4	0	339	0	5	395	39	23	9483	4	1	65	65	7	0	2	1	93	23	2	1	0	7
YOUNG	1202 9TH ST	93	629	4	0	328	0	5	395	37	58	16014	5	2	120	60	5	1	1	4	77	15	2	0	6	0
HABER	1202 9TH ST	93	630	4	0	522	0	8	667	59	66	18640	5	2	120	60	8	0	2	5	91	18	2	0	3	7
BONWIT	1202 9TH ST	93	631	4	0	390	0	6	532	38	21	12037	5	2	120	60	5	1	1	3	95	19	2	0	0	3
JACOBSON	1202 9TH ST	93	632	4	0	282	0	4	335	24	67	15867	2	2	120	60	5	1	2	3	47	24	2	0	0	2
DELANEY	1202 9TH ST	93	633	4	0	447	0	6	519	50	23	19249	6	1	65	65	5	1	1	3	105	18	2	0	0	0
MCGHEE	1202 9TH ST	93	634	4	0	544	0	7	571	57	46	17987	4	2	120	60	8	1	2	4	90	23	2	0	0	0
GARDNER	1202 9TH ST	93	635	4	0	574	0	8	637	67	13	20943	5	4	215	54	9	2	1	1	107	21	2	0	1	3
XAVIER	1202 9TH ST	93	636	4	0	511	0	9	728	84	61	21902	4	3	165	55	8	2	1	3	89	22	2	2	0	8
FRIEDMAN	1202 9TH ST	93	637	4	0	453	0	6	520	40	29	16917	6	2	120	60	6	1	2	3	136	23	2	0	0	0
WARSHAW	1202 9TH ST	93	638	4	0	438	0	5	365	36	15	18736	2	2	120	60	12	1	1	1	43	22	2	2	0	0
PEREZ	1202 9TH ST	93	639	4	0	521	0	8	621	64	61	17430	2	2	120	60	5	0	2	3	54	27	2	0	6	0
KARNBLUH	1202 9TH ST	93	640	4	0	606	0	8	661	65	80	26000	3	2	148	74	4	2	2	1	67	22	2	0	0	0
ROLTER	1202 9TH ST	93	641	4	0	583	0	9	714	71	39	30195	5	3	191	64	11	1	1	5	111	22	2	0	0	0
MASK	1202 9TH ST	93	642	4	0	486	0	7	579	58	69	21554	4	3	183	61	6	2	2	1	93	23	2	0	1	0
JAFFE	1202 9TH ST	93	643	4	0	499	0	6	486	46	16	27589	6	2	120	60	10	2	1	1	137	23	2	0	4	6
STADLER	1202 9TH ST	93	644	4	0	645	0	9	723	67	22	16730	4	1	65	65	4	1	2	1	106	27	2	0	0	0
PICKERING	1202 9TH ST	93	645	4	0	388	0	6	511	43	70	18836	4	1	65	65	6	1	1	4	73	18	2	2	0	0
KIMBLE	1202 9TH ST	93	646	4	0	609	0	9	743	82	92	22956	4	3	180	60	5	2	2	4	81	20	2	1	0	11
UBEDA	1201 10TH ST	93	647	4	0	535	0	9	750	71	38	23346	2	2	129	65	8	2	2	1	49	25	2	0	0	0
JAHN	1201 10TH ST	93	648	4	0	441	0	6	529	43	95	21242	3	2	138	69	3	0	1	4	76	25	2	0	0	0
FLORX	1201 10TH ST	93	649	4	0	407	0	6	472	50	16	18425	5	1	65	65	8	1	1	4	91	18	2	0	1	0
LEVENBERG	1201 10TH ST	93	650	4	0	397	0	6	439	48	69	25052	6	4	221	55	9	2	2	3	108	18	3	0	0	0
HALL	1201 10TH ST	93	651	4	0	440	0	6	502	46	33	27781	4	2	155	78	6	2	2	8	78	20	2	1	0	0
STIMSON	1201 10TH ST	93	652	4	0	499	0	6	481	41	42	22901	5	3	197	66	9	2	1	6	116	23	2	2	0	2
MANN	1201 10TH ST	93	653	4	0	458	0	6	444	45	21	28133	6	4	217	54	8	2	1	6	132	22	2	0	0	3
OCONNELL	1201 10TH ST	93	654	4	0	620	0	8	657	65	22	29640	4	3	177	59	4	1	1	3	111	28	2	0	0	0
GORDON	1201 10TH ST	93	655	4	0	411	0	6	522	44	12	23675	8	2	120	60	4	2	1	8	174	22	2	2	0	0
TYCHE	1201 10TH ST	93	656	4	0	447	0	6	467	50	20	29070	7	3	165	55	4	2	1	2	156	22	3	0	0	9
CROWN	1201 10TH ST	93	657	4	0	450	0	6	472	40	20	28604	4	4	214	54	2	0	1	1	94	24	2	0	0	9
HAMLYN	1201 10TH ST	93	658	4	0	528	0	8	657	67	38	22708	2	2	144	72	4	0	1	1	54	27	2	0	4	5
CUSSO	1201 10TH ST	93	659	4	0	507	0	8	653	73	62	28077	3	2	134	67	4	1	2	1	70	23	2	2	0	0
ALEA	1201 10TH ST	93	660	4	0	567	0	8	610	68	29	23927	5	3	194	65	2	1	1	1	126	25	2	2	0	0
ESSNER	1201 10TH ST	93	661	4	0	538	0	6	436	48	16	28116	8	2	120	60	8	2	2	6	156	20	2	0	0	11
FLETCHER	1201 10TH ST	93	662	4	0	348	0	5	448	38	20	28110	1	1	93	93	4	2	1	1	33	33	2	0	0	7
BOOKER	1201 10TH ST	93	663	4	0	424	0	6	495	46	37	27962	5	4	204	51	5	2	2	4	93	19	2	2	0	10
LUCE	1201 10TH ST	93	664	4	0	470	0	6	495	46	17	27946	5	4	208	52	4	1	2	4	103	21	2	0	0	0
TIMMS	1201 10TH ST	93	665	4	0	435	0	6	455	51	77	21450	3	2	133	67	9	2	2	3	70	23	2	0	0	0
KAPLAN	1201 10TH ST	93	666	4	0	376	0	6	415	44	22	27030	2	2	130	65	7	1	2	1	51	26	2	2	0	0
CROVELLA	1202 10TH ST	94	667	4	0	398	0	6	470	44	17	8495	7	1	65	65	7	0	1	0	164	23	2	0	0	1
BOOTHE	1202 10TH ST	94	668	4	0	447	0	6	453	44	94	12964	6	2	120	60	6	1	2	6	99	17	2	0	0	0
LUIS	1202 10TH ST	94	669	4	0	499	0	7	598	54	59	6892	4	1	65	65	7	0	1	0	77	19	2	0	1	0
PEREZ	1202 10TH ST	94	670	4	0	432	0	6	471	46	30	16470	7	1	65	65	3	0	2	7	161	23	2	0	0	0
OSSIP	1202 10TH ST	94	671	4	0	495	0	7	559	55	84	15948	4	1	65	65	9	0	2	2	67	17	2	2	0	0
LOWE	1202 10TH ST	94	672	4	0	353	0	5	379	39	19	11480	5	1	65	65	5	0	2	1	119	24	2	0	0	0
PEREZ	1202 10TH ST	94	673	4	0	275	0	5	427	41	56	17565	6	3	165	55	2	0	2	9	142	24	2	2	0	1
GIRO	1202 10TH ST	94	674	4	0	582	0	8	614	69	83	14408	3	2	120	60	7	0	2	0	70	23	2	0	0	0
PHELPS	1202 10TH ST	94	675	4	0	437	0	6	470	49	24	12580	7	1	65	65	4	0	1	3	169	24	1	2	0	6
SHEPHARD	1202 10TH ST	94	676	4	0	492	0	7	600	56	15	12755	5	1	65	65	6	0	2	5	108	22	2	0	0	0
HALLSTROM	1202 10TH ST	94	677	4	0	616	0	10	788	86	17	17854	3	2	120	60	8	1	2	3	74	25	2	0	0	5
CANDELL	1202 10TH ST	94	678	4	0	359	0	5	376	32	27	13347	5	2	120	60	7	1	2	1	95	19	2	0	4	0
SMITH	1202 10TH ST	94	679	4	0	501	0	7	577	54	90	15597	5	1	65	65	5	1	2	4	93	19	2	2	0	0
JONES	1202 10TH ST	94	680	4	0	489	0	7	548	54	19	14055	4	1	65	65	7	0	2	6	83	21	2	1	0	5
THOMPSON	1202 10TH ST	94	681	4	0	432	0	6	486	49	13	15504	3	2	120	60	7	1	1	4	63	21	2	0	0	0
MALM	1202 10TH ST	94	682	4	0	310	0	5	429	39	15	14864	4	2	120	60	11	1	2	4	76	19	2	0	1	0
HEMINGWAY	1202 10TH ST	94	683	4	0	434	0	5	341	37	11	15937	4	1	65	65	5	0	2	4	80	20	2	0	0	1
STEINHAM	1202 10TH ST	94	684	4	0	667	0	8	668	59	20	12539	5	1	65	65	4	1	2	0	117	23	3	0	4	0
ODELL	1202 10TH ST	94	685	4	0	325	0	6	511	47	35	16566	4	2	120	60	6	1	2	3	97	24	1	0	0	0

```
              0    0              0   000  0              0    1    1  1   1  1 112  2   2  2  2  2 2
              1    2              3   456  7      8        9 0  1    2  3   45 6 7 890 1   2  3  4  5 6 7
**********************************************************************************************************
TANNER     1202 10TH  ST  94  686 4 0 356   0  6  473  42  39  15272  7 2 120  60  8 1 1  5 139 20 2 0 5  0
COX        1202 10TH  ST  94  687 4 0 547   0  8  647  60  97  14967  5 1  65  65  5 0 1  4  92 18 2 0 0  1
AXELROD    1201 11TH  ST  94  688 4 0 572   0  8  618  62  16  16856  4 2 120  60  6 0 2  4  86 22 2 0 0  7
GFATTER    1201 11TH  ST  94  689 4 0 433   0  5  431  36  22  14369  7 1  65  65 10 0 1  3 151 22 2 0 0  7
HAINES     1201 11TH  ST  94  690 4 0 609   0  8  675  63  16  17302  4 2 120  60  6 0 1  2  87 22 2 2 0  2

JUNEAU     1201 11TH  ST  94  691 4 0 242   0  4  317  32  10  18078  1 2 120  60  4 0 2  1  24 24 2 1 0  0
LEIDERMAN  1201 11TH  ST  94  692 4 0 600   0  9  705  73  16  18489  4 1  65  65  7 1 1  4  79 20 2 2 0  4
DELACRUZ   1201 11TH  ST  94  693 4 0 225   0  3  303  22  20  19419  2 2 120  60  1 0 1  1  29 29 2 0 0  1
SHEADE     1201 11TH  ST  94  694 4 0 354   0  6  450  47  24  17601  6 1  65  65  5 0 2  2  70 18 3 0 3  3
LONGEN     1201 11TH  ST  94  695 4 0 400   0  6  485  52  17  14592  5 2 120  60  4 1 2  4  99 20 2 0 3  3

FLIEGEL    1201 11TH  ST  94  696 4 0 379   0  6  506  48  15  16149  7 2 120  60  5 1 2  3 140 20 2 0 0  0
HOLLADAY   1201 11TH  ST  94  697 4 0 376   0  5  422  32  73  15739  4 2 120  60  6 1 2  1  90 23 2 0 0  0
CANFIELD   1201 11TH  ST  94  698 4 0 454   0  5  379  35  19  12544  5 2 120  60  9 1 1  5 104 21 2 1 0  2
FLYNN      1201 11TH  ST  94  699 4 0 509   0  8  672  67  20  11102  4 2 120  60  9 0 1  1  77 19 2 0 0  0
QUIROS     1201 11TH  ST  94  700 4 0 323   0  4  288  33  20  14217  5 2 120  60 12 1 2  4 100 20 2 0 4 10

HAMILTON   1201 11TH  ST  94  701 4 0 321   0  5  416  37  16   5184  8 2 120  60 10 1 1  8 176 22 2 0 0  0
HOLKO      1201 11TH  ST  94  702 4 0 585   0  8  651  64  39  16006  4 1  65  65  9 0 1  2 105 26 2 0 3  0
NELSON     1201 11TH  ST  94  703 4 0 240   0  3  401  38  24  23862  3 2 122  61  4 2 1  4  65 22 2 0 0  0
ALVAREZ    1201 11TH  ST  94  704 4 0 429   0  6  471  43  85  13755  9 1  65  65  2 1 1  7 189 21 2 0 4  1
RODRIGUEZ  1201 11TH  ST  94  705 4 0 425   0  7  578  54  80  16726  4 1  65  65  6 1 1  4  74 19 2 0 0  0

TARRUELLA  1201 11TH  ST  94  706 4 0 404   0  7  590  57  71  12004  4 1  65  65  5 0 2  4  72 18 2 2 0  5
HOLLAND    1201 11TH  ST  94  707 4 0 488   0  6  428  46  84  21191  6 2 120  60 11 1 2  3 104 22 2 0 0  9
SWAN       1202 11TH  ST  95  708 4 0 292   0  3  380  32  80  15796  5 2 120  60  1 1 1  6 106 21 2 1 0  0
SWAN       1202 11TH  ST  95  709 4 0 402   0  5  392  35  15   9416  8 2 120  60  8 1 1  6 160 20 2 1 0  1
OTERO      1202 11TH  ST  95  710 4 0 409   0  6  416  51  11   9213  3 2 120  60  4 1 1  7  75 25 2 0 0  1

HINES      1202 11TH  ST  95  711 4 0 449   0  6  493  55  16  10961  6 1  65  65  6 0 2  3 143 24 2 0 0  0
WAGNER     1202 11TH  ST  95  712 4 0 462   0  6  475  42  18  13527  5 2 120  60  7 0 2  7 157 22 2 2 0  0
BRYANT     1202 11TH  ST  95  713 4 0 493   0  7  620  53  55   9759  3 2 120  60  4 1 2  2  66 22 2 2 0  1
GORDON     1202 11TH  ST  95  714 4 0 353   0  6  521  48  21  12078  6 2 120  60  5 1 1  4 110 18 1 0 1  0
APONTE     1202 11TH  ST  95  715 4 0 281   0  3  215  20  24  18864  2 2 120  60  7 0 2  1  45 23 2 0 2  0

ABLE       1202 11TH  ST  95  716 4 0 389   0  5  445  33  70  12364  6 3 165  55 10 1 1  8 128 21 3 0 0  0
EDDY       1202 11TH  ST  95  717 4 0 505   0  8  637  73  13  17484  5 1  65  65  1 0 2  1 111 22 2 0 1  5
KIMMEL     1202 11TH  ST  95  718 4 0 477   0  6  483  50  19  19359  5 2 120  60  6 0 1  1 130 26 2 0 3  3
SLACK      1202 11TH  ST  95  719 4 0 146   0  3  183  20  13  17045  3 2 120  60  3 0 1  1  63 21 2 0 3  0
HOFFMANN   1202 11TH  ST  95  720 4 0 455   0  5  380  39  77  16777  6 1  65  65  4 0 2  3 115 19 2 0 0  0

TOLEDO     1202 11TH  ST  95  721 4 0 202   0  3  262  22  34  13329  4 2 120  60  7 1 2  6  93 23 2 0 0  0
WATKINS    1202 11TH  ST  95  722 4 0 412   0  5  413  40  23  12872  4 1  65  65  7 1 2  5  77 19 2 2 0  1
MATTHIAS   1202 11TH  ST  95  723 4 0 564   0  8  638  69  68  13377  3 2 120  60  7 0 1  1  67 22 2 0 0  3
STOUFFER   1202 11TH  ST  95  724 4 0 518   0  8  596  62  20  17943  3 2 120  60  6 1 1  1  53 27 2 0 2  3
TOMAS      1202 11TH  ST  95  725 4 0 404   0  6  477  45 110  15310  5 2 120  60 10 1 1  5 119 24 2 0 2  3

POSEDA     1202 11TH  ST  95  726 4 0 467   0  7  565  59  26  16743  4 2 120  60  8 1 2  5  96 24 1 0 0  1
KING       1202 11TH  ST  95  727 4 0 341   0  5  439  44  53  15458  3 2 120  60  4 0 2  4  48 16 2 0 0  0
WILLIS     1202 11TH  ST  95  728 4 0 238   0  5  433  34  17  17288  6 2 120  60 10 1 2  4 107 27 2 0 0  5
GLOVER     1201 12TH  ST  95  729 4 0 498   0  6  486  49  24  13937  7 2 120  60  8 0 1  5 152 22 2 0 6  5
MATTHEW    1201 12TH  ST  95  730 4 0 168   0  3  247  20  19  16900  1 2 120  60  5 0 2  0  32 32 1 2 0  0

MCBRIDE    1201 12TH  ST  95  731 4 0 375   0  6  432  47  97  17309  8 1  65  65  6 0 1  8 162 20 2 0 1  2
BLUM       1201 12TH  ST  95  732 4 0 363   0  5  429  40  19  10795  9 2 120  60 11 1 2  2 188 21 1 2 0  6
YAEGER     1201 12TH  ST  95  733 4 0 381   0  5  414  37  25   5897  8 1  65  65  6 0 1  4 127 16 2 0 0  1
MORERA     1201 12TH  ST  95  734 4 0 435   0  6  482  47  54  20553  6 2 120  60  6 1 2  1 135 23 2 0 2  0
HABER      1201 12TH  ST  95  735 4 0 313   0  6  498  44  27  12741  8 1  65  65  6 0 2 10 162 20 2 2 2  4

ABBEY      1201 12TH  ST  95  736 4 0 307   0  5  392  38  14  18404  2 2 120  60 12 1 2  2  73 18 2 0 4  3
FLEMING    1201 12TH  ST  95  737 4 0 407   0  5  407  34  58  13397  4 1  65  65  3 1 1  3 135 23 2 0 0  0
COLLINS    1201 12TH  ST  95  738 4 0 445   0  6  489  45  21  12821  4 2 120  60  8 1 2  2  87 22 2 0 0  6
HECHT      1201 12TH  ST  95  739 4 0 514   0  7  526  53  34  11381  2 2 120  60  9 1 1  1  50 25 2 0 0 10
PHILLIPPE  1201 12TH  ST  95  740 4 0 536   0  7  523  62  12  11186  4 2 120  60  4 0 2  4  75 19 2 0 0 11

HORAN      1201 12TH  ST  95  741 4 0 564   0  7  570  62  86  16674  3 1  65  65  8 0 1  2  66 22 2 0 0  0
TONKIN     1201 12TH  ST  95  742 4 0 484   0  6  451  44  18  18317  3 1  65  65  8 1 2  3 133 22 2 0 0 11
WILLOUGHBY 1201 12TH  ST  95  743 4 0 647   0  9  741  73  16  16473  4 2 120  60  8 1 1  5  85 21 2 0 0  0
KAHN       1201 12TH  ST  95  744 4 0 330   0  5  414  38  11   4405  4 1  65  65  3 0 2  5  99 25 2 0 1  2
SMYTHE     1201 12TH  ST  95  745 4 0 380   0  5  356  39  77  25044  2 1 100 100  4 2 2  2  39 20 2 2 0  6

FLOWERS    1201 12TH  ST  95  746 4 0 561   0  8  652  69  29  12769  4 2 120  60  6 1 2  0  87 22 2 1 0  0
LAMM       1201 12TH  ST  95  747 4 0 315   0  5  420  34  46   7691  4 2 120  60  3 0 2  4 134 22 2 0 0  0
SODERHOLM  1201 12TH  ST  95  748 4 0 575   0  7  575  61  19  10380  4 2 120  60  5 0 1  2  44 22 2 1 0  0
MCBEE      1201 12TH  ST  96  749 4 0 513   0  6  517  41  17   8460  7 2 120  60  5 1 2  3 127 18 2 0 0  4
BROWN      1202 12TH  ST  96  750 4 0 314   0  5  380  37  33  10754  5 2 120  60  5 1 1  3 124 22 2 0 3  0

POTTER     1202 12TH  ST  96  751 4 0 478   0  6  461  48  20  10530  6 2 120  60  7 1 1  4 124 21 2 0 0  0
FIREMAN    1202 12TH  ST  96  752 4 0 307   0  5  393  38  21  11187  4 1  65  65  6 1 1  2  98 25 2 0 0  0
SINCLAIR   1202 12TH  ST  96  753 4 0 694   0 10  826  87  18  11643  6 1  65  65  6 1 1  2 100 25 2 0 0  3
MASI       1202 12TH  ST  96  754 4 0 357   0  5  422  39  27  13761  4 2 120  60  3 1 2  1  42 21 2 0 0  3
TOMAS      1202 12TH  ST  96  755 4 0 532   0  7  582  56  17  10039  4 2 120  60  5 0 2  2  83 21 2 0 0  1

PHELAN     1202 12TH  ST  96  756 4 0 270   0  5  427  40  39   9490  7 2 120  60  3 1 2  3 155 23 2 0 0  0
TORRES     1202 12TH  ST  96  757 4 0 496   0  6  501  42  66  12194  7 1  65  65  5 0 1  7 142 20 2 0 0  0
MONROE     1202 12TH  ST  96  758 4 0 381   0  6  489  44  16  10361  8 3 165  55  8 0 1  6 182 23 2 0 0  0
FLORIO     1202 12TH  ST  96  759 4 0 622   0  8  646  73  77  12717  4 2 120  60  8 0 1  3  75 18 2 0 0  7
BARREIRO   1202 12TH  ST  96  760 4 0 314   0  5  384  40  19  10268  8 1  65  65  5 0 2  8 162 20 2 1 0  0

KOPP       1202 12TH  ST  96  761 4 0 280   0  4  331  28  60  10045  2 2 120  60  5 0 1  1  38 19 2 0 0  3
BECK       1202 12TH  ST  96  762 4 0 455   0  7  577  54  28   8857  4 2 120  60  4 1 2  1  97 24 2 0 0  2
SCHUMACKER 1202 12TH  ST  96  763 4 0 126   0  3  292  21  21   8571  1 2 120  60  2 2 2  1  28 28 2 0 0  2
BAKER      1202 12TH  ST  96  764 4 0 143   0  3  245  19  81   8102  2 2 120  60  3 0 2  1  55 28 2 0 0  2
MARKS      1202 12TH  ST  96  765 4 0 369   0  5  383  37  25  18788  7 1  65  65  7 1 2  3 134 22 2 0 2  0

GRAVER     1202 12TH  ST  96  766 4 0 336   0  5  439  36  18   8772  7 2 120  60  6 0 1  7 125 18 2 1 0  7
CRAVEN     1202 12TH  ST  96  767 4 0 608   0  7  594  59  62   7809  4 2 120  60  6 0 1  3  86 22 2 0 1 10
MODER      1202 12TH  ST  96  768 4 0 276   0  4  365  27  37   5630  4 2 120  60  0 0 1  0  39 39 2 0 1 10
GRAVER     1202 12TH  ST  96  769 4 0 441   0  6  494  48  54   6038  6 2 120  60 10 1 1  1 122 20 3 0 0  0
STEWART    1201 12TH  ST  96  770 4 0 470   0  7  539  52  26  10831  4 1  65  65  5 1 2  1  80 20 2 0 0  4

ROSEN      1201 13TH  ST  96  771 4 0 467   0  6  535  44  11   7158  5 2 120  60  4 1 1  8 109 22 2 0 0  0
WURST      1201 13TH  ST  96  772 4 0 231   0  3  261  22  36   4322  2 2 120  60 10 0 1  1  71 18 2 0 0  0
ALLEN      1201 13TH  ST  96  773 4 0 294   0  4  284  18  26   8983  2 2 120  60  4 0 1  2  41 21 1 0 0  6
SHUMP      1201 13TH  ST  96  774 4 0 294   0  4  258  29  21   4886  3 2 120  60  3 0 1  1  61 20 2 0 0  6
WEISMAN    1201 13TH  ST  96  775 4 0 340   0  5  284  39  18   3168  8 1  65  65  3 1 1  8 177 22 2 0 1  1

CALHOUN    1201 13TH  ST  96  776 4 0 343   0  5  385  35  23   5830  4 1  65  65  3 1 1  3  87 22 2 0 0  0
STANFORD   1201 13TH  ST  96  777 4 0 533   0  8  691  63  19  17585  2 2 120  60  1 1 2  1  47 24 2 0 0  0
FOX        1201 13TH  ST  96  778 4 0 298   0  5  391  38  27   6476  2 2 120  60  2 0 1  3 138 23 1 0 0  0
BITTNER    1201 13TH  ST  96  779 4 0 390   0  6  418  38  20   7844  6 2 120  60  7 1 12  1 114 19 2 0 0  0
HAYES      1201 13TH  ST  96  780 4 0 325   0  5  418  40  41   6269  4 2 120  60  7 0 1  1  84 21 2 0 0  0

GOLDEN     1201 13TH  ST  96  781 4 0 272   0  6  443  43  18   8618  5 1  65  65  6 1 2  4 108 22 2 0 1  6
OWENS      1201 13TH  ST  96  782 4 0 327   0  6  466  46  10   8142  5 1  65  65  9 0 1  2 103 20 2 0 0  6
HERNANDEZ  1201 13TH  ST  96  783 4 0 332   0  5  390  39  13   4919  5 2 120  60  7 0 2  0  98 20 2 0 0  0
DAVIS      1201 13TH  ST  96  784 4 0 616   0  7  670  69  22   9763  3 1  65  65  4 1 2  1  73 24 2 0 4  3
KIRKPATRIK 1201 13TH  ST  96  785 4 0 403   0  7  525  56  38   9765  4 2 120  60  7 1 1  4  95 24 2 0 4  0
```

C Stat City data base

```
                                                    VARIABLES
   0           0         0     0 0 0 0   0   0   1   1   1     1   1 1 1   1   1 1 2   2   2 2 2   2 2 2
   1           2         3     4 5 6 7   8   9   0   1   2     3   4 5 6   7   8 9 0   1   2 3 4   5 6 7
*****************************************************************************************************

DODD       1201 13TH ST  96  786 4 0 413   0  6  530  42  88   9943  7 1  65 65  2 1 1  2 134 19 2  0 4 0
POWELL     1201 13TH ST  96  787 4 0 280   0  4  340  28  35   6241  5 2 120 60  3 0 1  4 114 23 2  1 0 3
IRWIN      1201 13TH ST  96  788 4 0 286   0  4  425  36  63   7548  7 2 120 60  3 1 2  3  99 14 2  0 1 1
JOHNSON    1201 13TH ST  96  789 4 0 279   0  4  285  31  54   2402  7 2 120 60  3 1 1  1  92 23 2  1 0 1
LONG       1102  9TH ST  97  790 4 0 416   0  4  494  51  13  14757  5 2 120 60  3 2 2  3 108 22 1  2 0 1

KRUGER     1102  9TH ST  97  791 4 0 558   0  7  574  57  20  12185  5 1  65 65  6 0 2  1  85 17 2  0 1 0
WOLFE      1102  9TH ST  97  792 4 0 528   0  7  582  53  19  12063  3 2 120 60  7 0 2  1  79 26 2  0 0 0
VELLANTI   1102  9TH ST  97  793 4 0 482   0  5  428  43  41  18811  7 1  65 65  1 1 2  5 153 22 2  0 0 10
NICHOLSON  1102  9TH ST  97  794 4 0 430   0  5  451  39  25  20323  5 3 198 66 11 1 1  1 112 22 2  0 0 0
HEYMAN     1102  9TH ST  97  795 4 0 333   0  5  374  40  73   9558  5 1  65 65  4 1 2  6 107 21 2  0 0 0

SPRINGER   1102  9TH ST  97  796 4 0 391   0  5  406  41  77  16200  6 2 120 60  6 0 1  4 109 18 2  0 1 10
PATTERSON  1102  9TH ST  97  797 4 0 278   0  4  392  30  19  13705  4 1  65 65  3 1 2  1  94 24 2  2 0 0
FIELDER    1102  9TH ST  97  798 4 0 384   0  5  447  38  50  19876  6 1  65 65  1 1 2 10 129 22 2  0 4 0
CAMPBELL   1102  9TH ST  97  799 4 0 252   0  3  232  21  74  19071  2 2 120 60  9 0 1  2  37 19 1  2 0 0
EARHART    1102  9TH ST  97  800 4 0 426   0  6  463  48  63  22691  6 3 165 55  7 2 1  1 129 22 2  2 0 0

NORBECK    1102  9TH ST  97  801 4 0 541   0  7  582  54 126  19261  4 1  65 65  4 1 2  2 102 26 2  0 0 0
HILL       1102  9TH ST  97  802 4 0 349   0  5  494  39  20  16371  2 2 120 60  8 0 2  1  51 26 2  0 0 0
TILLERMAN  1102  9TH ST  97  803 4 0 766   0 10  826  89  11  26377  3 2 159 80  6 2 1  1  60 20 2  0 0 0
MAST       1102  9TH ST  97  804 4 0 299   0  6  502  45  15  18096  7 1  65 65  5 0 1  5 130 19 2  0 0 0
FERNANDEZ  1102  9TH ST  97  805 4 0 366   0  5  435  37  14  22002  5 4 222 56  3 0 2  4 111 22 2  0 1 4

BARTON     1102  9TH ST  97  806 4 0 653   0  9  720  85 109  22557  4 3 176 59  5 2 2  3  89 22 2  0 5 0
HICKS      1102  9TH ST  97  807 4 0 540   0  8  690  69  78  24995  4 2 147 74  6 2 2  5  71 18 2  2 0 10
GILBERT    1102  9TH ST  97  808 4 0 411   0  6  521  46  16  17348  5 1  65 65  1 0 2  1  99 20 2  0 0 0
DRAGO      1102  9TH ST  97  809 4 0 546   0  7  538  63  35  19029  4 2 120 60  5 1 2  2  87 22 2  2 0 0
OAKES      1102  9TH ST  97  810 4 0 377   0  6  522  49  37  17715  4 2 120 60  2 0 2  2  90 23 2  1 0 0

FEY        1101 10TH ST  97  811 4 0 581   0  8  647  63  29  18795  3 2 120 60  5 0 1  4  82 27 2  0 0 0
RUSCH      1101 10TH ST  97  812 4 0 528   0  7  545  55  70  24504  4 3 168 56  9 0 2  4  85 21 2  0 0 5
PETRUCCI   1101 10TH ST  97  813 4 0 344   0  5  505  42  17  20118  5 4 201 50  5 2 2  4 101 20 2  2 0 1
ISGAR      1101 10TH ST  97  814 4 0 267   0  3  248  22  12  15322  2 2 120 60  5 0 2  0  46 23 2  2 7 7
JEFFERY    1101 10TH ST  97  815 4 0 362   0  4  281  31  11  13161  1 2 120 60  7 0 2  1  26 26 2  2 5 1

CAPMAN     1101 10TH ST  97  816 4 0 603   0  7  557  57  16  25718  4 3 179 60  8 1 1  7  97 24 2  0 5 0
KAPLAN     1101 10TH ST  97  817 4 0 449   0  7  570  60  17  21660  3 2 155 68  6 0 2  2  62 21 2  0 6 7
LEWIS      1101 10TH ST  97  818 4 0 482   0  7  562  52  24  21081  2 2 135 68  5 1 1  3  56 28 2  0 6 5
PABLO      1101 10TH ST  97  819 4 0 652   0  9  682  73  38  22992  3 2 155 78  6 1 2  3  58 19 2  0 0 0
INGRAM     1101 10TH ST  97  820 4 0 711   0  9  726  80  82   8473  5 2 120 60  7 0 1  3  89 18 2  0 0 0

MALLERY    1101 10TH ST  97  821 4 0 436   0  6  503  50  14  19498  7 1  65 65  6 1 2  9 137 20 2  0 0 5
DAVIDSON   1101 10TH ST  97  822 4 0 434   0  6  522  45  14  17777  5 1  65 65  2 0 2  3 118 24 2  0 0 0
ALEMAN     1101 10TH ST  97  823 4 0 587   0  8  644  65 101  21115  3 2 157 79  6 0 2  1  56 19 2  0 0 8
FLOOK      1101 10TH ST  97  824 4 0 460   0  6  444  44  25  18092  2 2 120 60  8 1 2  0  50 25 2  0 0 3
INGALLS    1101 10TH ST  97  825 4 0 389   0  6  473  44  14  23502  4 3 179 60  7 2 1  2  96 24 2  0 0 1

GAFF       1101 10TH ST  97  826 4 0 417   0  6  505  49  16  23257  6 2 120 60  5 2 1  6 122 20 2  0 0 0
PEXTON     1101 10TH ST  97  827 4 0 412   0  6  492  45  20  23872  6 4 222 56  7 1 2  8 125 21 1  0 1 3
OGLE       1101 10TH ST  97  828 4 0 361   0  6  456  42  80  25649  6 4 224 56  6 1 1  6 113 19 2  2 0 3
FISHER     1101 10TH ST  97  829 4 0 224   0  4  268  30  20  24567  1 1  65 65  3 2 2  2  26 26 3  2 0 0
RAMOS      1101 10TH ST  97  830 4 0 409   0  6  541  48  21  25456  7 2 120 60  8 2 1  3 164 23 2  0 0 2

JACKSON    1102 10TH ST  98  831 4 0 377   0  5  394  38  22  14573  5 1  65 65  3 1 1  5 117 23 2  0 4 0
SKINNER    1102 10TH ST  98  832 4 0 313   0  5  407  35  72  13669  6 1  65 65  5 1 2  1 120 20 2  0 2 4
NESBITT    1102 10TH ST  98  833 4 0 269   0  4  460  36  70   6095  5 1  65 65  5 1 1  1 116 19 2  0 0 0
GAINES     1102 10TH ST  98  834 4 0 670   0  9  702  73  76  10488  5 2 120 60  5 0 2  3 109 22 2  0 1 6
HINSLEY    1102 10TH ST  98  835 4 0 483   0  7  601  59  12  13661  3 2 120 60  7 0 2  5  69 23 2  0 1 6

SANCHEZ    1102 10TH ST  98  836 4 0 530   0  8  682  68  22  14572  4 1  65 65  3 0 2  1  91 23 2  0 3 0
WILLIAMS   1102 10TH ST  98  837 4 0 564   0  8  656  68  39  13276  3 2 120 60  3 0 2  2  79 26 2  0 1 0
LEVY       1102 10TH ST  98  838 4 0 329   0  5  391  33  19  10291  4 2 120 60  3 1 2  3 119 20 2  0 1 0
ALCORN     1102 10TH ST  98  839 4 0 254   0  5  422  41  24  14253  7 2 120 60  3 1 1  3 140 20 2  0 0 0
MILLER     1102 10TH ST  98  840 4 0 501   0  6  478  45  44  14305  7 1  65 65  1 1 1  3 132 19 2  2 0 1 10

HITT       1102 10TH ST  98  841 4 0 238   0  4  317  31  98  10857  1 2 120 60  7 1 1  0  26 26 2  0 5 0
CAMPBELL   1102 10TH ST  98  842 4 0 381   0  6  481  49  19  15166  5 2 120 60  6 1 1  5 109 22 2  0 3 3
MALLOY     1102 10TH ST  98  843 4 0 326   0  6  419  40  53  14832  5 2 120 60  6 1 2  6 105 21 2  1 0 1
BARRON     1102 10TH ST  98  844 4 0 525   0  7  556  54  59  16144  2 2 120 60  4 0 1  1  63 32 2  0 0 0
ENGLING    1102 10TH ST  98  845 4 0 379   0  6  487  45  38  11838  4 2 120 60  6 1 1  6  84 21 2  0 0 2

SPELLMAN   1102 10TH ST  98  846 4 0 418   0  5  374  41  15  16995  3 2 120 60  7 0 2  4  65 22 2  0 1 0
FOSTER     1102 10TH ST  98  847 4 0 402   0  6  482  47  15  15130  7 1  65 65  5 0 1  2 138 20 2  0 5 10
MCCARTNEY  1102 10TH ST  98  848 4 0 354   0  7  569  50  35  13721  6 1  65 65  5 0 1  2  77 19 3  1 0 0
HARRISON   1102 10TH ST  98  849 4 0 373   0  6  524  47  37  14868  7 1  65 65  3 1 2  3 120 17 2  1 0 0
STARR      1102 10TH ST  98  850 4 0 298   0  5  456  37  23  18348  5 2 120 60  8 0 1  0 115 23 2  0 0 0

LENNON     1102 10TH ST  98  851 4 0 485   0  6  510  50  68  12642  5 1  65 65  3 0 2  5 110 22 3  0 3 0
BERAN      1101 11TH ST  98  852 4 0 385   0  6  509  46  49  18609  5 2 120 60  3 1 1  0  93 19 2  0 0 4
DEAN       1101 11TH ST  98  853 4 0 320   0  5  345  31  41  15113  4 2 120 60  2 1 2  5  79 20 2  0 0 0
PORTER     1101 11TH ST  98  854 4 0 395   0  6  504  45  23  12097  7 1  65 65  5 0 1  5 150 21 2  0 0 8
DAY        1101 11TH ST  98  855 4 0 422   0  6  536  46  38  14460  4 1  65 65  5 0 1  4  90 23 2  0 0 0

TAYLOR     1101 11TH ST  98  856 4 0 418   0  6  518  44  21  16727  3 2 120 60  6 0 2  1  73 24 2  0 0 0
LEONE      1101 11TH ST  98  857 4 0 326   0  6  502  50  22  18919  6 2 120 60  4 0 1  4 143 24 2  0 2 0
DOYLE      1101 11TH ST  98  858 4 0 246   0  3  256  21  21  17052  2 2 120 60  6 0 1  1  26 22 2  0 2 6
BRUNNER    1101 11TH ST  98  859 4 0 332   0  6  455  40  15  18060  5 2 120 60  6 0 1  4 114 23 2  2 5 3
QUAT       1101 11TH ST  98  860 4 0 545   0  8  634  67  15  11499  3 2 120 60  5 1 2  0  62 21 2  0 6 3

KRAUSE     1101 11TH ST  98  861 4 0 395   0  6  467  48  17  15055  7 1  65 65  4 1 2  2 162 23 2  0 6 0
BELLOMY    1101 11TH ST  98  862 4 0 656   0  9  727  68  22  17522  3 2 120 60  7 1 1  6  56 19 2  0 0 0
BESANT     1101 11TH ST  98  863 4 0 341   0  5  335  36  25  16916  5 1  65 65  8 0 1  7 119 24 2  0 4 0
HARRIS     1101 11TH ST  98  864 4 0 386   0  5  416  35  20  15236  4 2 120 60  8 0 1  9  77 19 2  0 0 4
CALDWELL   1101 11TH ST  98  865 4 0 420   0  5  364  40  84  15200  5 1  65 65  2 0 1  4  90 18 2  0 0 1

BAINTON    1101 11TH ST  98  866 4 0 267   0  5  436  41  15  12507  4 1  65 65  5 0 2  4  90 23 2  0 0 0
IRVING     1101 11TH ST  98  867 4 0 377   0  6  454  44  41  15501  6 2 120 60  8 1 1  3 133 22 2  1 0 1
LAPPE      1101 11TH ST  98  868 4 0 450   0  6  450  44  66  16001  2 2 120 60  8 1 1  2  96 19 2  0 0 0
JAMES      1101 11TH ST  98  869 4 0 505   0  6  480  46  11  11257  4 2 120 60 10 0 1  2  79 20 2  0 4 9
COOPER     1101 11TH ST  98  870 4 0 361   0  6  445  47  11  23114  7 3 165 55  7 0 1  3 147 21 2  0 6 0

SINGER     1101 11TH ST  98  871 4 0 303   0  5  386  37  16  17569  5 2 120 60  4 0 2  8 113 23 1  0 0 0
HARRIS     1102 11TH ST  99  872 4 0 260   0  3  265  19  19   7465  3 1  65 65  8 0 1  5  59 22 2  0 2 0
BURROUGHS  1102 11TH ST  99  873 4 0 485   0  6  464  51  25  11509  6 1  65 65  7 0 1  8  97 16 2  0 3 0
JORDAN     1102 11TH ST  99  874 4 0 338   0  5  398  42  19  11410  6 2 120 60  6 0 2  2 116 19 2  1 0 0
MCKENDRICK 1102 11TH ST  99  875 4 0 335   0  5  395  37  63  13243  5 2 120 60  6 0 2  3  87 17 3  0 0 7

BURKE      1102 11TH ST  99  876 4 0 425   0  6  505  54  19   8308  5 2 120 60  9 0 2  1 116 23 2  0 1 1
HOWATCH    1102 11TH ST  99  877 4 0 239   0  5  405  36  21  12296  5 1  65 65  6 0 2  1 109 22 2  0 0 0
SANGER     1102 11TH ST  99  878 4 0 410   0  6  533  48  82  13466  5 2 120 60  5 1 1  1  50 17 2  0 0 0
MASSEE     1102 11TH ST  99  879 4 0 280   0  6  529  47  71  13903  5 1  65 65  4 0 2  4 101 20 2  0 0 0
WHITE      1102 11TH ST  99  880 4 0 373   0  6  408  40  13  11445  5 2 120 60  6 0 1  0  70 20 3  0 0 0

ALLEN      1102 11TH ST  99  881 4 0 689   0  9  760  75  78  10441  5 1  65 65  5 1 1  5  94 19 2  0 0 0
ETHAN      1102 11TH ST  99  882 4 0 416   0  6  486  44  20  18993  3 2 120 60  4 1 2  4 111 22 2  0 0 5
LUTHOR     1102 11TH ST  99  883 4 0 356   0  6  391  34  23  10885  6 2 120 60  4 0 1  6 125 21 2  0 0 0
CLARKE     1102 11TH ST  99  884 4 0 311   0  5  363  33 139  11445  4 2 120 60  6 1 1  3 107 18 2  0 0 0
WAYNE      1102 11TH ST  99  885 4 0 293   0  5  407  39  62  10873  4 2 120 60  6 1 1  3  70 18 2  0 0 11
```

Appendixes

VARIABLES

01	02	03	04	05	06	07	08	09	10	11	12	13	14	15	16	17	18	19	20	21	22	23	24	25	26	27
BALLARD	1102 11TH ST	99	886	4	0	458	0	6	488	49	18	13897	5	1	65	65	4	1	2	5	111	22	2	0	0	0
NORRIS	1102 11TH ST	99	887	4	0	372	0	5	417	40	80	16641	5	2	120	60	4	0	1	4	94	19	2	0	0	0
DIAMOND	1102 11TH ST	99	888	4	0	438	0	6	509	46	61	12870	7	2	120	60	7	1	2	2	131	19	2	0	1	2
BULL	1102 11TH ST	99	889	4	0	370	0	5	396	36	14	14590	5	1	65	65	3	0	1	6	90	18	2	0	4	3
PEZEL	1102 11TH ST	99	890	4	0	374	0	5	391	39	14	15787	3	2	120	60	7	0	1	2	69	23	2	0	3	1
ABRAMS	1102 11TH ST	99	891	4	0	145	0	4	369	27	16	10024	3	2	120	60	9	0	1	3	77	26	2	0	2	4
THOREAU	1102 11TH ST	99	892	4	0	571	0	8	638	68	36	15678	3	1	65	65	5	0	2	4	62	21	2	0	0	0
DOBSON	1101 12TH ST	99	893	4	0	341	0	5	394	41	94	10373	6	2	120	60	4	0	1	3	120	20	2	2	0	0
CAYCE	1101 12TH ST	99	894	4	0	448	0	6	483	53	20	16307	4	2	120	60	4	1	2	4	80	20	2	0	0	0
MELVILLE	1101 12TH ST	99	895	4	0	327	0	5	385	41	25	11960	7	2	120	60	4	1	1	3	161	23	2	0	0	0
POE	1101 12TH ST	99	896	4	0	290	0	4	329	29	17	15904	1	2	120	60	4	0	2	1	36	36	2	0	3	2
EMERSON	1101 12TH ST	99	897	4	0	340	0	5	373	39	26	11969	7	1	65	65	4	1	2	2	138	20	3	0	0	0
DODSON	1101 12TH ST	99	898	4	0	419	0	6	530	49	78	14765	5	2	120	60	6	1	2	3	101	20	2	1	0	0
HAWTHORNE	1101 12TH ST	99	899	4	0	221	0	3	246	22	21	15823	2	2	120	60	12	0	2	1	46	23	1	0	4	11
KAFKA	1101 12TH ST	99	900	4	0	362	0	6	492	50	15	14657	6	1	65	65	5	0	2	6	126	21	2	0	1	11
GUERAD	1101 12TH ST	99	901	4	0	510	0	7	520	53	35	15262	4	2	120	60	7	1	1	3	75	19	2	2	0	0
SETON	1101 12TH ST	99	902	4	0	223	0	4	334	30	39	18786	5	2	120	60	7	0	2	1	106	21	2	0	0	0
FREUD	1101 12TH ST	99	903	4	0	412	0	5	448	35	40	17922	7	2	120	60	5	0	1	10	114	16	2	2	0	0
HOFFMAN	1101 12TH ST	99	904	4	0	428	0	6	564	50	21	6095	2	2	120	60	10	1	1	1	53	27	2	0	0	0
WILLIAMS	1101 12TH ST	99	905	4	0	338	0	5	387	35	15	16141	7	2	120	60	2	1	2	5	149	21	2	2	0	0
KENDA	1101 12TH ST	99	906	4	0	460	0	8	644	57	30	17915	5	2	120	60	7	0	2	3	100	20	2	0	0	6
DURANT	1101 12TH ST	99	907	4	0	465	0	6	417	52	27	18539	6	1	65	65	3	0	2	4	105	18	2	2	0	9
AMES	1101 12TH ST	99	908	4	0	419	0	6	451	48	40	10614	6	2	120	60	4	1	1	1	104	17	2	0	0	0
RULE	1101 12TH ST	99	909	4	0	316	0	5	399	34	17	18430	3	1	65	65	2	0	2	1	59	20	2	0	0	6
PAULUS	1101 12TH ST	99	910	4	0	379	0	5	426	37	23	14508	7	2	120	60	7	1	2	3	156	22	2	0	1	9
SIMON	1101 12TH ST	99	911	4	0	411	0	6	501	44	16	5375	6	2	120	60	5	1	2	8	101	17	2	0	1	6
STONE	1101 12TH ST	99	912	4	0	487	0	7	600	58	57	16601	4	1	65	65	5	0	1	2	101	25	2	0	6	3
CROSBY	1102 12TH ST	100	913	4	0	332	0	5	456	33	45	7846	2	2	120	60	8	0	2	1	46	23	2	0	0	0
STILLS	1102 12TH ST	100	914	4	0	378	0	5	430	40	56	8372	3	2	120	60	3	0	1	1	70	23	2	0	5	5
NASH	1102 12TH ST	100	915	4	0	672	0	9	757	77	68	9527	4	2	120	60	5	1	2	4	87	22	3	2	0	7
COLLINS	1102 12TH ST	100	916	4	0	358	0	5	337	41	36	3714	6	2	120	60	7	1	2	1	113	19	2	1	0	3
BAEZ	1102 12TH ST	100	917	4	0	303	0	4	353	28	11	11300	4	1	65	65	4	0	1	3	69	17	2	0	0	0
DENVER	1102 12TH ST	100	918	4	0	342	0	4	333	27	17	6305	3	2	120	60	4	0	2	2	70	23	2	2	0	9
STRAUSS	1102 12TH ST	100	919	4	0	325	0	5	404	35	29	7806	6	2	120	60	6	0	2	4	120	20	2	0	0	0
GERSHWIN	1102 12TH ST	100	920	4	0	284	0	3	212	22	18	9239	4	1	65	65	9	0	2	3	96	24	2	1	0	0
LIGHTFOOT	1102 12TH ST	100	921	4	0	457	0	6	462	43	38	3976	6	2	120	60	5	0	2	0	123	21	2	0	0	0
TENILLE	1102 12TH ST	100	922	4	0	695	0	10	760	95	20	9648	4	2	120	60	7	0	2	4	79	20	2	0	0	0
CROFTS	1102 12TH ST	100	923	4	0	575	0	7	554	46	75	10483	4	1	65	65	6	0	1	5	84	21	2	0	0	0
SEALS	1102 12TH ST	100	924	4	0	439	0	5	449	31	23	9648	5	2	120	60	6	1	1	6	112	22	2	1	0	0
BATCOCK	1102 12TH ST	100	925	4	0	377	0	7	539	63	17	9194	2	2	120	60	6	0	2	1	48	24	2	0	0	0
SCARNE	1102 12TH ST	100	926	4	0	396	0	6	471	44	84	9438	7	2	120	60	7	0	1	9	172	25	2	0	1	0
GELENDER	1102 12TH ST	100	927	4	0	356	0	5	391	34	17	9449	4	2	120	60	6	0	2	4	75	19	2	0	0	7
MYER	1102 12TH ST	100	928	4	0	357	0	5	379	40	84	7711	4	2	120	60	4	1	1	2	75	19	2	1	0	1
BACH	1102 12TH ST	100	929	4	0	680	0	8	633	62	17	11873	3	2	120	60	4	1	1	1	70	23	2	0	1	0
HYMES	1102 12TH ST	100	930	4	0	380	0	6	526	45	19	7473	7	2	120	60	5	1	2	3	132	19	2	0	0	0
SORUM	1102 12TH ST	100	931	4	0	304	0	4	324	30	59	8987	4	2	120	60	7	0	2	4	99	25	2	0	0	5
MATSON	1102 12TH ST	100	932	4	0	416	0	6	468	48	24	4896	5	1	65	65	2	1	2	0	116	23	2	2	0	7
ALTHOUSE	1102 12TH ST	100	933	4	0	382	0	6	490	46	64	16051	5	1	65	65	4	0	2	5	114	23	2	1	0	3
WISE	1101 13TH ST	100	934	4	0	277	0	4	308	30	62	3509	3	1	65	65	7	0	2	1	58	19	2	0	0	3
BAUMAN	1101 13TH ST	100	935	4	0	385	0	5	349	37	48	2574	7	1	65	65	5	0	1	5	136	19	2	0	1	11
FULLER	1101 13TH ST	100	936	4	0	302	0	3	239	18	22	4842	2	2	120	60	4	0	2	1	43	22	2	1	0	0
GEORGE	1101 13TH ST	100	937	4	0	490	0	6	451	50	12	7869	8	2	120	60	6	1	2	6	126	16	2	2	0	0
JEFFERSON	1101 13TH ST	100	938	4	0	480	0	7	569	56	12	4322	5	1	65	65	5	0	2	4	104	21	2	0	4	6
MADISON	1101 13TH ST	100	939	4	0	602	0	9	740	67	16	11033	4	2	120	60	7	1	2	4	65	16	2	2	0	3
WILSON	1101 13TH ST	100	940	4	0	351	0	5	399	38	76	8637	4	2	120	60	5	1	1	3	74	19	2	0	4	4
KENNEDY	1101 13TH ST	100	941	4	0	569	0	7	585	55	33	9726	4	2	120	60	9	0	2	5	101	25	2	1	0	5
HUMPHREY	1101 13TH ST	100	942	4	0	497	0	6	475	47	10	3938	3	2	120	60	4	1	1	1	50	17	2	0	0	1
CARTER	1101 13TH ST	100	943	4	0	309	0	5	386	34	65	8197	7	1	65	65	11	1	2	7	145	21	1	0	0	0
NIXON	1101 13TH ST	100	944	4	0	463	0	6	481	46	67	8453	3	2	120	60	8	0	2	3	82	27	2	0	0	0
STEVENSON	1101 13TH ST	100	945	4	0	292	0	3	280	18	73	10898	2	2	120	60	3	1	2	2	29	15	2	0	0	0
WALKER	1101 13TH ST	100	946	4	0	263	0	4	298	31	12	4087	3	2	120	60	5	0	1	4	53	18	3	0	1	4
BACTZER	1101 13TH ST	100	947	4	0	344	0	5	388	37	25	4148	4	2	120	60	6	1	2	3	104	26	2	2	0	4
FENSKE	1101 13TH ST	100	948	4	0	555	0	6	452	47	75	10380	5	1	65	65	4	1	2	4	109	22	2	0	0	0
KIRKLEY	1101 13TH ST	100	949	4	0	359	0	5	393	39	14	11867	5	1	65	65	6	0	2	6	130	26	1	2	0	9
WEBB	1101 13TH ST	100	950	4	0	309	0	5	392	36	53	6162	4	1	65	65	5	1	2	4	89	22	2	0	0	0
GEIL	1101 13TH ST	100	951	4	0	400	0	6	457	49	19	7673	5	1	65	65	6	1	1	4	100	20	2	0	0	0
RANDLE	1101 13TH ST	100	952	4	0	512	0	7	593	62	21	8962	4	2	120	60	4	1	1	3	84	21	2	0	0	0
BATEY	1101 13TH ST	100	953	4	0	349	0	6	475	50	21	9245	6	1	65	65	8	0	1	4	121	20	2	0	0	0
BOLTON	1004 7TH ST	101	954	4	0	477	0	7	565	54	62	14832	4	1	65	65	4	1	2	2	88	22	2	0	0	0
COLON	1004 7TH ST	101	955	4	0	396	0	5	414	38	35	8055	3	2	120	60	7	1	2	1	68	23	2	0	5	3
ERICKSON	1004 7TH ST	101	956	4	0	442	0	6	498	45	11	11543	6	1	65	65	5	1	2	1	131	22	2	0	0	0
FINKE	1004 7TH ST	101	957	4	0	451	0	7	576	45	28	22379	5	3	184	61	4	2	2	4	93	19	2	0	0	0
PERKINS	1004 7TH ST	101	958	4	0	363	0	6	553	49	42	26105	7	3	165	55	11	1	2	2	150	21	2	0	0	0
SPRAGGETT	1004 7TH ST	101	959	4	0	600	0	8	681	66	11	24312	3	2	137	69	6	2	2	6	70	23	2	0	1	11
WORRALL	1004 7TH ST	101	960	4	0	473	0	6	428	45	16	19304	4	2	120	60	4	1	1	2	23	18	2	2	0	0
BIERSTEIN	1004 7TH ST	101	961	4	0	542	0	8	704	63	13	25657	4	3	178	59	9	1	2	3	86	22	3	0	1	4
ERRICO	1004 7TH ST	101	962	4	0	365	0	6	501	48	10	7852	6	2	120	60	3	0	2	3	139	23	2	2	0	1
FISKE	1004 7TH ST	101	963	4	0	263	0	5	451	38	20	28581	3	2	149	75	6	1	1	3	52	17	2	0	4	0
GEDDES	1004 7TH ST	101	964	4	0	374	0	6	503	48	22	24560	6	5	239	48	4	1	2	9	129	22	2	1	0	2
JONES	1003 8TH ST	101	965	4	0	453	0	5	400	37	43	35201	2	1	99	99	6	1	2	3	40	20	2	2	0	10
SLATER	1003 8TH ST	101	966	4	0	368	0	5	368	38	18	9963	3	2	120	60	8	0	1	1	59	20	2	0	0	0
TRIBBE	1003 8TH ST	101	967	4	0	413	0	6	430	47	21	26820	6	2	120	60	3	1	2	4	149	25	3	0	2	0
HEARST	1003 8TH ST	101	968	4	0	375	0	6	478	48	44	21518	5	4	209	52	6	1	1	4	109	22	2	2	0	4
TUBBS	1003 8TH ST	101	969	4	0	532	0	8	616	68	17	21071	2	2	126	63	9	2	1	2	44	22	3	0	4	9
BUBECK	1003 8TH ST	101	970	4	0	337	0	6	474	47	14	21258	6	4	223	56	12	1	2	9	126	21	3	0	0	3
FREEMAN	1003 8TH ST	101	971	4	0	571	0	8	689	70	92	23203	4	2	152	76	6	1	1	2	75	19	2	0	2	1
ELTON	1003 8TH ST	101	972	4	0	232	0	4	325	29	17	21765	1	1	88	88	8	2	2	1	33	33	2	1	0	0
DURBIN	1003 8TH ST	101	973	4	0	402	0	6	478	43	80	22127	4	3	190	63	11	2	2	2	74	19	2	0	4	4
OBRIEN	1003 8TH ST	101	974	4	0	471	0	6	433	52	17	21523	2	2	133	67	6	1	2	2	50	25	2	0	0	0
HOLES	1004 8TH ST	102	975	4	0	444	0	6	526	47	64	24134	5	4	205	51	9	1	2	4	118	24	2	1	0	0
HIGGINS	1004 8TH ST	102	976	4	0	475	0	6	448	45	38	28488	7	2	120	60	5	2	2	0	147	21	2	0	0	0
RAUSCHER	1004 8TH ST	102	977	4	0	264	0	5	418	37	109	21897	1	1	87	87	5	2	2	1	28	28	3	0	0	0
SWEENEY	1004 8TH ST	102	978	4	0	446	0	6	456	44	69	24772	6	4	218	55	6	2	1	6	121	20	2	0	5	10
WRIGHT	1004 8TH ST	102	979	4	0	320	0	5	369	37	14	18807	6	1	65	65	1	0	2	3	126	21	2	1	0	11
MORGAN	1004 8TH ST	102	980	4	0	682	0	9	748	77	12	22087	4	3	175	58	3	2	1	2	80	20	2	0	0	0
SHADBOLT	1004 8TH ST	102	981	4	0	647	0	9	721	77	14	21542	2	2	142	71	2	0	2	1	53	27	2	0	0	11
BISHOP	1004 8TH ST	102	982	4	0	242	0	4	332	26	20	22970	3	2	158	79	6	0	2	2	59	20	2	0	0	0
RATZLAFF	1004 8TH ST	102	983	4	0	454	0	7	540	53	65	18909	4	2	120	60	5	0	2	4	75	19	2	0	0	3
CHARI	1004 8TH ST	102	984	4	0	441	0	6	487	40	19	24413	6	2	120	60	5	1	2	1	110	18	3	0	0	0
PALMER	1004 8TH ST	102	985	4	0	584	0	8	667	63	18	23211	3	2	128	64	6	1	1	3	63	21	2	0	0	2

```
                                                    VARIABLES
         0          0           0     0 0 0   0       0 1  1   1   1 1 1   1   1 1 2   2   2 2 2 2   2 2
         1          2           3     4 5 6   7       8 9  0   1   2   3   4 5 6   7   8 9 0   1   2 3 4 5   6 7
****************************************************************************************************************
STEINER    1003  9TH ST   102   986 4 0   569   0 7  533 57  26  23533  4 2 156  78  9 0 2  1   89 22 2  0 0 0
JACOBS     1003  9TH ST   102   987 4 0   437   0 6  503 48  16  20488  5 2 185  62  5 1 1  6   89 18 2  0 0 4
ROSE       1003  9TH ST   102   988 4 0   461   0 6  443 50  23  19205  6 2 120  60  8 1 1  6  129 22 1  0 0 0
MURPHY     1003  9TH ST   102   989 4 0   590   0 8  676 60  21  22009  5 3 186  62  5 0 2  1  100 20 2  0 5 0
HORN       1003  9TH ST   102   990 4 0   474   0 8  676 67  21  19681  6 2 120  60  6 1 1  3   73 18 2  0 3 11

PUGH       1003  9TH ST   102   991 4 0   580   0 8  644 68  18  20249  4 4 207  52  4 1 1  1   88 22 2  0 4 1
MATTHEWS   1003  9TH ST   102   992 4 0   519   0 6  502 47  15  33790  5 4 211  53  7 2 2  3  104 21 2  0 0 0
WAGNER     1003  9TH ST   102   993 4 0   557   0 9  700 74  46  25260  3 2 122  61 10 2 2  3   69 23 2  0 0 0
WARD       1003  9TH ST   102   994 4 0   388   0 5  406 36  14  34964  1 1  65  65  8 2 2  0   23 23 2  0 6 0
EVANSTON   1003  9TH ST   102   995 4 0   483   0 7  585 52  69  23331  5 3 190  63  6 0 2  0   95 19 2  1 0 0

KRISTOL    1004  9TH ST   103   996 4 0   491   0 7  598 56  10  18157  4 1  65  65  4 1 2  4  102 26 2  0 0 1
SINGER     1004  9TH ST   103   997 4 0   355   0 5  429 36  16  17201  3 2 120  60  5 1 1  1   51 17 2  0 1 0
ENGELBARDT 1004  9TH ST   103   998 4 0   338   0 6  463 46  48  21591  6 4 225  56  7 2 1  1  111 19 2  0 0 0
WILL       1004  9TH ST   103   999 4 0   339   0 5  397 38 120  17691  6 2 120  60  7 1 1  6  127 21 2  0 3 0
WERNICK    1004  9TH ST   103  1000 4 0   333   0 5  362 38  24  20070  2 1  99  99  4 1 1  2   34 17 2  1 0 9

GILBERT    1004  9TH ST   103  1001 4 0   321   0 6  506 49  19  20197  8 2 120  60  8 0 2  2  151 19 2  1 0 0
PARSONS    1004  9TH ST   103  1002 4 0   514   0 8  647 63  78  20517  3 2 162  81  8 2 2  5   81 20 2  0 0 6
LOGAN      1004  9TH ST   103  1003 4 0   318   0 5  393 35  58  19710  6 2 120  60  9 1 1  3  141 24 2  0 4 0
EISELEY    1004  9TH ST   103  1004 4 0   467   0 6  429 45  21  36149  5 4 201  50  4 2 2  8  103 21 1  0 4 10
KISSINGER  1004  9TH ST   103  1005 4 0   406   0 6  453 48  12  20075  3 2 128  64  6 2 1  1   64 21 2  0 0 1

THURBER    1004  9TH ST   103  1006 4 0   462   0 6  445 45  42  21581  7 2 120  60  8 1 1  0  136 19 2  2 0 0
CRAWFORD   1004  9TH ST   103  1007 4 0   388   0 6  518 46  18  25900  5 4 207  52  4 1 1  4  111 22 2  0 0 0
PARTON     1004  9TH ST   103  1008 4 0   431   0 7  552 54  20  20216  4 3 181  60  8 2 1  1   95 24 2  0 5 0
HOLMES     1003 10TH ST   103  1009 4 0   270   0 3  254 21 134  19477  3 1  65  65  7 0 1  1   62 21 2  0 1 4
ERNST      1003 10TH ST   103  1010 4 0   360   0 6  518 50  24  17010  7 1  65  65  2 0 1  5  152 22 2  0 0 5

BATSON     1003 10TH ST   103  1011 4 0   270   0 5  396 36  78  16007  5 1  65  65  6 0 2  6   98 20 3  0 0 0
HAUSER     1003 10TH ST   103  1012 4 0   371   0 6  504 45  78  23411  5 4 205  51  9 2 2  6  109 22 2  0 0 0
GHANDI     1003 10TH ST   103  1013 4 0   352   0 5  420 36  16  20760  4 3 167  56  8 1 2  0   98 25 2  0 5 6
VASHOLZ    1003 10TH ST   103  1014 4 0   615   0 9  799 82  32  24059  4 2 157  79  8 1 2  3   86 22 2  0 5 5
BLANDINA   1003 10TH ST   103  1015 4 0   412   0 6  504 46  22  18330  5 2 120  60  6 1 1  6   99 23 2  0 0 0

GREENFELD  1003 10TH ST   103  1016 4 0   396   0 6  512 46  19  24800  5 3 185  62  5 2 1  9  105 21 2  0 0 0
FUNK       1004 10TH ST   104  1017 4 0   403   0 5  391 41  16  10581  3 2 120  60  6 0 1  4   46 15 2  0 1 11
BOLCH      1004 10TH ST   104  1018 4 0   404   0 7  616 53  46  14985  4 1  65  65  5 0 2  2   78 22 2  1 0 5
GELDER     1004 10TH ST   104  1019 4 0   522   0 7  549 54  87  15404  3 2 120  60  5 0 2  1   60 20 2  0 0 1
PORTER     1004 10TH ST   104  1020 4 0   565   0 8  653 69  31  10569  5 2 120  60  6 1 1  3   97 19 2  0 0 0

HOWARD     1004 10TH ST   104  1021 4 0   310   0 4  315 28  12  14160  1 2 120  60  5 1 2  1   36 36 3  0 3 0
WINCHESTER 1004 10TH ST   104  1022 4 0   579   0 9  735 82  18  15789  1 2 120  60  5 1 2  3  111 28 2  2 0 0
INGE       1004 10TH ST   104  1023 4 0   291   0 4  301 28  19  18889  3 2 120  60  6 1 1  3   70 23 2  0 5 0
KRESS      1004 10TH ST   104  1024 4 0   458   0 5  568 53  28  12822  2 2 120  60  7 1 2  1   59 30 2  0 0 5
ADAMS      1004 10TH ST   104  1025 4 0   455   0 6  510 42  12  17580  6 2 120  60  3 1 1  9  115 19 1  1 0 0

SHEERAN    1004 10TH ST   104  1026 4 0   393   0 6  486 43  74  13041  3 1  65  65  5 0 1  4   75 25 2  0 0 0
COLE       1004 10TH ST   104  1027 4 0   607   0 8  605 67  24  21026  3 2 152  76  5 1 2  4   83 21 1  0 0 0
WARDER     1003 11TH ST   104  1028 4 0   270   0 6  315 30  26  15240  1 2 120  60  3 0 2  1   23 23 2  0 4 11
LETULI     1003 11TH ST   104  1029 4 0   618   0 8  668 65  78  12212  4 2 120  60  6 0 1  3   78 22 2  0 0 0
ATCHLEY    1003 11TH ST   104  1030 4 0   641   0 8  678 64  20  15273  4 2 120  60  4 0 2  4   67 17 2  2 0 6

BIXLER     1003 11TH ST   104  1031 4 0   326   0 5  404 35  11  14461  5 1  65  65  6 1 2  4   96 19 2  0 5 0
GROTH      1003 11TH ST   104  1032 4 0   429   0 6  471 40  37  18651  5 1  65  65  6 1 2  4  110 16 1  0 0 0
SIDNEY     1003 11TH ST   104  1033 4 0   417   0 5  408 41  20  14215  6 2 120  60  6 1 2  6  138 21 1  0 0 7
MAROONE    1003 11TH ST   104  1034 4 0   425   0 6  487 43  16  15387  6 2 120  60  7 1 2  3   77 19 2  0 0 2
MEYENDORFF 1003 11TH ST   104  1035 4 0   392   0 6  472 42  92  17928  4 2 120  60 12 0 2  2   80 20 1  0 3 3

BOYLE      1003 11TH ST   104  1036 4 0   550   0 6  443 50  17  16847  7 1  65  65  7 0 2  3  182 26 2  0 0 5
WINFREY    1003 11TH ST   104  1037 4 0   380   0 6  497 50  92  16336  5 2 120  60  6 0 2  4   99 20 2  0 0 0
TAYLOR     1004 11TH ST   105  1038 4 0   444   0 5  437 38  79  15340  6 2 120  60  3 0 2  1  149 19 2  0 3 6
SHERHAG    1004 11TH ST   105  1039 4 0   263   0 3  200 21  78   7228  5 1  65  65  3 0 1  3   79 16 2  0 3 6
RUSSELL    1004 11TH ST   105  1040 4 0   446   0 6  519 49  23  13837  7 1  65  65  3 1 1  9  146 21 2  0 8 6

BRENNAN    1004 11TH ST   105  1041 4 0   425   0 6  503 47  21  10987  6 1  65  65 10 0 1  6  114 19 2  2 0 0
NIVEN      1004 11TH ST   105  1042 4 0   399   0 6  460 46  16   9930  5 2 120  60  6 0 2  4   97 19 2  0 3 0
COCO       1004 11TH ST   105  1043 4 0   531   0 6  626 69  36   8987  4 2 120  60  7 1 1  2   99 17 2  0 0 0
FALK       1004 11TH ST   105  1044 4 0   435   0 6  460 47  63  11892  5 2 120  60  4 1 2  1  106 21 2  0 3 0
BAILEY     1004 11TH ST   105  1045 4 0   507   0 6  490 48  14   9780  5 1  65  65  5 0 2  8  101 20 1  0 2 0

NEUMAN     1004 11TH ST   105  1046 4 0   556   0 7  532 59  98  11424  3 1  65  65  4 1 1  4   73 24 2  2 0 11
GOULEY     1004 11TH ST   105  1047 4 0   250   0 3  236 21  25  10659  2 2 120  60  2 1 1  4   34 17 2  0 0 0
WAYNE      1004 11TH ST   105  1048 4 0   566   0 8  618 63  14   9099  5 1  65  65  5 1 2  1  103 21 2  0 4 0
HEPBURN    1003 12TH ST   105  1049 4 0   368   0 8  411 41 117   9417  6 1  65  65  7 1 2  3  119 20 2  0 0 1
WOODWARD   1003 12TH ST   105  1050 4 0   527   0 8  575 60  16   8664  3 2 120  60  4 1 2  3   59 20 2  2 0 0

UGGAMS     1003 12TH ST   105  1051 4 0   571   0 8  571 63  12  12125  3 1  65  65  8 1 2  4   70 23 2  0 0 0
DENNIS     1003 12TH ST   105  1052 4 0   343   0 5  423 37  59   8618  2 2 120  60  5 1 2  4  124 21 2  0 0 0
PECK       1003 12TH ST   105  1053 4 0   355   0 6  490 49  67  12661  4 1  65  65  6 1 2  4   86 22 1  0 4 2
HAYES      1003 12TH ST   105  1054 4 0   302   0 5  424 39  14   9089  8 2 120  60  6 1 2  6  158 20 2  1 0 0
BENNY      1003 12TH ST   105  1055 4 0   404   0 5  369 37  15  13371  6 1  65  65  5 1 2  8  138 23 2  0 5 10

MARX       1003 12TH ST   105  1056 4 0   294   0 4  342 27  18   6620  4 1  65  65  7 1 2  1   87 22 2  0 0 0
PRESLEY    1003 12TH ST   105  1057 4 0   227   0 3  279 29  10  19842  4 2 120  60  3 0 2  1   36 18 2  0 1 11
ALLEN      1003 12TH ST   105  1058 4 0   251   0 3  252 19  17  19917  2 2 120  60  7 1 2  2   50 25 2  0 1 0
BENEDICT   1004 12TH ST   106  1059 4 0   300   0 5  415 35  12   5853  8 1  65  65  5 1 1  4   41 21 2  0 1 0
POMEROX    1004 12TH ST   106  1060 4 0   395   0 6  484 54  21   7288  8 1  65  65  5 1 1  4  160 20 3  0 0 7

MCMANUS    1004 12TH ST   106  1061 4 0   430   0 5  403 40  26   1399  4 2 120  60  3 0 2  7   98 25 2  0 1 0
SADAT      1004 12TH ST   106  1062 4 0   299   0 4  300 27  16   7921  4 2 120  60  4 1 2  4  101 20 2  0 0 0
KROLL      1004 12TH ST   106  1063 4 0   298   0 5  370 34 107   7717  5 2 120  60 10 1 2  9   98 20 2  0 0 0
MCCAULEY   1004 12TH ST   106  1064 4 0   433   0 6  648 80  16   7941  4 2 120  60  4 1 2  9  102 20 2  0 6 0
PETERS     1004 12TH ST   106  1065 4 0   182   0 3  202 20  18   5469  3 2 120  60  9 1 1  3   75 25 2  0 0 0

RESTON     1004 12TH ST   106  1066 4 0   466   0 7  577 55  49   8048  5 1  65  65  5 1 1  3   95 19 2  0 0 0
MCKAY      1004 12TH ST   106  1067 4 0   179   0 3  301 32  44  10811  3 2 120  60  5 1 2  2   32 32 2  0 5 1
FROELICH   1004 12TH ST   106  1068 4 0   154   0 3  239 20  19   6273  3 2 120  60  3 1 2  2   72 24 2  0 0 0
MORRIS     1004 12TH ST   106  1069 4 0   506   0 8  632 60  22   8148  3 2 120  60  2 1 2  2   67 22 2  0 0 0
DEANE      1004 12TH ST   106  1070 4 0   307   0 5  361 30  27   7936  1 2 120  60  4 1 2  1   35 35 2  0 0 0

BURGESS    1003 13TH ST   106  1071 4 0   549   0 6  446 44  40   5813  5 1  65  65 11 0 2  4  110 22 2  0 4 9
AMANS      1003 13TH ST   106  1072 4 0   478   0 6  693 64  20   9005  6 2 120  60  6 0 1  4  109 22 2  2 0 0
RYAN       1003 13TH ST   106  1073 4 0   359   0 5  381 38  71   7306  6 2 120  60  7 0 1  6  138 23 2  0 1 7
GREENWOOD  1003 13TH ST   106  1074 4 0   377   0 6  466 44  18  11770  6 1  65  65  6 1 2  4  116 19 2  0 1 0
PARKER     1003 13TH ST   106  1075 4 0   435   0 6  499 44  26   5175  6 1  65  65  6 1 2  4  123 21 2  0 1 10

HOBSON     1003 13TH ST   106  1076 4 0   467   0 6  413 48  20  27032  6 4 218  55  5 1 1  3  112 19 2  0 0 0
MCMINN     1003 13TH ST   106  1077 4 0   434   0 6  422 38  66   6920  4 2 120  60  9 1 1  3  169 24 2  0 0 0
CAMPBELL   1003 13TH ST   106  1078 4 0   283   0 5  387 43  91   6894  6 1  65  65  5 1 1  2  111 19 2  0 0 0
MILLS      1003 13TH ST   106  1079 4 0   377   0 5  367 39  62   6246  6 1  65  65  5 0 1  4   74 19 2  0 0 0
MALONEY    1002  7TH ST   107  1080 4 0   352   0 5  451 35  17  13113  6 1  65  65  5 1 2  3  132 22 1  0 0 4

LAWRENCE   1002  7TH ST   107  1081 4 0   503   0 7  568 53  17  19787  3 2 120  60  6 1 2  3   70 23 2  1 0 0
PENN       1002  7TH ST   107  1082 4 0   305   0 5  393 32  28  19814  3 2 120  60  4 1 2  3   94 16 2  0 1 0
KNIGHT     1002  7TH ST   107  1083 4 0   560   0 8  688 68  14  31908  5 3 196  65  5 2 1  2  111 22 1  0 0 3
TROY       1002  7TH ST   107  1084 4 0   339   0 5  362 38  51  49410  5 3 117 117  3 2 1  2   40 20 2  0 0 2
VIGUERIE   1002  7TH ST   107  1085 4 0   389   0 5  427 37  14  28679  5 3 189  63  5 2 1  2   94 19 2  0 0 9
```

VARIABLES

01	02	03	04	05	06	07	08	09	10	11	12	13	14	15	16	17	18	19	20	21	22	23	24	25	26	27
CHARLES	1002 7TH ST	107	1086	4	0	222	0	3	293	20	23	20539	5	3	191	64	5	1	2	4	88	18	2	0	0	0
SIMS	1002 7TH ST	107	1087	4	0	519	0	8	700	65	20	19207	4	2	120	60	8	0	1	4	88	22	2	0	0	0
OLSON	1002 7TH ST	107	1088	4	0	426	0	6	428	47	12	32365	6	2	120	60	6	2	1	6	125	21	2	2	0	0
KROHNFELDT	1002 7TH ST	107	1089	4	0	434	0	6	441	47	56	30697	6	4	218	55	4	2	2	4	127	21	2	0	5	1
SULK	1002 7TH ST	107	1090	4	0	412	0	6	478	46	21	24361	6	2	120	60	8	2	1	6	143	24	2	2	0	2
CLARK	1001 8TH ST	107	1091	4	0	436	0	6	466	46	21	35832	4	2	148	74	9	0	1	8	75	19	2	0	0	0
VOELL	1001 8TH ST	107	1092	4	0	525	0	7	592	56	13	26576	4	2	149	75	6	2	2	5	78	20	2	0	0	0
PEALE	1001 8TH ST	107	1093	4	0	328	0	6	445	39	71	25152	6	3	165	55	5	1	2	4	119	20	2	0	5	0
WILBUR	1001 8TH ST	107	1094	4	0	372	0	5	465	40	14	20876	4	2	156	78	6	1	1	6	80	20	2	2	0	6
SELLECK	1001 8TH ST	107	1095	4	0	405	0	6	465	45	23	27968	7	2	120	60	6	2	2	5	143	20	2	0	0	0
MEDOR	1001 8TH ST	107	1096	4	0	499	0	7	495	57	58	23083	5	3	185	62	3	1	2	5	80	16	2	2	0	1
WILLS	1001 8TH ST	107	1097	4	0	241	0	4	283	29	79	19691	2	2	120	60	5	1	2	2	39	20	2	0	6	2
COHEN	1001 8TH ST	107	1098	4	0	354	0	6	475	44	19	22100	7	3	165	55	10	1	1	7	132	19	2	2	0	5
MANDELL	1001 8TH ST	107	1099	4	0	666	0	9	719	76	38	29134	4	3	190	63	8	0	2	3	73	18	2	0	0	2
MARGOLIS	1001 8TH ST	107	1100	4	0	432	0	6	498	43	82	20653	5	4	206	52	6	1	2	3	105	21	2	2	0	0
CANE	1002 8TH ST	108	1101	4	0	358	0	6	471	48	70	17005	5	1	65	65	5	1	1	3	105	21	3	0	4	2
WORCHESTER	1002 8TH ST	108	1102	4	0	640	0	9	749	85	14	19518	4	2	120	60	7	1	1	6	94	24	2	2	0	0
TAFT	1002 8TH ST	108	1103	4	0	526	0	6	514	41	14	21607	6	2	120	60	6	1	1	4	138	23	2	0	0	0
LINCOLN	1002 8TH ST	108	1104	4	0	468	0	7	583	62	20	24259	4	2	162	81	9	2	1	1	80	20	2	2	0	2
EVSLIN	1002 8TH ST	108	1105	4	0	569	0	8	648	68	15	25132	5	3	199	66	6	1	2	5	106	21	2	0	1	11
COFFRIN	1002 8TH ST	108	1106	4	0	354	0	5	405	36	17	19336	5	1	65	65	6	0	1	1	119	24	2	2	0	2
PLESHAW	1002 8TH ST	108	1107	4	0	677	0	9	709	77	19	31397	4	2	158	79	4	2	2	0	70	18	2	1	0	8
DAHLIN	1002 8TH ST	108	1108	4	0	552	0	8	673	62	19	19621	2	2	120	60	6	1	1	2	49	25	2	2	0	4
DEWITT	1002 8TH ST	108	1109	4	0	550	0	8	636	68	76	20826	4	2	157	79	10	2	1	8	82	21	2	0	0	0
HOWARD	1002 8TH ST	108	1110	4	0	432	0	6	470	44	13	23902	7	3	165	55	5	0	2	7	134	19	2	1	0	0
KRAFT	1001 9TH ST	108	1111	4	0	504	0	8	613	67	94	23180	5	3	197	66	3	2	1	3	93	19	1	0	6	10
WELSH	1001 9TH ST	108	1112	4	0	469	0	6	503	45	34	22603	8	2	120	60	3	1	2	2	144	18	2	0	4	5
CHESTERTON	1001 9TH ST	108	1113	4	0	433	0	6	517	48	19	19245	5	1	65	65	4	0	1	4	112	22	3	1	0	0
QUADE	1001 9TH ST	108	1114	4	0	334	0	6	500	39	66	18105	5	1	65	65	8	1	1	6	95	19	2	0	0	0
DILLON	1001 9TH ST	108	1115	4	0	358	0	6	509	52	90	22379	7	2	120	60	7	2	1	5	142	20	2	0	4	1
COHEN	1001 9TH ST	108	1116	4	0	423	0	6	452	48	16	34432	4	3	165	55	11	0	2	1	98	25	2	0	0	1
STUART	1001 9TH ST	108	1117	4	0	478	0	6	465	41	82	30937	5	3	189	63	7	2	2	0	104	21	2	0	0	9
FUDIM	1001 9TH ST	108	1118	4	0	344	0	5	404	34	72	20851	4	3	173	58	10	2	1	6	86	22	1	0	6	0
HUGO	1001 9TH ST	108	1119	4	0	539	0	7	596	54	64	23802	4	3	168	56	8	2	2	2	85	21	2	0	1	5
FERNANDEZ	1001 9TH ST	108	1120	4	0	373	0	5	360	42	16	4519	3	2	120	60	8	0	1	1	78	26	2	0	0	0
SILER	1001 9TH ST	108	1121	4	0	373	0	6	528	45	21	22564	6	4	216	54	5	2	1	6	114	19	2	0	0	0
ETONS	1002 9TH ST	109	1122	4	0	389	0	6	466	47	84	14900	5	1	65	65	4	1	2	3	110	22	2	0	3	3
TIMOTHY	1002 9TH ST	109	1123	4	0	265	0	4	310	29	102	16711	1	2	120	60	7	0	2	1	29	29	2	0	0	0
PEREZ	1002 9TH ST	109	1124	4	0	357	0	5	432	38	28	14333	2	2	120	60	4	0	1	2	57	29	2	0	3	0
GREGORY	1002 9TH ST	109	1125	4	0	423	0	6	482	51	16	17724	5	2	120	60	9	1	1	4	128	26	2	2	0	9
SHOUMATOFF	1002 9TH ST	109	1126	4	0	325	0	6	476	47	19	22005	4	2	162	81	6	2	1	4	87	22	2	0	1	0
HEFFRON	1002 9TH ST	109	1127	4	0	517	0	8	673	52	17	18341	4	2	120	60	7	1	2	2	73	18	2	0	5	1
DEFARIA	1002 9TH ST	109	1128	4	0	609	0	8	678	68	24	26511	5	3	199	66	7	2	2	1	114	23	2	0	4	2
PETERSON	1002 9TH ST	109	1129	4	0	416	0	7	543	57	13	18771	4	1	65	65	6	1	2	6	66	17	2	0	0	0
FRITTS	1002 9TH ST	109	1130	4	0	344	0	6	521	42	21	19307	6	2	120	60	4	0	2	6	124	21	2	2	0	5
MCCABE	1002 9TH ST	109	1131	4	0	334	0	5	384	35	49	24604	2	1	105	105	5	1	2	1	39	20	2	2	0	0
ALLEN	1002 9TH ST	109	1132	4	0	233	0	4	363	28	18	19179	5	2	120	60	7	1	2	4	108	22	2	0	0	11
KAPLAN	1001 10TH ST	109	1133	4	0	495	0	6	526	50	15	20791	7	2	120	60	2	1	2	5	129	18	2	0	0	5
HOOVER	1001 10TH ST	109	1134	4	0	468	0	6	492	37	17	17395	8	1	65	65	4	1	2	6	163	20	2	2	0	0
TURNER	1001 10TH ST	109	1135	4	0	327	0	4	340	26	28	21870	1	1	65	65	6	0	2	0	23	23	2	1	0	0
OGRADY	1001 10TH ST	109	1136	4	0	424	0	6	458	43	105	23923	8	2	120	60	5	2	2	8	162	20	2	0	5	1
STORM	1001 10TH ST	109	1137	4	0	351	0	6	506	47	38	27784	5	4	206	52	6	2	2	5	102	20	2	2	0	0
GREGG	1001 10TH ST	109	1138	4	0	369	0	6	497	50	14	18481	5	1	65	65	6	0	2	4	116	23	2	0	0	10
WIRTA	1001 10TH ST	109	1139	4	0	421	0	6	469	46	28	18924	5	2	120	60	4	0	2	3	95	19	2	0	1	4
ROWLAND	1001 10TH ST	109	1140	4	0	519	0	6	420	51	19	20883	5	4	208	52	3	1	2	5	122	24	1	0	1	1
LUBITZ	1001 10TH ST	109	1141	4	0	610	0	8	644	63	39	25317	3	2	130	65	8	2	2	1	69	23	1	2	0	4
KOGER	1001 10TH ST	109	1142	4	0	441	0	6	506	47	21	20648	3	2	158	79	4	2	1	1	60	20	2	0	0	0
LONIEWSKI	1002 10TH ST	110	1143	4	0	268	0	3	229	21	16	12770	4	2	120	60	6	1	1	0	74	19	2	0	5	0
HALBROOKS	1002 10TH ST	110	1144	4	0	345	0	6	482	46	63	12597	5	1	65	65	8	1	1	5	126	25	2	1	0	9
CALDWELL	1002 10TH ST	110	1145	4	0	526	0	7	560	59	22	13043	3	2	120	60	7	1	1	1	61	20	2	1	0	1
SUSSEX	1002 10TH ST	110	1146	4	0	233	0	4	384	26	26	18919	1	2	120	60	5	1	1	1	26	26	2	2	0	0
VICKSLEY	1002 10TH ST	110	1147	4	0	270	0	5	415	34	25	14790	6	2	120	60	7	0	1	8	119	20	2	0	4	8
JOSEPHS	1002 10TH ST	110	1148	4	0	525	0	8	609	64	60	18718	4	2	120	60	9	0	1	5	70	18	2	0	0	0
GORMAN	1002 10TH ST	110	1149	4	0	438	0	8	651	69	27	19670	4	2	120	60	5	0	2	4	87	22	2	0	0	4
RUIZ	1002 10TH ST	110	1150	4	0	491	0	8	632	68	12	12653	3	2	120	60	5	1	1	4	61	20	2	0	0	0
PURCELL	1002 10TH ST	110	1151	4	0	403	0	5	425	36	81	18639	8	2	120	60	4	0	1	6	161	20	2	2	0	0
HINDS	1002 10TH ST	110	1152	4	0	412	0	6	529	42	40	11171	7	1	65	65	4	0	2	7	154	22	2	2	0	0
HOROWITZ	1002 10TH ST	110	1153	4	0	323	0	5	433	33	25	12575	5	1	65	65	2	0	1	3	120	24	1	0	2	0
NORTON	1001 11TH ST	110	1154	4	0	551	0	9	762	65	38	19199	4	1	65	65	4	0	1	4	94	24	2	0	1	0
MIDDLETON	1001 11TH ST	110	1155	4	0	445	0	7	555	61	22	14018	3	1	65	65	7	1	2	1	73	24	2	0	0	0
BASSETT	1001 11TH ST	110	1156	4	0	493	0	6	484	50	13	27039	6	4	224	56	8	2	1	4	130	22	2	0	3	9
FOSTER	1001 11TH ST	110	1157	4	0	402	0	6	393	44	29	17098	3	1	65	65	5	0	2	3	66	22	2	2	0	0
KAHN	1001 11TH ST	110	1158	4	0	261	0	4	260	29	89	7531	4	1	65	65	8	1	2	4	98	25	2	0	4	0
DUNLAN	1001 11TH ST	110	1159	4	0	417	0	6	465	49	19	18141	4	2	120	60	7	1	1	5	79	20	2	0	0	0
PRAEFKE	1001 11TH ST	110	1160	4	0	237	0	4	338	28	21	17381	3	1	65	65	6	0	2	4	81	27	2	0	5	0
OLESON	1001 11TH ST	110	1161	4	0	388	0	6	489	47	16	11075	6	2	120	60	6	0	1	0	136	23	2	2	0	4
MENDOZA	1001 11TH ST	110	1162	4	0	527	0	8	623	63	24	20610	3	2	151	76	6	2	2	1	60	20	2	2	0	8
ACEVEDO	1001 11TH ST	110	1163	4	0	583	0	8	615	64	99	19334	5	2	120	60	4	1	2	6	97	19	2	2	0	6
CLARKSON	1002 11TH ST	111	1164	4	0	206	0	3	263	22	82	7369	2	2	120	60	5	0	2	1	41	21	2	2	0	2
NEUMAN	1002 11TH ST	111	1165	4	0	282	0	3	229	24	96	11763	2	2	120	60	3	1	1	1	53	27	2	2	0	0
KATZMAN	1002 11TH ST	111	1166	4	0	448	0	6	488	46	46	9336	5	2	120	60	6	0	2	5	117	23	2	0	0	2
BAGWELL	1002 11TH ST	111	1167	4	0	372	0	5	382	34	24	17790	2	2	120	60	9	1	1	3	50	25	2	0	0	0
GREEN	1002 11TH ST	111	1168	4	0	341	0	5	452	38	22	8998	8	2	120	60	9	0	2	2	186	23	2	0	0	1
SELAWRY	1002 11TH ST	111	1169	4	0	230	0	4	279	30	51	11561	1	2	120	60	6	0	1	1	25	25	2	0	0	9
NAGEL	1002 11TH ST	111	1170	4	0	346	0	6	522	45	55	9609	4	1	65	65	4	1	2	2	84	21	2	0	5	0
HENDRICKS	1002 11TH ST	111	1171	4	0	270	0	5	412	40	81	7406	5	2	120	60	6	1	1	1	112	22	2	1	0	5
WILLIAMS	1002 11TH ST	111	1172	4	0	563	0	6	525	38	31	7983	7	2	120	60	3	1	1	9	113	16	2	0	0	7
ROOTH	1002 11TH ST	111	1173	4	0	402	0	5	429	40	13	9844	5	2	120	60	4	1	2	3	121	24	2	1	0	0
BALOUGH	1002 11TH ST	111	1174	4	0	336	0	5	412	41	68	9755	6	2	120	60	8	0	2	12	132	22	2	0	0	4
WEBB	1001 12TH ST	111	1175	4	0	339	0	5	417	34	14	10227	5	2	120	60	7	1	2	5	117	23	2	0	4	0
PRICE	1001 12TH ST	111	1176	4	0	475	0	7	523	55	85	15981	4	1	65	65	8	0	2	4	81	20	2	2	0	5
FERNANDEZ	1001 12TH ST	111	1177	4	0	513	0	7	568	51	41	19315	3	2	120	60	5	1	2	6	59	20	2	0	0	0
CARLSON	1001 12TH ST	111	1178	4	0	481	0	7	545	56	18	13035	4	1	65	65	7	0	1	3	81	20	2	2	0	0
ARMINIO	1001 12TH ST	111	1179	4	0	556	0	8	628	61	18	13622	4	2	120	60	7	0	2	2	84	21	2	0	1	9
HERNANDEZ	1001 12TH ST	111	1180	4	0	608	0	8	666	71	19	11156	3	2	120	60	7	1	2	5	76	25	3	0	4	9
DURAN	1001 12TH ST	111	1181	4	0	569	0	7	608	55	22	18194	2	2	120	60	10	1	1	2	53	27	2	0	4	4
FERGUSON	1001 12TH ST	111	1182	4	0	351	0	6	479	43	45	15918	6	2	120	60	3	1	1	8	133	22	2	0	0	0
COHEN	1001 12TH ST	111	1183	4	0	463	0	7	559	51	88	14584	5	2	120	60	5	1	2	1	81	16	2	2	0	0
HARRIS	1001 12TH ST	111	1184	4	0	424	0	5	433	35	13	6549	6	2	120	60	5	0	1	3	132	22	2	0	0	0
DELEON	1002 12TH ST	112	1185	4	0	287	0	5	350	33	13	4581	3	2	120	60	5	0	2	4	58	19	2	0	4	0

C Stat City data base

331

VARIABLES

01	02	03	04	05	06	07	08	09	10	11	12	13	14	15	16	17	18	19	20	21	22	23	24	25	26	27
MURPHY	1002 12TH ST	112	1186	4	0	379	0	6	480	45	14	6677	6	2	120	60	6	1	1	3	130	22	2	0	0	0
RASKOSKY	1002 12TH ST	112	1187	4	0	307	0	4	312	31	18	6083	2	2	120	60	5	1	1	1	35	18	2	0	0	0
OSTROW	1002 12TH ST	112	1188	4	0	343	0	5	411	38	107	7504	7	2	120	60	8	1	2	5	159	23	1	2	0	5
CORONA	1002 12TH ST	112	1189	4	0	449	0	6	500	47	25	1807	5	2	120	60	6	0	2	5	125	25	2	0	1	1
BILLINGS	1002 12TH ST	112	1190	4	0	474	0	5	434	36	19	8126	3	2	120	60	4	1	2	1	66	22	2	0	0	0
PARKS	1002 12TH ST	112	1191	4	0	337	0	5	352	37	15	7520	5	2	120	60	6	1	2	4	104	21	2	0	6	6
FISHER	1002 12TH ST	112	1192	4	0	405	0	6	466	52	16	5786	7	1	65	65	2	1	2	5	124	18	2	0	0	6
DELGADO	1002 12TH ST	112	1193	4	0	266	0	5	365	42	22	7865	6	2	120	60	8	1	1	9	114	19	2	0	0	0
REIZEN	1002 12TH ST	112	1194	4	0	214	0	3	220	19	15	8841	1	2	120	60	6	1	1	1	28	28	2	2	0	0
TERCILLA	1002 12TH ST	112	1195	4	0	367	0	5	384	35	13	8256	3	2	120	60	6	0	1	4	71	24	1	2	0	4
ESCOBAR	1001 13TH ST	112	1196	4	0	281	0	4	373	26	26	5985	4	2	120	60	5	0	1	8	81	20	2	0	4	0
OWENS	1001 13TH ST	112	1197	4	0	405	0	5	423	32	91	5532	8	2	120	60	8	1	1	4	176	22	2	2	0	2
LENHARDT	1001 13TH ST	112	1198	4	0	308	0	4	344	31	39	5881	5	2	120	60	5	1	2	4	96	19	2	1	0	0
MALZONE	1001 13TH ST	112	1199	4	0	379	0	6	487	49	20	8189	5	2	120	60	2	1	2	5	93	19	2	0	0	11
TANNER	1001 13TH ST	112	1200	4	0	386	0	6	458	49	30	7723	5	1	65	65	7	0	2	4	126	25	2	0	0	0
STREETER	1001 13TH ST	112	1201	4	0	213	0	4	299	29	25	2974	5	1	65	65	2	0	2	3	87	17	2	0	0	0
BERKOWITZ	1001 13TH ST	112	1202	4	0	267	0	4	335	29	105	4739	3	1	65	65	3	0	1	4	85	28	2	0	4	4
FONTANA	1001 13TH ST	112	1203	4	0	441	0	6	504	45	21	5302	5	2	120	60	3	0	2	3	122	24	2	0	5	4
SINGLETON	1001 13TH ST	112	1204	4	0	243	0	3	210	22	38	5846	4	1	65	65	6	0	1	4	85	21	2	0	1	4
ANDERSON	1001 13TH ST	112	1205	4	0	264	0	5	428	45	89	5429	6	2	120	60	6	0	1	4	126	21	2	0	6	3
WALKER	904 9TH ST	113	1206	4	0	466	0	6	458	56	19	13222	7	1	65	65	6	1	1	2	113	16	2	0	3	0
HIROKAWA	904 9TH ST	113	1207	4	0	468	0	7	624	60	22	12451	4	1	65	65	3	0	2	4	95	24	2	0	0	0
MACHT	904 9TH ST	113	1208	4	0	504	0	8	608	65	37	14475	4	2	120	60	6	1	2	2	89	22	2	0	5	2
JIMENEZ	904 9TH ST	113	1209	4	0	379	0	5	384	34	34	10477	3	2	120	60	6	1	2	3	70	23	2	0	0	0
LINDSAY	904 9TH ST	113	1210	4	0	539	0	8	629	65	19	16986	3	2	120	60	7	0	2	1	81	27	3	0	0	10
WALLACH	904 9TH ST	113	1211	4	0	403	0	6	495	46	15	21285	5	4	202	51	5	2	2	4	110	22	2	0	1	0
SIBLEX	904 9TH ST	113	1212	4	0	487	0	7	542	49	77	22910	5	3	187	62	8	2	2	1	94	19	2	0	4	0
HILL	904 9TH ST	113	1213	4	0	382	0	6	499	42	20	21847	6	4	218	55	7	2	1	6	137	23	2	1	0	6
PACCIONE	904 9TH ST	113	1214	4	0	162	0	3	243	21	30	13461	2	2	120	60	8	0	2	1	45	23	2	2	0	0
LEHMAN	904 9TH ST	113	1215	4	0	538	0	6	487	48	20	21094	5	4	207	52	7	0	1	4	123	25	2	2	0	1
TORRES	904 9TH ST	113	1216	4	0	690	0	10	817	94	74	29141	4	3	183	61	10	0	1	4	79	20	2	0	0	0
JORGE	903 10TH ST	113	1217	4	0	267	0	5	384	39	27	20175	1	1	77	77	5	2	2	1	31	31	2	0	1	3
NATHANSON	903 10TH ST	113	1218	4	0	435	0	5	423	35	94	40836	1	1	65	65	5	2	2	1	26	26	1	0	5	6
EARLY	903 10TH ST	113	1219	4	0	361	0	7	521	53	18	19757	4	1	65	65	5	1	1	5	82	21	2	0	5	1
MACIEL	903 10TH ST	113	1220	4	0	561	0	8	669	64	50	27651	2	2	120	60	7	1	2	1	53	27	2	0	4	4
ZUCKERMAN	903 10TH ST	113	1221	4	0	423	0	6	490	45	72	24703	5	5	233	47	6	2	2	6	111	22	2	0	1	0
SANDUSKY	903 10TH ST	113	1222	4	0	365	0	5	402	36	15	20390	1	1	85	85	6	0	1	0	39	39	2	0	0	0
FERRE	903 10TH ST	113	1223	4	0	420	0	6	500	51	19	29086	6	2	120	60	3	1	2	8	105	18	2	2	0	0
MARQUEZ	903 10TH ST	113	1224	4	0	446	0	6	488	46	38	25525	4	3	169	56	2	0	1	4	79	20	2	0	0	2
GRABOWSKI	903 10TH ST	113	1225	4	0	618	0	8	666	66	53	26071	3	2	128	64	6	2	1	1	74	25	2	1	0	3
CORBETT	903 10TH ST	113	1226	4	0	364	0	5	418	40	79	19567	5	1	65	65	2	0	2	5	91	18	2	0	4	0
BAUER	904 10TH ST	114	1227	4	0	395	0	7	560	51	38	10312	4	2	120	60	10	0	1	4	109	27	2	0	0	0
DORSEY	904 10TH ST	114	1228	4	0	433	0	6	433	48	21	7967	6	1	65	65	8	1	1	3	123	21	1	0	4	11
ALVAREZ	904 10TH ST	114	1229	4	0	441	0	5	442	38	72	16968	5	1	65	65	3	0	2	3	89	18	2	0	0	3
SENN	904 10TH ST	114	1230	4	0	322	0	4	302	25	19	11648	2	2	120	60	7	0	2	3	36	18	2	0	2	3
VARGAS	904 10TH ST	114	1231	4	0	440	0	7	586	59	22	19576	3	1	65	65	3	0	2	2	68	23	2	0	4	3
BASSETT	904 10TH ST	114	1232	4	0	511	0	7	571	56	72	14331	4	1	65	65	3	0	1	4	79	20	2	0	0	0
ROSENBLUM	904 10TH ST	114	1233	4	0	656	0	10	832	79	71	13477	4	1	65	65	4	0	2	1	97	24	2	1	0	3
COOLIDGE	904 10TH ST	114	1234	4	0	475	0	6	477	49	13	17580	3	2	120	60	7	1	2	0	88	29	2	0	5	6
TERRELL	904 10TH ST	114	1235	4	0	282	0	4	295	28	16	15042	1	2	120	60	8	0	1	1	37	37	2	0	1	3
PRIETO	904 10TH ST	114	1236	4	0	469	0	7	544	52	13	19745	3	2	120	60	4	1	1	3	58	19	2	0	0	0
ADAMSON	904 10TH ST	114	1237	4	0	576	0	8	643	59	21	21549	3	2	155	78	5	1	1	3	50	17	3	1	0	6
SCHWARTZ	903 11TH ST	114	1238	4	0	453	0	6	487	42	15	25459	6	2	120	60	4	2	2	8	146	24	2	2	0	3
PROULX	903 11TH ST	114	1239	4	0	278	0	5	430	41	22	13326	3	2	120	60	7	1	1	1	74	25	2	0	2	0
WINNICK	903 11TH ST	114	1240	4	0	498	0	8	637	68	76	25241	4	3	165	55	3	2	2	2	93	23	2	0	0	4
RUBIN	903 11TH ST	114	1241	4	0	438	0	7	572	58	75	18150	5	2	120	60	8	0	2	3	109	22	2	0	1	0
CORDOVA	903 11TH ST	114	1242	4	0	411	0	6	471	47	22	24743	4	3	168	56	8	2	2	3	73	18	2	2	0	2
LO	903 11TH ST	114	1243	4	0	377	0	5	394	37	15	17542	6	1	65	65	4	1	1	3	109	18	2	2	0	1
KIRSCH	903 11TH ST	114	1244	4	0	352	0	5	420	39	23	16095	3	1	65	65	2	1	1	2	78	26	2	2	0	7
TALMADGE	903 11TH ST	114	1245	4	0	485	0	6	466	45	77	22413	4	3	180	60	7	2	2	4	96	24	2	1	0	0
MELLA	903 11TH ST	114	1246	4	0	432	0	6	494	47	58	24855	4	3	195	65	7	2	2	3	85	21	2	0	0	0
LESTER	903 11TH ST	114	1247	4	0	377	0	6	494	45	78	21735	5	4	212	53	6	2	2	3	120	24	2	1	0	1
SUTHERLAND	904 11TH ST	115	1248	4	0	437	0	6	432	48	14	5780	8	1	65	65	6	1	1	12	178	22	2	0	0	0
FOLTMAN	904 11TH ST	115	1249	4	0	113	0	2	164	13	20	5660	4	2	120	60	5	1	2	3	84	21	2	0	0	0
MENDOZA	904 11TH ST	115	1250	4	0	422	0	7	554	60	107	10145	5	1	65	65	7	0	2	0	96	19	2	2	0	11
KINSEY	904 11TH ST	115	1251	4	0	416	0	6	455	46	25	10803	5	2	120	60	10	1	2	0	94	19	2	2	0	3
HORNER	904 11TH ST	115	1252	4	0	393	0	6	460	48	24	10116	5	2	120	60	8	0	1	1	123	25	2	0	6	7
SENGUPTA	904 11TH ST	115	1253	4	0	337	0	5	419	43	93	11296	2	2	120	60	8	0	2	1	51	26	2	0	0	1
NOTARIO	904 11TH ST	115	1254	4	0	370	0	6	480	48	20	13827	7	1	65	65	2	1	2	10	117	17	2	2	0	1
JACOBSON	904 11TH ST	115	1255	4	0	353	0	5	434	40	25	9187	4	2	120	60	5	0	2	3	80	20	1	2	0	2
FOLLETT	904 11TH ST	115	1256	4	0	270	0	4	338	29	24	7492	3	2	120	60	3	0	2	1	55	18	2	2	0	7
HWANG	904 11TH ST	115	1257	4	0	549	0	8	667	63	21	14385	3	1	65	65	3	1	2	1	80	27	2	0	0	0
SCHNEIDER	904 11TH ST	115	1258	4	0	370	0	5	395	37	22	6473	2	2	120	60	7	1	1	1	46	23	3	0	1	1
MIDDLETON	903 12TH ST	115	1259	4	0	567	0	7	608	55	18	16406	5	2	120	60	7	1	1	0	101	20	2	0	5	5
TYLER	903 12TH ST	115	1260	4	0	512	0	7	569	58	18	10392	5	2	120	60	6	0	1	4	121	24	2	0	0	7
MCKENRY	903 12TH ST	115	1261	4	0	524	0	7	561	53	39	10925	5	2	120	60	8	1	1	3	106	21	1	0	1	1
KAPROVE	903 12TH ST	115	1262	4	0	291	0	3	256	21	21	11557	1	2	120	60	5	1	2	0	33	33	2	2	0	3
MORGAN	903 12TH ST	115	1263	4	0	352	0	5	406	35	16	18385	4	2	120	60	6	1	1	3	85	21	2	0	1	2
VASQUEZ	903 12TH ST	115	1264	4	0	514	0	7	559	58	42	12860	3	2	120	60	4	1	1	2	55	18	2	0	6	8
JACKSON	903 12TH ST	115	1265	4	0	689	0	9	700	84	38	23745	4	2	150	75	11	0	1	2	74	19	2	1	0	1
TANNER	903 12TH ST	115	1266	4	0	431	0	7	547	53	23	20685	3	2	140	70	9	1	1	4	75	25	3	0	0	0
PROVENZO	903 12TH ST	115	1267	4	0	288	0	5	408	35	47	8466	5	1	65	65	6	0	1	3	115	23	2	0	0	0
LEVERMORE	903 12TH ST	115	1268	4	0	458	0	7	586	52	24	24872	5	3	185	62	7	1	2	0	80	16	2	0	6	2
SEYMOUR	904 12TH ST	116	1269	4	0	398	0	6	490	48	18	4877	7	1	65	65	10	0	2	3	142	20	2	0	5	4
OYEMURA	904 12TH ST	116	1270	4	0	377	0	5	412	36	62	9179	6	1	65	65	2	0	1	3	136	23	2	0	0	0
ROSS	904 12TH ST	116	1271	4	0	405	0	5	461	33	13	6494	3	2	120	60	3	0	2	2	65	22	2	0	0	0
GREGORY	904 12TH ST	116	1272	4	0	356	0	5	422	36	20	13933	5	1	65	65	5	0	2	4	104	21	2	0	0	0
LIPMAN	904 12TH ST	116	1273	4	0	396	0	5	411	32	20	6798	4	2	120	60	6	1	2	1	68	17	2	2	0	3
VICTOR	904 12TH ST	116	1274	4	0	454	0	7	586	59	20	11148	4	1	65	65	6	1	2	2	96	24	2	0	0	0
FONT	904 12TH ST	116	1275	4	0	520	0	7	586	59	26	16418	4	2	120	60	7	1	1	5	83	21	1	1	0	0
DUVENHAGE	904 12TH ST	116	1276	4	0	489	0	6	515	43	37	3342	5	1	65	65	2	0	2	4	112	22	2	0	1	1
LAYTON	904 12TH ST	116	1277	4	0	445	0	5	470	37	24	6590	6	1	65	65	3	1	1	9	128	21	2	0	0	0
SIMMONS	904 12TH ST	116	1278	4	0	309	0	4	318	29	71	4664	4	2	120	60	7	0	2	4	77	19	2	0	1	0
PACUCH	904 12TH ST	116	1279	4	0	313	0	4	290	26	36	5555	2	2	120	60	6	1	2	1	45	23	2	0	1	0
FRANKLIN	903 13TH ST	116	1280	4	0	354	0	5	404	36	17	6670	5	2	120	60	8	1	1	4	110	22	2	0	1	0
GWYN	903 13TH ST	116	1281	4	0	372	0	5	401	37	45	6785	4	1	65	65	5	0	2	2	93	23	2	2	0	2
POOLEY	903 13TH ST	116	1282	4	0	301	0	5	451	41	11	5326	6	2	120	60	4	1	2	6	111	19	2	0	0	0
MONOSA	903 13TH ST	116	1283	4	0	271	0	4	303	27	16	7080	2	2	120	60	6	0	1	2	48	24	2	0	0	0
SANCHEZ	903 13TH ST	116	1284	4	0	311	0	6	529	52	20	7920	6	1	65	65	5	1	2	9	113	19	2	0	0	0
TOLEDO	903 13TH ST	116	1285	4	0	211	0	4	317	24	27	6671	1	2	120	60	6	1	2	1	30	30	2	0	0	0

Appendixes

332

VARIABLES

01	02	03	04	05	06	07	08	09	10	11	12	13	14	15	16	17	18	19	20	21	22	23	24	25	26	27
HERNANDEZ	903 13TH ST	116	1286	4	0	462	0	6	442	49	12	8521	6	2	120	60	6	0	1	6	140	23	2	0	5	0
LYNCH	903 13TH ST	116	1287	4	0	394	0	6	472	50	19	3102	9	1	65	65	6	1	1	7	180	20	2	0	1	7
GRAHAM	903 13TH ST	116	1288	4	0	316	0	6	487	42	50	3383	4	1	65	65	4	0	2	3	91	23	2	0	0	1
CORTINA	903 13TH ST	116	1289	4	0	453	0	6	461	48	13	8726	5	2	120	60	6	0	2	5	109	22	2	0	1	6
MEDINA	902 9TH ST	117	1290	4	0	353	0	5	382	35	15	16555	6	1	65	65	7	0	2	3	130	22	2	0	5	4
SMITH	902 9TH ST	117	1291	4	0	402	0	5	470	35	23	14120	3	2	120	60	5	1	2	2	55	18	2	2	0	0
ROHRBACK	902 9TH ST	117	1292	4	0	415	0	6	503	48	40	7480	6	1	65	65	6	0	2	1	130	22	2	2	0	1
GLASSFORD	902 9TH ST	117	1293	4	0	163	0	3	271	19	69	13723	2	2	120	60	8	0	2	2	43	22	2	2	0	7
BERGER	902 9TH ST	117	1294	4	0	432	0	6	503	40	14	14656	9	2	120	60	7	1	1	9	154	17	2	0	0	0
BOLD	902 9TH ST	117	1295	4	0	499	0	5	414	37	19	31619	4	2	152	76	4	0	2	2	70	18	2	1	0	3
HANSEN	902 9TH ST	117	1296	4	0	335	0	5	382	38	34	20703	4	3	177	59	5	1	2	2	99	25	2	0	4	2
PLANTE	902 9TH ST	117	1297	4	0	583	0	8	647	70	15	18922	5	1	65	65	6	0	1	4	109	22	2	1	0	0
KIRSCHEN	902 9TH ST	117	1298	4	0	230	0	3	221	21	75	19654	1	2	120	60	8	0	2	1	31	31	3	0	5	9
SKINNER	902 9TH ST	117	1299	4	0	305	0	4	346	29	76	29004	4	2	160	80	7	2	1	2	85	21	2	0	6	0
GOYNE	902 9TH ST	117	1300	4	0	332	0	5	418	37	18	23132	2	1	110	110	6	2	2	1	35	18	3	0	0	0
SANCHEZ	901 10TH ST	117	1301	4	0	479	0	7	566	56	17	20073	4	3	170	57	8	2	1	5	88	22	2	0	0	0
TEASLEY	901 10TH ST	117	1302	4	0	297	0	5	304	36	17	24761	2	1	95	95	4	1	2	1	34	17	2	2	0	3
GOLDFARB	901 10TH ST	117	1303	4	0	476	0	6	497	41	25	18216	6	2	120	60	7	1	1	4	122	20	2	1	0	0
LARKIN	901 10TH ST	117	1304	4	0	524	0	8	603	67	26	24422	5	3	199	66	7	2	1	4	107	21	2	0	3	9
PALMER	901 10TH ST	117	1305	4	0	439	0	6	494	53	56	22516	6	2	120	60	7	2	2	1	107	18	2	0	0	10
SOCORRO	901 10TH ST	117	1306	4	0	600	0	9	688	74	68	29034	5	3	190	63	8	0	1	9	108	22	2	1	0	0
VEIGA	901 10TH ST	117	1307	4	0	405	0	6	463	48	21	27158	6	4	218	55	5	1	2	4	127	21	2	0	5	0
FEINMEL	901 10TH ST	117	1308	4	0	386	0	6	484	43	18	28058	2	2	135	68	8	2	1	1	60	30	2	0	6	0
HAMPTON	901 10TH ST	117	1309	4	0	312	0	5	403	38	26	24029	4	2	145	73	12	1	1	1	78	20	2	0	3	3
CRANE	901 10TH ST	117	1310	4	0	325	0	5	414	38	91	28180	4	3	184	61	9	0	1	5	109	27	2	1	0	0
LAVIN	902 10TH ST	118	1311	4	0	439	0	7	589	54	60	16962	4	1	65	65	3	0	1	3	88	22	2	2	0	0
ROTH	902 10TH ST	118	1312	4	0	277	0	4	340	31	107	11936	2	2	120	60	4	1	1	3	32	16	2	0	0	0
LO	902 10TH ST	118	1313	4	0	170	0	3	226	19	21	6971	5	1	65	65	3	0	2	6	90	18	2	0	4	0
CONWAY	902 10TH ST	118	1314	4	0	438	0	6	475	47	72	14298	6	2	120	60	6	0	1	1	91	15	2	0	0	0
ABDO	902 10TH ST	118	1315	4	0	377	0	5	385	38	14	19456	7	1	65	65	7	0	2	9	155	22	2	0	6	0
HASTINGS	902 10TH ST	118	1316	4	0	261	0	3	220	19	18	12494	1	2	120	60	4	0	1	1	22	22	2	1	0	0
WULF	902 10TH ST	118	1317	4	0	452	0	5	423	40	23	26335	1	1	65	65	2	1	1	1	25	25	2	1	0	2
SHAPIRO	902 10TH ST	118	1318	4	0	350	0	5	424	38	20	14303	6	2	120	60	4	0	1	8	100	17	2	0	1	0
GUERRIER	902 10TH ST	118	1319	4	0	289	0	5	437	36	18	10349	6	1	65	65	3	0	2	4	125	21	2	1	0	5
SINGLETON	902 10TH ST	118	1320	4	0	450	0	6	470	46	10	19193	7	2	120	60	4	1	1	2	142	20	2	0	0	0
HERNANDEZ	902 10TH ST	118	1321	4	0	350	0	6	481	47	81	25493	4	4	204	51	7	2	2	3	99	25	2	0	6	0
CANETA	901 11TH ST	118	1322	4	0	599	0	8	670	66	13	16204	5	2	120	60	8	1	1	4	98	20	2	0	0	1
MILLS	901 11TH ST	118	1323	4	0	375	0	6	550	43	16	19772	6	1	65	65	2	1	1	3	145	24	2	0	0	0
KAHN	901 11TH ST	118	1324	4	0	292	0	5	456	38	33	20395	2	1	100	100	3	2	2	3	39	20	2	0	0	0
WATKINS	901 11TH ST	118	1325	4	0	407	0	6	455	46	20	13694	4	1	65	65	7	1	1	2	95	24	2	0	0	0
ROGOW	901 11TH ST	118	1326	4	0	408	0	6	501	44	56	23242	5	3	191	64	7	2	2	5	87	17	2	1	0	0
BLOCK	901 11TH ST	118	1327	4	0	400	0	5	452	31	70	18355	5	2	120	60	6	1	1	4	117	23	2	0	1	0
GUTIERREZ	901 11TH ST	118	1328	4	0	423	0	6	503	47	25	18497	5	2	120	60	6	1	2	6	88	18	2	0	3	0
PEREZ	901 11TH ST	118	1329	4	0	437	0	6	446	42	31	26012	2	2	137	69	6	1	2	1	58	29	2	0	1	0
ESPIRITO	901 11TH ST	118	1330	4	0	486	0	7	578	55	36	18369	4	2	120	60	5	1	1	5	90	23	2	0	5	0
BOLINGER	901 11TH ST	118	1331	4	0	461	0	7	564	55	73	19738	5	2	120	60	3	1	1	1	105	21	2	1	0	0
SAGE	902 11TH ST	119	1332	4	0	156	0	3	222	20	22	8999	2	2	120	60	2	0	1	1	42	21	2	0	0	1
CATTERY	902 11TH ST	119	1333	4	0	321	0	5	440	42	50	8807	2	2	120	60	4	1	2	2	49	25	2	0	1	5
FOLEY	902 11TH ST	119	1334	4	0	329	0	5	388	38	25	9257	7	2	120	60	5	0	2	3	145	21	2	0	0	5
BALDWIN	902 11TH ST	119	1335	4	0	286	0	5	398	36	39	8907	6	1	65	65	5	1	1	4	130	22	2	0	1	3
CANCELA	902 11TH ST	119	1336	4	0	234	0	4	262	27	97	10960	4	1	65	65	5	0	2	3	78	20	2	0	1	5
LEIBOWITZ	902 11TH ST	119	1337	4	0	553	0	8	629	64	19	14338	4	2	120	60	3	1	1	1	101	25	2	2	0	3
RAMIREZ	902 11TH ST	119	1338	4	0	375	0	7	542	56	23	11897	3	1	65	65	5	0	2	3	69	23	2	0	1	2
KIMBLER	902 11TH ST	119	1339	4	0	431	0	5	443	34	21	12668	6	2	120	60	5	0	2	0	141	24	2	0	0	0
DUPUIS	902 11TH ST	119	1340	4	0	408	0	7	538	60	89	10032	4	1	65	65	4	1	1	2	85	21	2	2	0	2
MORENO	902 11TH ST	119	1341	4	0	362	0	5	400	38	16	24789	4	2	149	75	7	2	1	4	83	21	2	0	6	1
QUINTERO	902 11TH ST	119	1342	4	0	407	0	6	485	51	15	16392	6	2	120	60	10	1	1	3	132	22	2	0	0	0
SMITH	901 12TH ST	119	1343	4	0	309	0	6	474	48	14	23260	7	2	120	60	5	2	2	0	159	23	2	0	4	0
KUTUN	901 12TH ST	119	1344	4	0	568	0	8	646	61	15	21308	3	2	144	72	7	0	2	3	51	17	2	0	4	1
PRIEST	901 12TH ST	119	1345	4	0	284	0	5	342	38	13	12639	6	1	65	65	2	1	2	0	112	19	2	0	6	4
HAWTHORNE	901 12TH ST	119	1346	4	0	389	0	5	412	35	26	15875	6	3	165	55	7	1	1	12	115	19	2	2	0	6
TURNER	901 12TH ST	119	1347	4	0	447	0	7	486	48	18	13580	3	1	65	65	6	1	1	0	62	21	2	0	0	0
MACMASTER	901 12TH ST	119	1348	4	0	611	0	8	629	65	19	19695	4	1	65	65	3	0	1	1	72	18	2	2	0	5
RICCI	901 12TH ST	119	1349	4	0	425	0	6	518	44	82	22455	7	2	120	60	4	2	1	9	164	23	3	0	0	0
ALANIS	901 12TH ST	119	1350	4	0	483	0	7	580	54	21	23560	4	3	167	56	4	2	2	3	74	19	2	1	0	6
DOROSKI	901 12TH ST	119	1351	4	0	257	0	4	335	29	49	20432	4	2	164	82	8	2	1	7	85	21	2	0	0	0
REEVES	901 12TH ST	119	1352	4	0	387	0	5	388	38	71	5305	2	2	120	60	6	1	1	1	47	24	2	2	0	1
LAZARUS	902 12TH ST	120	1353	4	0	539	0	8	713	69	32	15445	3	1	65	65	2	1	1	2	82	27	2	0	0	0
MARTINEZ	902 12TH ST	120	1354	4	0	464	0	6	466	49	28	8312	5	1	65	65	5	0	2	4	98	20	2	0	0	0
COTO	902 12TH ST	120	1355	4	0	347	0	5	468	38	16	6444	3	2	120	60	6	0	2	1	68	23	2	0	0	0
FEINBURG	902 12TH ST	120	1356	4	0	440	0	5	486	37	15	8544	5	2	120	60	5	1	2	6	110	22	2	2	0	0
VALDES	902 12TH ST	120	1357	4	0	381	0	6	483	47	15	3826	6	2	120	60	6	1	1	6	164	27	2	0	0	0
DEITZ	902 12TH ST	120	1358	4	0	436	0	5	410	37	29	9207	5	1	65	65	6	1	2	3	103	21	2	2	0	0
OVINNIO	902 12TH ST	120	1359	4	0	423	0	6	500	43	11	5073	6	2	120	60	6	1	2	3	136	23	2	0	4	1
WALKER	902 12TH ST	120	1360	4	0	246	0	4	291	30	49	5451	2	2	120	60	9	0	2	1	54	27	2	0	0	0
PINERO	902 12TH ST	120	1361	4	0	304	0	5	405	35	25	8998	3	1	65	65	6	1	2	2	72	24	2	2	0	0
SWANSON	902 12TH ST	120	1362	4	0	388	0	6	487	45	16	5743	8	1	65	65	4	0	2	12	182	23	2	1	0	9
COWART	902 12TH ST	120	1363	4	0	382	0	6	482	43	74	5556	8	2	120	60	7	1	2	10	160	20	2	2	0	3
PARCELL	901 13TH ST	120	1364	4	0	414	0	6	495	42	47	7145	3	2	120	60	4	1	1	1	79	26	2	0	0	7
ILNITSKY	901 13TH ST	120	1365	4	0	425	0	6	484	46	62	9715	2	2	120	60	6	1	2	2	55	28	2	0	6	7
BENBON	901 13TH ST	120	1366	4	0	374	0	6	533	44	95	8967	4	2	120	60	6	0	2	5	74	19	2	2	0	6
ROBBINS	901 13TH ST	120	1367	4	0	349	0	5	444	39	11	5503	5	1	65	65	2	0	1	0	110	22	2	0	1	3
BLEDSOE	901 13TH ST	120	1368	4	0	416	0	6	511	43	15	8603	5	2	120	60	3	0	2	4	128	26	2	0	0	4
OLSON	901 13TH ST	120	1369	4	0	153	0	3	333	22	20	6398	1	2	120	60	3	0	2	1	24	24	2	2	0	0
REDFORD	901 13TH ST	120	1370	4	0	260	0	4	355	30	40	8648	1	2	120	60	3	1	1	1	23	23	2	2	0	4
FORD	901 13TH ST	120	1371	4	0	308	0	5	413	39	16	3655	5	2	120	60	5	0	2	6	115	23	2	0	1	2
NOVACK	901 13TH ST	120	1372	4	0	269	0	5	442	37	45	7732	5	1	65	65	5	0	2	6	79	16	2	0	1	4
PRICE	901 13TH ST	120	1373	4	0	363	0	5	374	38	19	4279	6	1	65	65	7	1	2	4	123	21	2	1	0	0

INDEX

This book has been set VIP, in 10 and 9 point Times Roman, leaded 2 points. Part numbers are 18 point City Light and chapter titles are 24 point City Light. The size of the type page is 42 by 58 picas.

LEGEND FOR THE STAT CITY MAP

Two types of dwelling units appear on the Stat City map, houses and apartment buildings. The information recorded for each house is explained below:

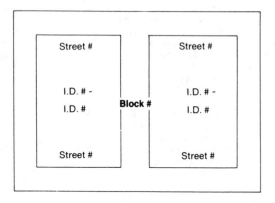

The information recorded for each apartment building is explained below:

Please note that apartment buildings in Stat City have entrances on two streets; consequently, two street numbers appear for each apartment building.

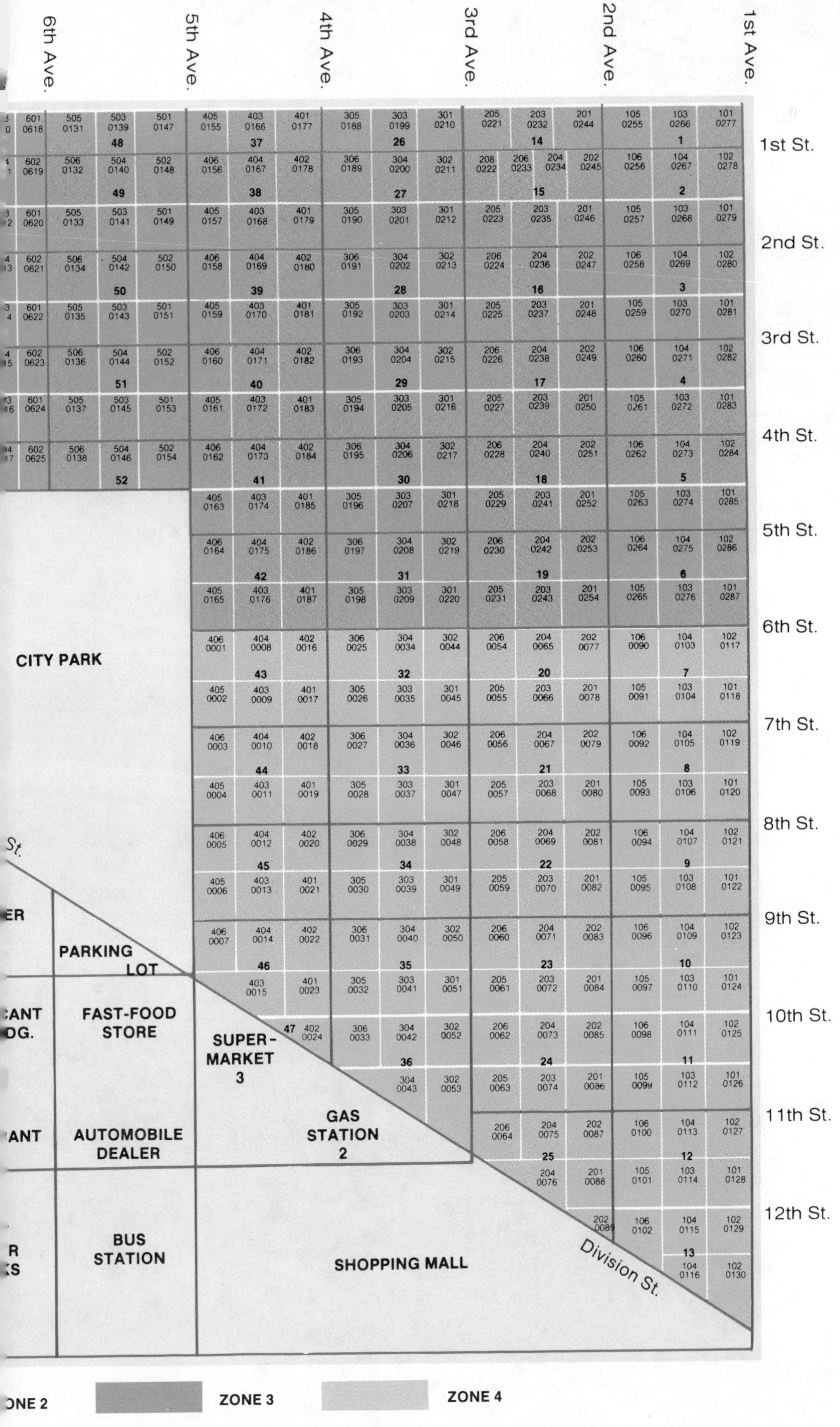

City street map — address grid

Avenues (top, left to right): 6th Ave. · 5th Ave. · 4th Ave. · 3rd Ave. · 2nd Ave. · 1st Ave.

Streets (right side, top to bottom): 1st St. · 2nd St. · 3rd St. · 4th St. · 5th St. · 6th St. · 7th St. · 8th St. · 9th St. · 10th St. · 11th St. · 12th St.

Left-side features: CITY PARK · PARKING LOT · FAST-FOOD STORE · SUPER-MARKET 3 · AUTOMOBILE DEALER · GAS STATION 2 · BUS STATION · SHOPPING MALL · Division St.

Legend: ZONE 2 · ZONE 3 · ZONE 4

Grid cells (top address number / lower parcel number), listed row by row. Bold block numbers appear below cell pairs.

Row (above 1st St.): 601/0618 · 505/0131 · 503/0139 · 501/0147 · 405/0155 · 403/0166 · 401/0177 · 305/0188 · 303/0199 · 301/0210 · 205/0221 · 203/0232 · 201/0244 · 105/0255 · 103/0266 · 101/0277
Blocks: **48** · **37** · **26** · **14** · **1**

Row: 602/0619 · 506/0132 · 504/0140 · 502/0148 · 406/0156 · 404/0167 · 402/0178 · 306/0189 · 304/0200 · 302/0211 · 208/0222 · 206/0233 · 204/0234 · 202/0245 · 106/0256 · 104/0267 · 102/0278
Blocks: **49** · **38** · **27** · **15** · **2**
— 1st St.

Row: 601/0620 · 505/0133 · 503/0141 · 501/0149 · 405/0157 · 403/0168 · 401/0179 · 305/0190 · 303/0201 · 301/0212 · 205/0223 · 203/0235 · 201/0246 · 105/0257 · 103/0268 · 101/0279

Row: 602/0621 · 506/0134 · 504/0142 · 502/0150 · 406/0158 · 404/0169 · 402/0180 · 306/0191 · 304/0202 · 302/0213 · 206/0224 · 204/0236 · 202/0247 · 106/0258 · 104/0269 · 102/0280
Blocks: **50** · **39** · **28** · **16** · **3**
— 2nd St.

Row: 601/0622 · 505/0135 · 503/0143 · 501/0151 · 405/0159 · 403/0170 · 401/0181 · 305/0192 · 303/0203 · 301/0214 · 205/0225 · 203/0237 · 201/0248 · 105/0259 · 103/0270 · 101/0281

Row: 602/0623 · 506/0136 · 504/0144 · 502/0152 · 406/0160 · 404/0171 · 402/0182 · 306/0193 · 304/0204 · 302/0215 · 206/0226 · 204/0238 · 202/0249 · 106/0260 · 104/0271 · 102/0282
Blocks: **51** · **40** · **29** · **17** · **4**
— 3rd St.

Row: 601/0624 · 505/0137 · 503/0145 · 501/0153 · 405/0161 · 403/0172 · 401/0183 · 305/0194 · 303/0205 · 301/0216 · 205/0227 · 203/0239 · 201/0250 · 105/0261 · 103/0272 · 101/0283

Row: 602/0625 · 506/0138 · 504/0146 · 502/0154 · 406/0162 · 404/0173 · 402/0184 · 306/0195 · 304/0206 · 302/0217 · 206/0228 · 204/0240 · 202/0251 · 106/0262 · 104/0273 · 102/0284
Blocks: **52** · **41** · **30** · **18** · **5**
— 4th St.

(CITY PARK occupies the 6th/5th Ave. area from here down)

Row: 405/0163 · 403/0174 · 401/0185 · 305/0196 · 303/0207 · 301/0218 · 205/0229 · 203/0241 · 201/0252 · 105/0263 · 103/0274 · 101/0285
— 5th St.

Row: 406/0164 · 404/0175 · 402/0186 · 306/0197 · 304/0208 · 302/0219 · 206/0230 · 204/0242 · 202/0253 · 106/0264 · 104/0275 · 102/0286
Blocks: **42** · **31** · **19** · **6**

Row: 405/0165 · 403/0176 · 401/0187 · 305/0198 · 303/0209 · 301/0220 · 205/0231 · 203/0243 · 201/0254 · 105/0265 · 103/0276 · 101/0287
— 6th St.

Row: 406/0001 · 404/0008 · 402/0016 · 306/0025 · 304/0034 · 302/0044 · 206/0054 · 204/0065 · 202/0077 · 106/0090 · 104/0103 · 102/0117
Blocks: **43** · **32** · **20** · **7**

Row: 405/0002 · 403/0009 · 401/0017 · 305/0026 · 303/0035 · 301/0045 · 205/0055 · 203/0066 · 201/0078 · 105/0091 · 103/0104 · 101/0118
— 7th St.

Row: 406/0003 · 404/0010 · 402/0018 · 306/0027 · 304/0036 · 302/0046 · 206/0056 · 204/0067 · 202/0079 · 106/0092 · 104/0105 · 102/0119
Blocks: **44** · **33** · **21** · **8**

Row: 405/0004 · 403/0011 · 401/0019 · 305/0028 · 303/0037 · 301/0047 · 205/0057 · 203/0068 · 201/0080 · 105/0093 · 103/0106 · 101/0120
— 8th St.

Row: 406/0005 · 404/0012 · 402/0020 · 306/0029 · 304/0038 · 302/0048 · 206/0058 · 204/0069 · 202/0081 · 106/0094 · 104/0107 · 102/0121
Blocks: **45** · **34** · **22** · **9**

Row: 405/0006 · 403/0013 · 401/0021 · 305/0030 · 303/0039 · 301/0049 · 205/0059 · 203/0070 · 201/0082 · 105/0095 · 103/0108 · 101/0122
— 9th St.

Row: 406/0007 · 404/0014 · 402/0022 · 306/0031 · 304/0040 · 302/0050 · 206/0060 · 204/0071 · 202/0083 · 106/0096 · 104/0109 · 102/0123
Blocks: **46** · **35** · **23** · **10**

(PARKING LOT, FAST-FOOD STORE on left)

Row: 403/0015 · 401/0023 · 305/0032 · 303/0041 · 301/0051 · 205/0061 · 203/0072 · 201/0084 · 105/0097 · 103/0110 · 101/0124
— 10th St.

Row: **47** 402/0024 · 306/0033 · 304/0042 · 302/0052 · 206/0062 · 204/0073 · 202/0085 · 106/0098 · 104/0111 · 102/0125
Blocks: **36** · **24** · **11**
(SUPER-MARKET 3 on left)

Row: 304/0043 · 302/0053 · 205/0063 · 203/0074 · 201/0086 · 105/0099 · 103/0112 · 101/0126
— 11th St.
(GAS STATION 2, AUTOMOBILE DEALER on left)

Row: 206/0064 · 204/0075 · 202/0087 · 106/0100 · 104/0113 · 102/0127
Blocks: **25** · **12**

Row: 204/0076 · 201/0088 · 105/0101 · 103/0114 · 101/0128
— 12th St.

Row: 202/0089 · 106/0102 · 104/0115 · 102/0129
Block: **13**

Row: 104/0116 · 102/0130

(BUS STATION, SHOPPING MALL, Division St. at bottom)

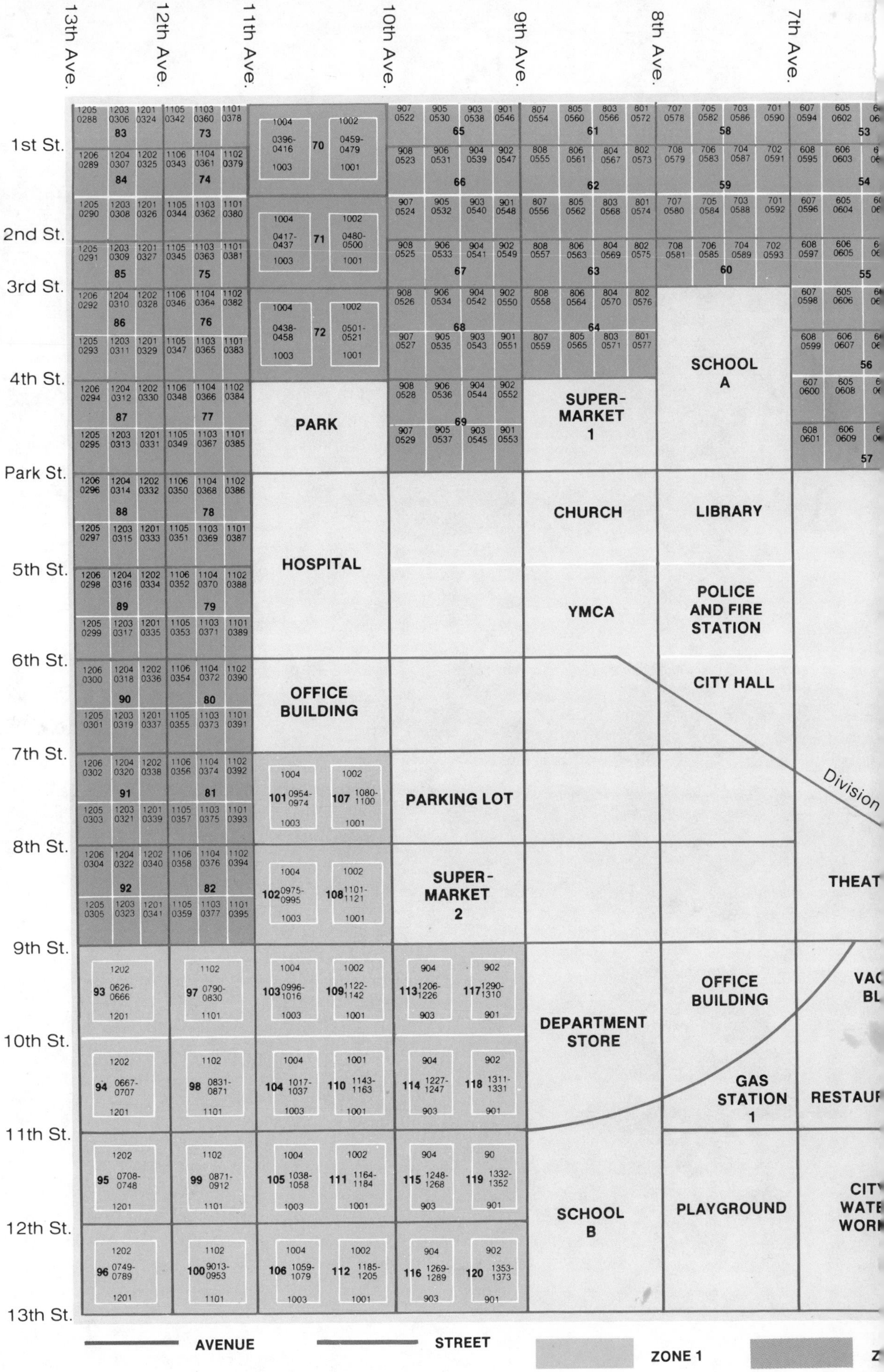

Map (city grid)

Avenues (top, left to right): 13th Ave. · 12th Ave. · 11th Ave. · 10th Ave. · 9th Ave. · 8th Ave. · 7th Ave.

Streets (left side, top to bottom): 1st St. · 2nd St. · 3rd St. · 4th St. · Park St. · 5th St. · 6th St. · 7th St. · 8th St. · 9th St. · 10th St. · 11th St. · 12th St. · 13th St.

Left grid blocks (13th–11th Ave.)

Block 83: 1205/0288 · 1203/0306 · 1201/0324 | 1105/0342 · 1103/0360 · 1101/0378 (73)
Block 84: 1206/0289 · 1204/0307 · 1202/0325 | 1106/0343 · 1104/0361 · 1102/0379 (74)
Block 85: 1205/0290 · 1203/0308 · 1201/0326 | 1105/0344 · 1103/0362 · 1101/0380 ; 1205/0291 · 1203/0309 · 1201/0327 | 1105/0345 · 1103/0363 · 1101/0381 (75)
Block 86: 1206/0292 · 1204/0310 · 1202/0328 | 1106/0346 · 1104/0364 · 1102/0382 ; 1205/0293 · 1203/0311 · 1201/0329 | 1105/0347 · 1103/0365 · 1101/0383 (76)
Block 87: 1206/0294 · 1204/0312 · 1202/0330 | 1106/0348 · 1104/0366 · 1102/0384 ; 1205/0295 · 1203/0313 · 1201/0331 | 1105/0349 · 1103/0367 · 1101/0385 (77)
Block 88: 1206/0296 · 1204/0314 · 1202/0332 | 1106/0350 · 1104/0368 · 1102/0386 ; 1205/0297 · 1203/0315 · 1201/0333 | 1105/0351 · 1103/0369 · 1101/0387 (78)
Block 89: 1206/0298 · 1204/0316 · 1202/0334 | 1106/0352 · 1104/0370 · 1102/0388 ; 1205/0299 · 1203/0317 · 1201/0335 | 1105/0353 · 1103/0371 · 1101/0389 (79)
Block 90: 1206/0300 · 1204/0318 · 1202/0336 | 1106/0354 · 1104/0372 · 1102/0390 ; 1205/0301 · 1203/0319 · 1201/0337 | 1105/0355 · 1103/0373 · 1101/0391 (80)
Block 91: 1206/0302 · 1204/0320 · 1202/0338 | 1106/0356 · 1104/0374 · 1102/0392 ; 1205/0303 · 1203/0321 · 1201/0339 | 1105/0357 · 1103/0375 · 1101/0393 (81)
Block 92: 1206/0304 · 1204/0322 · 1202/0340 | 1106/0358 · 1104/0376 · 1102/0394 ; 1205/0305 · 1203/0323 · 1201/0341 | 1105/0359 · 1103/0377 · 1101/0395 (82)

Lower-left blocks

93: 1202 / 0626–0666 / 1201
94: 1202 / 0667–0707 / 1201
95: 1202 / 0708–0748 / 1201
96: 1202 / 0749–0789 / 1201
97: 1102 / 0790–0830 / 1101
98: 1102 / 0831–0871 / 1101
99: 1102 / 0871–0912 / 1101
100: 1102 / 9013–0953 / 1101
103: 1004 / 0996–1016 / 1003
104: 1004 / 1017–1037 / 1003
105: 1004 / 1038–1058 / 1003
106: 1004 / 1059–1079 / 1003
109: 1002 / 1122–1142 / 1001
110: 1001 / 1143–1163
111: 1002 / 1164–1184 / 1001
112: 1002 / 1185–1205 / 1001
113: 904 / 1206–1226 / 903
114: 904 / 1227–1247 / 903
115: 904 / 1248–1268 / 903
116: 904 / 1269–1289 / 903
117: 902 / 1290–1310 / 901
118: 902 / 1311–1331 / 901
119: 90 / 1332–1352 / 901
120: 902 / 1353–1373 / 901

Center blocks

70: 1004 / 0396–0416 / 1003 | 1002 / 0459–0479 / 1001
71: 1004 / 0417–0437 / 1003 | 1002 / 0480–0500 / 1001
72: 1004 / 0438–0458 / 1003 | 1002 / 0501–0521 / 1001
101: 1004 / 0954–0974 / 1003
107: 1002 / 1080–1100 / 1001
102: 1004 / 0975–0995 / 1003
108: 1002 / 1101–1121 / 1001

Right grid blocks (10th–7th Ave.)

Block 65: 907/0522 · 905/0530 · 903/0538 · 901/0546
Block 66: 908/0523 · 906/0531 · 904/0539 · 902/0547
Block 67: 907/0524 · 905/0532 · 903/0540 · 901/0548 ; 908/0525 · 906/0533 · 904/0541 · 902/0549
Block 68: 908/0526 · 906/0534 · 904/0542 · 902/0550 ; 907/0527 · 905/0535 · 903/0543 · 901/0551
Block 69: 908/0528 · 906/0536 · 904/0544 · 902/0552 ; 907/0529 · 905/0537 · 903/0545 · 901/0553

Block 61: 807/0554 · 805/0560 · 803/0566 · 801/0572
Block 62: 808/0555 · 806/0561 · 804/0567 · 802/0573
Block 63: 807/0556 · 805/0562 · 803/0568 · 801/0574 ; 808/0557 · 806/0563 · 804/0569 · 802/0575
Block 64: 808/0558 · 806/0564 · 804/0570 · 802/0576 ; 807/0559 · 805/0565 · 803/0571 · 801/0577

Block 58: 707/0578 · 705/0582 · 703/0586 · 701/0590
Block 59: 708/0579 · 706/0583 · 704/0587 · 702/0591
Block 60: 707/0580 · 705/0584 · 703/0588 · 701/0592 ; 708/0581 · 706/0585 · 704/0589 · 702/0593

Block 53: 607/0594 · 605/0602 · 6../06..
Block 54: 608/0595 · 606/0603 · 6../06..
Block 55: 607/0596 · 605/0604 ; 608/0597 · 606/0605
Block 56: 607/0598 · 605/0606 ; 608/0599 · 606/0607
Block 57: 607/0600 · 605/0608 ; 608/0601 · 606/0609

Named features

PARK · HOSPITAL · OFFICE BUILDING · SCHOOL A · SUPER-MARKET 1 · CHURCH · LIBRARY · YMCA · POLICE AND FIRE STATION · CITY HALL · Division · PARKING LOT · SUPER-MARKET 2 · THEAT... · DEPARTMENT STORE · OFFICE BUILDING · VAC... BL... · GAS STATION 1 · RESTAUR... · SCHOOL B · PLAYGROUND · CIT... WATE... WORK...

Legend

———— AVENUE ———— STREET · ZONE 1 · Z...